pushing rubber tubing over a glass tube; a severe explosion and fire may result from attempting to distill a substance in a completely closed system.

Procedures to be followed in case of accidents are given in the Appendix and also are printed on the inside back cover of this manual.

Toxic Chemicals and Mutagens

Many common organic chemicals, particularly those containing nitrogen, are toxic or even lethal when ingested in amounts as small as a few tenths of a gram. A major rule of safety laboratory practice is *never* taste any compound.

Certain organic compounds, when ingested or inhaled over a long period of time, even in minute doses, may eventually induce tumors—some benign and some cancerous. An attempt has been made in this manual to avoid use of such compounds, but this is difficult because of the limited information available in most cases. It seems probable that many relatively safe compounds have been included on the "cancer suspect" list, whereas many others that should be included are not. Unfortunately, simplicity of structure or lack of active functional groups does not guarantee safety. In the absence of specific information, it is wise to minimize exposure to the vapors of all organic solvents; even pleasant-smelling vapors may be quite toxic.

Laboratory Apparel

It is desirable to protect street clothing from being soiled and damaged by chemicals and accidents of various sorts by wearing an inexpensive laboratory coat or a rubber apron. For freedom of arm movement, tight sleeves should be avoided; loose and bulky sleeves may cause overturning of fragile apparatus. Unprotected light-weight, flammable apparel constitutes a serious fire hazard in the organic laboratory. Many synthetic fabrics are soluble in acetone and other common organic solvents.

Shoes that cover at least the toes and instep must be worn in the laboratory.

Shoulder-length (or longer) hair should be tied back, especially if a burner is being used.

In experiments requiring transfer of corrosive chemicals, it is desirable to wear some type of resistant gloves. Inexpensive, disposable gloves made from polyethylene are available. Rubber gloves afford better protection, although they make laboratory operations more cumbersome.

EXPERIMENTAL ORGANIC CHEMISTRY
Theory and Practice

Charles F. Wilcox, Jr.
Cornell University

MACMILLAN PUBLISHING COMPANY
NEW YORK
Collier Macmillan Publishers
London

Experimental Organic Chemistry
is dedicated to the memory of
JOHN R. JOHNSON
scholar and teacher.

Copyright © 1984, Macmillan Publishing Company,
a division of Macmillan, Inc.

Printed in the United States of America

All rights reserved. No part of this book may be reproduced or
transmitted in any form or by any means, electronic or mechanical,
including photocopying, recording, or any information storage and
retrieval system, without permission in writing from the Publisher.

A portion of this book is reprinted from *Laboratory Experiments* in
Organic Chemistry, Seventh Edition, by Roger Adams,
John R. Johnson, and Charles F. Wilcox, Jr., copyright © 1979 by
Macmillan Publishing Co., Inc.

Macmillan Publishing Company
866 Third Avenue, New York, New York 10022

Collier Macmillan Canada, Inc.

Library of Congress Cataloging in Publication Data

Wilcox, Charles F., Date:
   Experimental organic chemistry.
   Includes index.
QD261.W498 1984   547'.0028   83-11253
ISBN 0-02-427600-6

Printing: 1 2 3 4 5 6 7 8   Year: 4 5 6 7 8 9 0 1 2

ISBN 0-02-427600-6

Preface

For more than 50 years, through seven editions, the text *Laboratory Experiments in Organic Chemistry* by Adams, Johnson, and (for the last three editions) Wilcox has been widely used. As the teaching of organic laboratory work evolved, the original brief manual was expanded to incorporate new needs. Now, with Roger Adams and Jack Johnson deceased, it was felt that the time had come to make a fresh start. The present text is that effort. Readers familiar with the seventh edition will recognize much of it in the new text, but closer examination will also reveal significant deletions, changes, and additions.

A new title, *Experimental Organic Chemistry: Theory and Practice*, was chosen to express the author's conviction that in learning how to do experimental organic chemistry students can and should understand the theoretical basis for both the reactions they study and the separation methods used to purify the products. It is self-evident that if students are to carry out experiments above a blind, mechanical level they need to understand the underlying chemistry. In the same vein, it is argued that a thorough grasp of the fundamentals of separation methods is essential because more effort is usually spent in isolating products than in running reactions. Another essential aspect of organic chemistry is the identification of materials and the evaluation of their purity.

With these several needs in mind, the book has been divided into three parts. Part I describes the more common separation methods and the theories underlying them. In this author's view, a powerful unifying theme that correlates the various separation methods is the concept of vapor pressure and how it is affected by molecular interactions. For this reason distillation is presented first. The basic ideas presented in the chapter on simple and fractional distillation are developed and extended in the following chapters.

Part II presents the identification of organic compounds by both classical chemical-test methods and modern spectrometric techniques. The section on classical methods does not attempt to present anything like a full identification scheme, but restricts itself to the common functional groups. At Cornell, our introductory organic laboratory course uses a blend of both approaches. The chemical methods, especially when done on the small scale described here, reinforce good laboratory technique. They also encourage clear chemical thinking. Spectrometry has become an integral part of research, and students should be exposed to it in a first course in order to develop an early understanding of its power and limitations. Although at Cornell the material of this chapter is introduced only after some preparative experiments taken from Part III have been performed, its placement in Part II simplifies the organization.

Part III provides a number of preparative experiments. The primary objective in their selection has been to provide syntheses and transformations that illustrate general reactions or mechanistic principles and furnish products in satisfactory yields. In a few cases a preparation has been included because the product is particularly interesting. Preference has been given to the use of relatively inexpensive reagents and a scale has been chosen that is practical in the hands of beginning students but minimizes cost and waste disposal. The author is keenly aware of the heightened sense of chemical toxicity and the challenges it raises. An effort has been made to eliminate, or at least sharply reduce, exposure to potentially toxic substances. With regret, therefore, several otherwise attractive preparations have been deleted. Where the risks were significant, but felt not to be severe enough to require deletion, appropriate warnings have been included.

A list of apparatus suitable for the laboratory work and of the chemicals needed for each experiment is given in the *Instructor's Manual*.

The author acknowledges gratefully his indebtedness to many colleagues and instructors throughout the country for suggestions and friendly criticism, including Macmillan's reviewers—W. J. Burke, Dennis McMinn, and Bradford P. Mundy. Particular thanks go to Drs. J. B. Ellern and Forrest Sheffy. Thanks also go to the many students who have suffered through the problems of early versions of the experiments and have contributed so much to their resolution.

C. F. W., Jr.

Contents

1 Introduction — 1
    1.1 General Precautions   1
    1.2 Apparatus   3
    1.3 Weighing and Measuring Reagents   7
    1.4 Heat Sources   9
    1.5 Stirring   11
    1.6 Laboratory Notes   12

I SEPARATION AND PURIFICATION OF ORGANIC COMPOUNDS

2 Simple and Fractional Distillation — 17
    2.1 Principles of Distillation   17
    2.2 Fractional Distillation   25
    2.3 Laboratory Practice   31
    2.4 Representative Simple and Fractional Distillations   39
    Questions   45

3 Vacuum Distillation — 45
    3.1 Principles of Vacuum Distillation   45
    3.2 Vacuum Pumps   46

    3.3 Laboratory Practice   51
    3.4 Representative Vacuum Distillations   53
    Questions   54

4 Steam Distillation   55

    4.1 Principles of Steam Distillation   55
    4.2 Distillation Temperature and Composition of Distillate   56
    4.3 Laboratory Practice   58
    4.4 Representative Steam Distillations   60
    Questions   61

5 Melting Points, Crystallization, and Sublimation   62

    5.1 Melting Points   62
    5.2 Crystallization   68
    5.3 Sublimation   74
    5.4 Representative Procedures   75
    Questions   78

6 Extraction with Solvents   79

    6.1 Extraction of Solids   79
    6.2 Extraction of Solutions   80
    6.3 Multiple Extraction   82
    6.4 Laboratory Practice   86
    6.5 Representative Extractions   88
    Questions   91

7 Chromatography   92

    7.1 Liquid–Solid Chromatography   92
    7.2 Ion-Exchange Chromatography   94
    7.3 Liquid–Liquid Chromatography   96
    7.4 Gas–Liquid Chromatography   97
    7.5 Laboratory Practice   100
    7.6 Representative Chromatographic Separations   111
    Questions   114

8 Accessory Laboratory Operations   116

    8.1 Drying Agents   116
    8.2 Cooling Baths   118
    8.3 Refluxing   118

8.4 Gas Absorption Traps   120
8.5 Mechanical and Magnetic Stirring   121
8.6 Rotary Evaporation   122

II IDENTIFICATION OF ORGANIC COMPOUNDS

9 Identification by Chemical Methods — 125
9.1 Introduction   125
9.2 Preliminary Examination   126
9.3 Purification of the Unknown Sample   127
9.4 Physical Constants   128
9.5 Element Identification   129
9.6 Solubility Classification   133
9.7 Functional Group Identification   138
9.8 Derivatization of Functional Groups   154
Questions   181

10 Identification of Structure by Spectrometric Methods — 182
10.1 Mass Spectrometry   183
10.2 Infrared Spectroscopy   186
10.3 Nuclear Magnetic Resonance   197
10.4 Ultraviolet and Visible Spectroscopy   211
Questions   213

III PREPARATIONS AND REACTIONS OF TYPICAL ORGANIC COMPOUNDS

11 General Remarks — 217
11.1 Preparation Before the Laboratory   217
11.2 Calculation of Yields   220
11.3 Laboratory Directions   222
11.4 In the Laboratory   223
11.5 Samples and Reports   223

12 Free-Radical Halogenation — 225
12.1 Mechanism of Free-Radical Chlorination   225
12.2 Chlorination by Means of Sulfuryl Chloride and AIBN   226

12.3 Energetics of Halogenation   227
12.4 Selectivity in Halogenations   228
12.5 Substituent Effects   229
12.6 Preparations and Reactions   230
    (A) Photochemical chlorination of 2,3-dimethylbutane   230
    (B) Substituent effects in free-radical chlorination   231
Questions   232

13 Conversion of Alcohols to Alkyl Halides — 233

13.1 Preparation of Alkyl Halides   233
13.2 Reactions of Alkyl Halides   235
13.3 Preparations   236
    (A) *n*-Butyl bromide   236
    (B) *sec*-Butyl chloride   237
    (C) *t*-Butyl chloride   238
Questions   239

14 Second-Order Nucleophilic Substitution — 241

14.1 Replacement Reactions   241
14.2 Stereochemistry and Kinetics   242
14.3 Nucleophilicity   242
14.4 Substrate Structure   242
14.5 Solvent   243
14.6 Preparation of *n*-Butyl Iodide   243
Questions   244

15 Chemical Kinetics: Solvolysis of *t*-Butyl Chloride — 245

15.1 First-Order Kinetics   245
15.2 Laboratory Practice   247
15.3 Measurement of the S_N1 Reaction Rate of *t*-Butyl Chloride   248
Questions   249

16 Alkenes — 250

16.1 Sources of Alkenes   256
16.2 Carbocation Rearrangements   251
16.3 Dimerization of Isobutylene (2-Methylpropene)   252
16.4 Reactions of Alkenes   253

CONTENTS

- 16.5 Preparations 254
  - (A) Methylpentenes 254
  - (B) Cyclohexene 255
  - (C) 2,4,4-Trimethyl-1- and -2-pentenes (diisobutylenes) 256
- *Questions* 257

17 A Multiple-Step Synthesis — 258
- 17.1 From *n*-Butyl Alcohol to 2-Methylhexenes 258
- 17.2 Grignard Synthesis of Alcohol 260
- 17.3 Preparation of 2-Methyl-1-hexene and 2-Methyl-2-hexene 261
- *Questions* 265

18 Hydration of Alkenes and Alkynes — 267
- 18.1 Hydration of Double Bonds 267
- 18.2 Oxymercuration–Demercuration of Alkenes 268
- 18.3 Hydration of Alkynes 269
- 18.4 Reactions and Preparations 270
  - (A) Oxymercuration–demercuration of 1-hexene 270
  - (B) 2-Heptanone by hydration of 1-heptyne 270
- *Questions* 271

19 Glaser–Eglinton–Hayes Acetylene Coupling — 272
- 19.1 Introduction 272
- 19.2 Mechanism of Acetylene Coupling 273
- 19.3 Preparation of 1-Ethynylcyclohexanol 273
- *Questions* 275

20 Preparation of Aldehydes and Ketones by Oxidation — 276
- 20.1 Chromic Acid Oxidation of Alcohols 276
- 20.2 Other Oxidation Methods 278
- 20.3 Preparations 278
  - (A) Methyl *n*-propyl ketone (2-pentanone) 278
  - (B) Cyclohexanone 279
- *Questions* 280

21 Reactions of Aldehydes and Ketones — 281
- 21.1 Carbonyl Addition Reactions 281
- 21.2 Reduction of Carbonyl Compounds 282

21.3 Reactions of Carbonyl Compounds 283
 (A) Equilibria and rates in carbonyl reactions: formation of 2-furaldehyde and cyclohexanone semicarbazones 283
 (B) Reduction by sodium borohydride: diphenylmethanol 284
Questions 286

22 A Modified Wittig Synthesis — 287
22.1 The Wittig Reaction 287
22.2 Preparation of *p*-Methyoxystilbene 289
Questions 290

23 The Canizzaro Reaction — 291
23.1 Reactions of Aromatic Aldehydes 291
23.2 Preparations and Reactions 292
 (A) Benzyl alcohol 292
 (B) Benzoic acid 293
Questions 294

24 Esters — 295
24.1 Esterification and Saponification 295
24.2 Glyceryl Esters—Fats and Fatty Oils 298
24.3 Detergents and Wetting Agents 299
24.4 Preparations and Reactions 299
 (A) *n*-Butyl acetate—esterification of acetic acid 299
 (B) Methyl benzoate 300
Questions 302

25 Ionization of Carboxylic Acids — 303
25.1 Introduction 303
25.2 Inductive Effects 304
25.3 Analysis of pH/Titer Data for pK 305
25.4 Measurement of the pK of a Carboxylic Acid 307
Questions 308

26 Side-Chain Oxidation of Aromatic Compounds — 309
26.1 Oxidation of Side Chains 309
26.2 Preparations 310
 (A) *p*-Nitrobenzoic acid 310
 (B) *o*-Nitrobenzoic acid 311
Questions 311

CONTENTS

27 Friedel–Crafts Reactions 312
  27.1 Alkylation of Benzene and Related Hydrocarbons 312
  27.2 Friedel–Crafts Acylation 313
  27.3 Preparation of 4-Acetylbiphenyl 315
  Questions 316

28 Nitration of Aromatic Compounds 318
  28.1 Mechanism of Nitration 318
  28.3 Preparations 320
    (A) *m*-Dinitrobenzene 320
    (B) *p*-Bromonitrobenzene 321
    (C) Methyl *m*-nitrobenzoate 322
  Questions 322

29 Nitration of Anilines: Use of a Protecting Group 324
  29.1 Protecting Groups 324
  29.2 Acetylation of Aniline 325
    (A) Acetylation in water—Lumière–Barbier method 325
    (B) Acetylation in acetic acid 325
    (C) Direct acetylation with acetic acid 326
  29.3 Nitration of Acetanilide and Deacetylation 326
    (A) *p*-Nitroacetanilide 328
    (B) *p*-Nitroaniline 328
  Questions 329

30 Compounds of Medicinal and Biological Interest 331
  30.1 Acetylsalicylic Acid (Aspirin) 332
    Preparation of acetylsalicylic acid 332
  30.2 *p*-Ethoxyacetanilide (Phenacetin) 333
    Preparation of *p*-ethoxyacetanilide 333
  30.3 *p*-Ethoxyphenylurea (Dulcin) 334
    (A) Preparation by cyanate method 335
    (B) Preparation by urea method 335
  30.4 *p*-Aminobenzoic Acid (PABA) and Esters 336
    (A) Preparation of PABA 336
    (B) Esterification of PABA 337
  30.5 Sulfanilamide 338
    Preparation of sulfanilamide 339
  Questions 341

31 Heterocyclic Aromatics: 3-Phenylsydnone — 342
31.1 Mesoionic Compounds — 342
31.2 Preparation of 3-Phenylsydnone — 343
Questions — 344

32 Aldol Condensation — 345
32.1 Introduction — 345
32.2 Preparation of Dibenzalacetone — 347
Questions — 347

33 The Benzoin Condensation — 348
33.1 Introduction — 348
33.2 Vitamin B_1 Catalysis — 350
33.3 Preparation and Reactions of Benzoin — 351
33.4 Preparation and Reactions of Benzil — 353
    (A) Oxidation of benzoin by cupric salts — 353
    (B) Oxidation of benzoin by nitric acid — 354
Questions — 357

34 The Benzilic Acid Rearrangement — 358
34.1 Introduction — 358
34.2 Preparation of Benzilic Acid — 359
    (A) From benzil — 359
    (B) From benzoin — 360
34.3 Reactions of Benzilic Acid — 360
    (A) Benzophenone from benzilic acid — 361
    (B) Acetylbenzilic acid (α-acetoxydiphenylacetic acid) — 361
    (C) Methyl benzilate — 361
Questions — 362

35 Triphenylmethanol — 363
35.1 Triarylmethanols — 363
35.2 Preparations and Reactions — 365
    (A) Grignard synthesis of triphenylmethanol — 365
    (B) β,β,β-Triphenylpropionic acid from triphenylmethanol — 367
Questions — 367

36 Pheromones and Insect Repellents — 369
36.1 Chemical Communication — 369

36.2 Insect Repellents   370
36.3 Preparation of *N*,*N*-Diethyl-*m*-toluamide   371
Questions   372

37 The Pinacol–Pinacolone Rearrangement   373
37.1 Introduction   373
37.2 Preparations   375
  (A) Benzopinacol by photochemical reduction   375
  (B) Benzopinacolone   376
Questions   377

38 Polycyclic Quinones   378
38.1 Quinones   378
38.2 Preparations   380
  (A) Anthraquinone   380
  (B) Phenathrenequinone   381
Questions   382

39 Enamine Synthesis of a Diketone: 2-Acetylcyclohexanone   383
39.1 The Enamine Reaction   383
39.2 Preparation of 2-Acetylcyclohexanone   386
Questions   386

40 Wagner–Meerwein Rearrangements: Camphor from Camphene   387
40.1 Introduction   387
40.2 Preparation of Camphor   388
Questions   389

41 The Diels–Alder Reaction   390
41.1 Introduction   390
41.2 Preparations   393
  (A) *N*-Phenylmaleimide   393
  (B) *N*-Phenylamaleimide adducts   393
  (C) Maleic anhydride adducts   394
Questions   395

42 Benzoquinone and Dihydroxytriptycene   396
42.1 Diels–Alder Reactions of Benzoquinone   396

42.2 Preparations and Reactions   398
    (A) *p*-Benzoquinone   398
    (B) Dihydroxytriptycene   400
Questions   400

43 Ferrocene and Acetylferrocene   **402**

43.1 Metallocenes   402
43.2 Preparations   403
    (A) Ferrocene   403
    (B) Acetylferrocene   406
Questions   407

44 Dyes and Indicators   **408**

44.1 Diazonium-Coupling Reactions   408
44.2 Preparations of Azo Dyes   410
    (A) Methyl orange   410
    (B) Para red   411
44.3 Phthalein and Sulfonphthalein Indicators   412
44.4 Preparation of *o*-Cresol Red   415
Questions   416

45 Solvatochromic Dyes   **417**

45.1 Merocyanin Dyes   417
45.2 Theoretical Basis for Solvatochromism   418
45.3 Synthesis of Merocyanin Dyes   418
45.4 Preparation and measurements   419
Questions   420

46 Sugars   **422**

46.1 Introduction   422
46.2 Monosaccharide and Disaccharide Tests   423
    (A) Test for reducing sugars   423
    (B) Osazone test   424
    (C) Acetylation of glucose   426
    (D) Benzoylation of glucose   427
Questions   427

47 Biosynthesis of Alcohols   **428**

47.1 Fermentation of Sugars   428
47.2 Ethanol by Fermentation   430
Questions   431

48 Peptides — 432

48.1 Structure 432
48.2 Biological Function 433
48.3 Synthesis of Polypeptides 433
48.4 Preparation of Phthaloylglycylglycine 435
Questions 436

Appendix — 437

Tables of Physical Data 437
In Case of Accident 442

Index — 443

1 Introduction

Laboratory work that accompanies an introductory course in organic chemistry is intended primarily to acquaint you with the principles and practice of laboratory operations used in this field and to reinforce the theoretical aspects of the subject. The experience of working with a variety of typical organic materials and observing their characteristic properties and transformations gives a sense of reality to the structural formulas and scientific names of organic molecules. Laboratory experiments can stimulate intellectual curiosity and develop powers of observation, in addition to giving training in careful and skillful manipulation. It is important to realize that none of these objectives will be achieved if the work is performed in an unenlightened routine fashion, if you follow one instruction and proceed to the next without having beforehand a general knowledge of the whole sequence of laboratory operations and the underlying principles.

1.1 General Precautions

The organic laboratory contains many hazards. Fortunately, the risks can be minimized if certain basic safety practices, described in the following paragraphs, are followed faithfully.

▶ **Safety Glasses.** Eyes are particularly vulnerable to injury from splashing droplets of corrosive chemicals and flying particles of glass or other solid fragments. Safety glasses with side shields or goggles must be worn *at all* *times* in the laboratory. Prescription glasses are not a substitute for safety glasses and must be supplemented with a pair of plastic goggles that fit over them. Goggles provide even more protection than safety glasses, and some laboratories specifically require the wearing of goggles.

Contact lenses, too, should *never* be worn in the laboratory without safety glasses or goggles. Contact lenses cannot be removed rapidly enough to prevent damage from reagents splashed in the eyes, and they offer no protection against shrapnel.

Fire Hazards. One of the chief dangers of organic laboratory work is the fire hazard associated with the manipulation of volatile, flammable, organic liquids. With few exceptions, organic liquids and vapors catch fire readily, and many organic vapors form explosive mixtures with air. Obviously, organic liquids must not be manipulated near an open flame, and precautions must be taken to avoid the escape of organic vapors into the laboratory. For general safety, you should form the habit of scanning the adjacent laboratory bench space for lighted burners before working with flammable solvents, and it is good practice to look around for fire hazards to yourself and adjacent workers before lighting a match or a burner.

The degree of flammability of organic compounds varies widely. The vapors of diethyl ether, petroleum ether, acetone, and ethanol catch fire quite readily, and the manipulation of these liquids requires careful attention at all times to fire hazards. Methylene chloride (bp 40°) is a much safer solvent. Carbon disulfide is so readily ignited (even by a hot steam pipe) that it should *never* be used by an inexperienced worker.

Chemical Burns and Cuts. Specific precautions for handling particularly dangerous chemicals are noted in the directions for procedures in which they are used, but any ordinary chemical or piece of apparatus can be dangerous if manipulated carelessly. It is important to develop a general awareness of dangers and accidents that can arise from carelessness in simple routine, operations. To cite two examples, a severe cut or laceration may result from carelessness in pushing rubber tubing over a glass tube; a severe explosion and fire may result from attempting to distill a substance in a completely closed system.

Procedures to be followed in case of accidents are given in the Appendix and also are printed on the inside back cover of this manual.

Toxic Chemicals and Mutagens. Many common organic chemicals, particularly those containing nitrogen, are toxic or even lethal when ingested in amounts as small as a few tenths of a gram. A major rule of safe laboratory practice is *never* taste any compound.

Certain organic compounds, when ingested or inhaled over a long period of time, even in minute doses, may eventually induce tumors—some benign and some cancerous. An attempt has been made in this manual to avoid use of such compounds, but this is difficult because of the limited information available in most cases. It seems probable that many relatively safe compounds have been included on the "cancer suspect" list whereas many others that should be included are not. Unfortunately, simplicity of structure or lack of active functional groups does not guarantee safety. In the absence of specific information, it is wise to minimize exposure to the vapors of all organic solvents; even pleasant-smelling vapors may be quite toxic.

Laboratory Apparel. It is desirable to protect street clothing from being soiled and damaged by chemicals and accidents of various sorts by wearing an inexpensive laboratory coat or a rubber apron. For freedom of arm movement, tight sleeves should be avoided; loose and bulky sleeves may cause overturning of fragile apparatus. Unprotected light-weight, flammable apparel constitutes a serious fire hazard in the organic laboratory. Many synthetic fabrics are soluble in acetone and other common organic solvents.

Shoes that cover at least the toes and instep must be worn in the laboratory.

Shoulder-length (or longer) hair should be tied back, especially if a burner is being used.

In experiments requiring transfer of corrosive chemicals, it is desirable to wear some type of resistant gloves. Inexpensive, disposable gloves made from polyethylene are available.[1] Rubber gloves, afford better protection, although they make laboratory operations more cumbersome.

1.2 Apparatus

One of your first activities in the laboratory will be to check in and determine if your assigned complement of glassware and hardware is present and undamaged. The rules of the laboratory governing the degree of financial responsibility for at the end of the term will vary from school to school, and you should understand your liability clearly before your begin. Carefully check the glassware for cracks and chips. It is a good idea to clean any dirty glassware to be certain that it can pass inspection. Determine that the thermometers are not broken and that the mercury thread is continuous.

Drawings of some of the more common organic laboratory apparatus are given in Figure 1.1.

---

[1] Poly gloves, distributed by Cole-Parmer, come in large and medium sizes; similar gloves are avilable from Will Scientific, Inc., in large, medium, and small sizes.

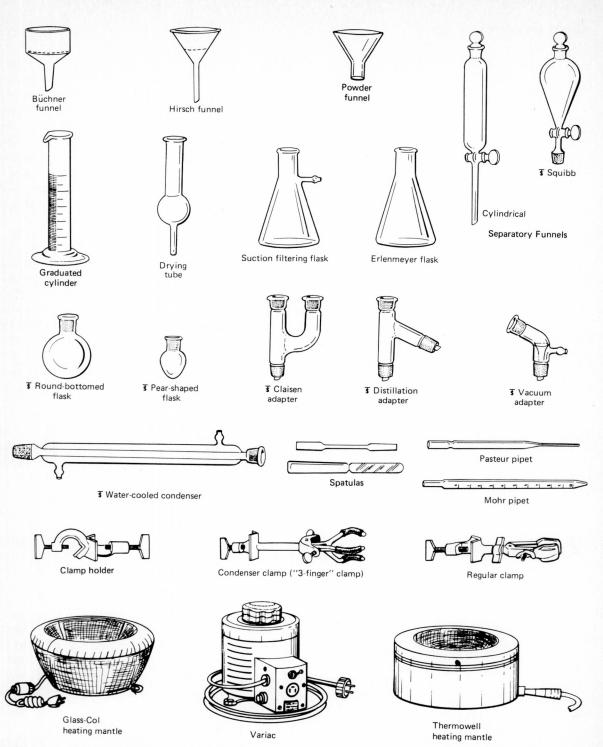

FIGURE 1.1 *Common Organic Laboratory Apparatus*

Cleaning and Drying Glassware. It is advantageous to clean laboratory glassware immediately after use, since tars and gummy matter are most easily removed before they harden. Much time is saved by having glassware clean and dry, ready for use at once. Many water-insoluble organic compounds and gums can be removed quickly and economically with scouring powder, a brush, and warm water. The use of strong acids, such as concentrated sulfuric or sulfuric–chromic acid cleaning solution, is dangerous and messy. *Nitric acid is particularly* dangerous as a cleaning agent because it reacts explosively with many organic compounds.

To remove resins and gummy material from glassware, first pour or scrape out as much material is possible, directly into a labeled waste container; *never put organic tars, paper, or other solid wastes into the sink.* Next, try to remove or loosen the resin by using a *small amount* of acetone or toluene (10–20 mL) and allowing the solvent to stand in contact with the material for 5 or 10 min. The solvent action may be hastened by warming the glassware on a steam bath (not over a flame) with care to avoid accidental ignition of the flammable solvent vapor. Do not expect tars and gums to dissolve quickly; allow ample time for the organic solvent to act. Toluene and acetone are usually good solvents for tars; ethanol is generally not effective.

To remove the remaining small amounts of tars and dirt, use scouring powder and a large test tube brush. If the brush is bent properly, it will reach the inner surfaces of flasks. The use of a little washing powder or liquid detergent followed by a good water rinse will give glassware a clean, brilliant sparkle when it dries. The best way to dry apparatus is to allow it to stand overnight on the laboratory desk. Beakers and flasks should be inverted to permit drainage; test tubes and small funnels may be inverted over crumpled paper placed in the bottom of a large beaker.

If wet glassware must be dried quickly for immediate use, it may be rinsed with one or two small portions, *not over 10 mL* of acetone, allowed to drain, and the last traces of acetone removed by drawing or passing a current of dry air through the apparatus. Methanol or ethanol may be used instead of acetone, but they evaporate less quickly. Ordinary compressed air is not suitable for drying purposes unless a good drying train is used, since the air is apt to be nearly saturated with water and may even contain suspended droplets of water or oil. It is more convenient to draw a stream of air through the apparatus by means of a glass tube connected to a suction pump.

Apparatus with Interchangeable Ground-Glass Joints. Most contemporary organic laboratory work uses apparatus having interchangeable ground-glass joints (standard taper joints, ⟁, rather than the older, but less expensive, unground glassware. The principal advantage of ground-jointed apparatus is that the joints are not affected by corrosive liquids and vapors that attack corks and rubber stoppers (chlorosulfonic acid, phosphorus trichloride, bromine, nitric acid, etc.) Reaction mixtures containing such corrosive materials may be distilled or refluxed without contamination of the product or loss of material through leakage at the joints. Dimensions of the joints have been standardized[2] ($\text{\$}$ 12/18, 14/20, etc.), so that a variety of assemblies can be set up from a small stock of standard taper flasks, condensers, and adapters.

The foremost rule for assembling ground-jointed apparatus is that the ground surfaces *must be free* of any gritty material that might score them when they are mated. It is good practice to wipe each surface gently with a lint-free cloth or tissue before assembling them.

Because of the highly precise grinding of standard taper apparatus, it is not necessary in most ordinary laboratory work to apply grease to the ground surfaces before assembly (this does not apply to stopcocks or other apparatus with ground surfaces that must be rotated during use). Grease must be applied

1. When the apparatus will be heated above 150°.[3]
2. When there is any possibility that the joint will come in contact with strongly alkaline solutions.
3. When the apparatus will be required to hold a vacuum.
4. When the surfaces will be rotated during use.

As an additional precaution, joints that have been exposed to base should be disassembled immediately after use, especially if they have been heated.

Although careful application of the foregoing rules will prevent mated joints fusing ("frozen") together, many workers prefer to apply grease to the ground surfaces in every situation. The trade-off is the risk of losing the apparatus versus the risk of contaminating the reaction with grease leached from the joint during the reaction.

A good method for applying grease is to place several small dabs on one of the two surfaces to be mated. The joint members are then placed together and rotated back and forth gently until the grease has formed a thin, continuous film between the two ground surfaces. Care should be taken to avoid excess grease, since the excess will gradually flow out of the bottom of the joint and contaminate any material with which it comes in contact. It should be remembered also that if a liquid is to be poured out of a flask having a greased joint, the surface must first be wiped clean.

Greased joints should always be cleaned thoroughly when the apparatus is disassembled. Hydrocarbon greases are readily removed by acetone,

---

[2] The joints are ground to a standard taper ($\text{\$}$) of 1-mm taper for a length of 10 mm; the maximum diameter and length in millimeters are given in that order by the designation 12/18, etc.

[3] Temperatures in this book are degrees Celsius unless otherwise specified.

carbon tetrachloride, and many other organic solvents. Silicone grease has the advantage over hydrocarbon greases of lower vapor pressure and lower solubility in most organic solvents, but if left on glass surfaces it tends to oxidize to an unsightly white film that cannot be removed and gives the appearance of a very dirty flask. At check-out time, an apparatus with such a film is likely to be rejected. To avoid the problem, it is essential that the minimum amount of silicone grease be used and that it be removed from the apparatus as soon as the experiment is completed; methylene chloride, CH_2Cl_2, is a good solvent.

A limitation of ground-jointed apparatus is that, unlike glassware assembled with corks or rubber stoppers, it has no mechanical flexibility. Special care must be exercised when a clamp is tightened onto a portion of an assembly that is jointed to another clamped member.

1.3 Weighing and Measuring Reagents

In performing laboratory experiments, it is important to weigh or measure the amounts of materials carefully and to use the *exact* quantities called for in the directions. When reagents need not be accurately measured (as in certain tests), the laboratory directions indicate approximate quantities. When an approximate quantity is indicated, as 1–2 mL, you should use a quantity within the specified limits. At first it will be advisable actually to measure the quantity used, in order to learn to judge such quantities, but after some experience you should be able to estimate approximate quantities.

In many cases the success of an operation depends on using the starting materials and certain reagents in definite amounts. It is usually advantageous and sometimes necessary to know at least fairly accurately the amount of each reagent that is present. A careful laboratory worker will acquire a habit of using solutions of known strength and of weighing or measuring the reagents and solutions used. The strengths of the common laboratory desk reagents are listed in Table 1.1 (for additional data on these reagents see the Appendix).

TABLE 1.1 Desk Reagents

Reagent	Density g/mL	Reagent concentration		
		g/100 g	g/100 mL	mole/L (approx.)
Acetic acid (glacial)	1.06	99.5	105.5	17.5
Hydrochloric acid (conc)	1.18	35.4	42	12
Nitric acid (conc)	1.42	70	100	17
Sulfuric acid (conc)	1.84	96	176	18
Sodium hydroxide solution (dil)	1.11	10	11.1	3
Ammonia solution (conc)	0.90	29	26	15

Several types of balances are used in undergraduate organic laboratories. The principal requirement is that masses as large as 100 g can be measured with an accuracy of about 0.1–0.01 g. All balances should be treated as delicate instruments and anything spilled on one should be removed immediately. Many balances use knife edges to support the balance beam, and these can be damaged easily by misuse.

Needless accuracy in weighing should be avoided, since preparative organic laboratory work is done with a precision of only about 1%. With this in mind, it is helpful to know how much weight is required to displace the balance beam pointer by a small amount from the null mark. A few minutes spent in gaining familiarity with the balance and its responsiveness can save much time later.

The balance has a zero adjustment screw, which will be adjusted before the laboratory starts. It should not be altered for the purpose of taring a weighing paper.

Interconversion of Weights and Volumes. In laboratory practice it is often necessary or desirable to convert weight measures into volume measures, and vice versa. These conversions may be made by use of the follow relationships.

$$\text{Weight (g)} = \text{volume (mL) at } t° \times \text{density (g/mL) at } t°$$

$$\text{Volume (mL) at } t° = \frac{\text{weight (g)}}{\text{density (g/mL) at } t°}$$

The numerical values of the density and the specific gravity of a particular liquid (at a given temperature) are usually so nearly equal that they may be used interchangeably for approximate calculations. Nevertheless, you should bear in mind the following accurate definitions.

The common laboratory volume unit is the milliliter (mL), which is $\frac{1}{1000}$ of a liter (L). A liter is the volume of a cube 10 cm on a side. Thus the milliliter and cubic centimeter (cm^3 or cc) are identical.

The density of a liquid is equal to the mass of a unit volume of the substance. An accurate density value includes a statement of the temperature and the units. For example, the density of water at 20° is 0.9982 g/mL, which may also be expressed by the notation $d_4^{20} = 0.9982$.

When solutions such as hydrochloric acid, concentrated sulfuric acid, and 95% ethanol are used, it is necessary to calculate the weight of the solute present.

$$\text{Weight of solute (g)} = \text{weight of solution} \times \frac{\text{g of solute}}{\text{g of solution}}$$

$$= \text{weight of solution} \times \frac{\% \text{ of solute by weight}}{100}$$

Another general statement of the equation is

$$\text{Weight of solute} = \text{volume (mL) at } t° \times \text{density (g/mL) at } t° \times \text{concentration by weight}$$

For convenience, the physical constants of solutions of the common acids and of ethanol are included in the Appendix. More complete tables may be found in chemical handbooks and in reference works.[4]

Example. Let us suppose that we wish to find the weight of pure ethanol present in 30 mL of ordinary ethanol (95% by volume or 92.5% by weight). By reference to an "alcohol table" the density of 92.5% (by weight) ethanol at 20° is found to be 0.8112 (d_4^{20}). Using the preceding equations, we have

$$\text{Weight of 30 mL of 92.5\% ethanol} = 30 \text{ mL} \times 0.8112 \text{ g/mL} = 24.34 \text{ g}$$

$$\text{Weight of pure ethanol (100\%)} = 24.34 \text{ g} \times \frac{92.5}{100} = 22.5 \text{ g}$$

1.4 Heat Sources

Many different sources of heat are used in the organic laboratory. Unfortunately, each has its limitations so that no uniform practice can be described or recommended.

The gas burner is the traditional means of heating and has the unique advantage of providing a wide range of heating that can be altered quickly if the situation requires it. Gas burners have the obvious disadvantage of an open flame, which requires constant awareness of any flammable vapors nearby. In practice burners are not as dangerous as they might seem because the fire hazard is so apparent that proper precautions are taken instinctively.

Modern gas burners allow control of both the gas flow (needle value at the burner base) and air flow (rotatable ring at the bottom of the burner barrel). To a first approximation the gas flow controls the flame size and the air flow controls the flame temperature (more air gives a hotter flame; more gas gives a bigger flame).

---

[4] Such as the Chemical Tables in the *Handbook of Chemistry and Physics* (Boca Raton, FL: CRC Press, Inc.); *Physical Properties of Chemical Compounds*, Vols. I–III (cumulative name index in Vol. III); American Chemical Society, *Advances in Chemistry Series*, Vols. 15, 22, and 29.

Proper operation of a burner is simple. To ignite the burner, turn off the air flow; turn the gas on part way and ignite with a lighted match held just below the top of the chimney barrel. Admitting air causes the large luminous flame to become increasingly blue and eventually form the desired blue "flame cone," which is resistant to drafts. If too much air is admitted, the flame may "flash back" into the barrel and burn inside at the needle jet. If that happens, turn the gas off, and relight the burner using either less air or more gas. Too much gas or air may cause the flame to extinguish.

Steam baths are suitable for heating or distilling when the required temperature is less than 100°. Vessels placed on top of the bath will usually reach only a temperature of about 90° because of heat losses from the sides of the vessel. A general precaution to be observed in using a steam bath is to avoid excess steam. Not only are clouds of steam unnecessary and annoying, but condensation of escaping steam may introduce moisture into the reaction vessel, which could be deleterious for some reactions.

For temperatures above 90°, an oil bath heated either electrically or by flame works well. Because of the high heat capacity of the oil, an oil bath provides a very steady source of heat. For the same reason the temperature cannot be lowered quickly if required, so that some care in increasing the bath temperature is called for. A solution is to mount the bath on a "Lab-Jack," which allows rapid adjustment of height. A less expensive, but more hazardous, solution is to use wooden blocks cut from "2 × 4s" or plywood. When a flame is used to heat the bath particular care must be taken not to ignite the hot oil vapors. In many laboratories oil baths are heated by a coil of resistance wire placed directly in the bath. This method, because of the electrical shock hazard, has been banned by OSHA.[5] Use of electrically insulated immersion heaters is legal and practical.

For very high temperatures, a sand bath heated by a flame has many desirable features. In addition to being inexpensive and a steady source of heat, sand, unlike oil, is not subject to fire hazards and does not decompose on prolonged heating.

Electrical hot plates, particularly those with built-in magnetic stirring motors, are widely used in research laboratories. Caution in their use is called for because solvent vapors can ignite if they come in contact with the hot surface. A hidden danger for the unwary is the use in many hot plates of "on-off" temperature regulators that produce sparks as the contacts open and close. Unless the contacts are properly sealed, the sparks can ignite combustible vapors.

Another widely used heating device is the Glas-Col heating mantle,[6]

---

[5] The Occupational Safety and Health Administration is a federal body charged with the responsibility of setting safety and health standards in the workplace and has the authority to enforce its regulations.

[6] Glas-Col mantles are available from most laboratory supply houses.

constructed by weaving nichrome wire heating elements into fiberglass cloth to form a hemispherical jacket that fits snugly against the lower part of the flask. Loose-fitting mantles tend to overheat, which is dangerous; a different mantle must be used for each size of flask. Other shapes are available for special types of apparatus. Current OSHA regulations require that the outside case be grounded.

Closely related to the Glas-Col mantle is the Thermowell[7] heating mantle, in which the heating element is imbedded in a rigid hemispherical shell fabricated from a refractory material. In concept, the Thermowell unit is intermediate between a hot plate and a Glas-Col mantle. A snug fit is not required so that one unit can be used with several sizes and shapes of flasks.

Some electrically heated devices contain their own current controllers. More frequently a separate external controller is required. Usually, a Variac or Powerstat is used; these are autotransformers capable of converting 110-V alternating current into any voltage in the range of 0–130 V by the twist of a dial on the top of the case. Autotransformers, when kept clean, do not spark and do not pose a fire hazard.

1.5 Stirring

In the organic laboratory, stirring is often needed to hasten solution or reaction by bringing a solid and liquid, or two immiscible liquids, into good contact. Stirring of a homogeneous solution is advantageous only when one desires to bring the liquid into good contact with the walls of the vessel to render external heating or cooling more effective and to insure uniform temperature throughout the liquid. Often the movement of a boiling liquid is sufficient to give good mixing.

For ordinary laboratory preparations, satisfactory mixing is generally accomplished by intermittent shaking by hand; a circular swirling motion is most effective and reduces the danger of splashing. Occasionally a rather violent shaking is required, as when a heavy solid like metallic zinc or iron must be brought into contact with an organic liquid. For material in a beaker, sufficient mixing can usually be obtained by stirring by hand with a glass rod or wooden paddle.

Mechanical stirring is usually needed for large-scale preparations and in any operation where continuous stirring is required for a long time. Descriptions of stirring assemblies with motor-driven stirrers may be found in laboratory manuals for advanced work (see also Section 8.5).

---

[7] Thermowell units are available only from Laboratory Craftsmen, P.O. Box 148, Beloit, WI, 53511.

1.6 Laboratory Notes[8]

It is essential to have a suitable notebook in which to record directly the observations made during experiments and to assemble information that will aid in their performance. For this purpose, obtain a stiff-covered bound notebook, about 8×10 in., preferably with cross-ruled paper (to make the preparation of tables of physical constants required in the later experiments easier). It is convenient to use a notebook with numbered pages and tear-out carbon-copy duplicate pages. The use of spiral or loose-leaf notebooks for laboratory records is not satisfactory, and the recording of observations on loose sheets or scraps of paper is not permissible. Notebook entries should be made in ink, and any corrections that are necessary should be made by adding notes rather than erasing.

Before you come to the laboratory, the following steps should be carried out for the exercises on separation and purification. The schedule of exercises will be announced beforehand, so that you will have an opportunity to prepare your notebook and be ready to start laboratory work at the beginning of the laboratory period.

1. Read the descriptive pages concerning the laboratory operation to be carried out (these are found immediately preceding each experiment). In your notebook, write a title and general statement of the process to be studied.
2. Read the laboratory directions *for the entire experiment*, note particularly the cautions for handling materials, and consider the reasons for the procedure to be followed. In your notebook jot down any points that require special observation or reminders of specific details.
3. Write the names and formulas of the compounds to be used, and where chemical tests are to be made, write equations for the reactions.

Careful planning of laboratory work is essential. Effective use of laboratory time requires that you know in advance just what you are going to do in the laboratory. Instead of watching idly while a liquid is being heated for an hour or more, you can use periods *when full attention is not required* to conduct another experiment, to clean apparatus, or to prepare for subsequent work.

When you come to the laboratory, proceed as follows.

1. Arrange the apparatus for the experiment and secure the approval of the laboratory instructor for the setup.

---

[8] The following specific directions for the preparation of notebooks and the general laboratory procedure are based on those that have been used in the elementary courses in organic chemistry at Cornell University. For the particular conditions that pertain in other laboratories, the instructor may wish to alter these directions or substitute others.

2. Perform the experiment according to the laboratory directions and record observations directly in your notebook. When the exercise has been completed, dismantle the setup and immediately clean the glassware and apparatus, as described in Section 1.5.
3. Write answers to the questions given at the end of the chapter. Make complete statements in answering the questions.

I SEPARATION AND PURIFICATION OF ORGANIC COMPOUNDS

2 Simple and Fractional Distillation

Since organic compounds do not usually occur in a pure condition in nature and are accompanied by impurities when synthesized, the purification of materials forms an important part of laboratory work in chemistry. Four general separation procedures are used frequently in organic work in university laboratories and in industry: distillation, chromatography, crystallization, and extraction. Sublimation is used occasionally, and various special techniques such as electrophoresis and zone-refining are available for advanced work. The process used in any particular case depends upon the characteristics of the substance to be purified and the impurities to be removed. To select the most appropriate process and to employ it effectively, you must understand the principles involved as well as the correct methods of manipulation.

Simple and fractional distillations have been placed first because they depend clearly on the physical concept of vapor pressure, which is needed to understand the other separation techniques to be described. Distillation also happens to be one of the most commonly used purification methods.

The objective of distillation is to separate a mixture of two or more materials that differ in ease of vaporization. The next section explores that process.

2.1 Principles of Distillation

Boiling. In a liquid the molecules are in constant motion and have a tendency to escape from the surface and become gaseous molecules, even at temperatures far below the boiling point. When a liquid is placed in an

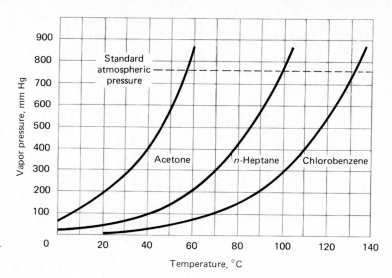

FIGURE 2.1
Temperature Variation of Vapor Pressure

enclosed space, the pressure exerted by the gaseous molecules increases until it reaches the equilibrium value for that particular temperature. The equilibrium pressure is known as the vapor pressure and is a constant characteristic of the material at a specific temperature. Although vapor pressures vary widely with different materials, vapor pressure always increases as the temperature increases (Figure 2.1). The vapor pressure is commonly expressed as the height, in millimeters, of a mercury column that produces an equivalent pressure (mm Hg). The addition of soluble substances to a pure compound alters the measured vapor pressure.

In ordinary glass flasks, there are microscopic pockets of air trapped in the pores and crevices of the walls. With a liquid in the container the pockets are filled with vapor of the liquid at its equilibrium vapor pressure. When the temperature of the liquid is raised, the vapor remains compressed until the vapor pressure exceeds the applied pressure (the pressure at the liquid surface plus the hydrostatic pressure), whereupon the trapped vapor rapidly expands to form bubbles that rise to the surface and expel their vapor. The resulting agitation (boiling) churns more air bubbles into the liquid where they continue the process after receiving new charges of vapor. Liquids heated in containers that have been degased do not boil, although they vaporize explosively if heated to a sufficiently high temperature. To avoid the hazards associated with sudden, irregular boiling (bumping) a dependable source of bubbles should always be introduced into a flask before its contents are heated to boiling. When a liquid is boiled at atmospheric pressure, the bubble source is customarily a boiling chip (see Section 2.3); with vacuum distillations boiling chips do not work as reliably and other sources are frequently used (see Chapter 3).

Boiling Point and Boiling Temperature. The boiling point of a liquid is defined as the temperature at which its vapor pressure equals the atmospheric pressure. By convention, boiling points are reported in the scientific literature at a pressure of 1 atm except when otherwise specified. The boiling temperature is the actual observed temperature when boiling occurs and is generally a few hundredths to a few degrees above the boiling point because of experimental difficulties involved in the measurement.

Distillation of a Pure Compound. Distillation consists of boiling a liquid and condensing the vapor in such a manner that the condensate (distillate) is collected in a separate container. A simple apparatus assembly for this operation is shown in Figure 2.10.

When a pure substance is distilled at constant pressure, the temperature of the distilling vapor will remain constant throughout the distillation, provided that sufficient heat is supplied to insure a uniform rate of distillation and superheating is avoided. In actual practice these ideal conditions are not realized; drafts in the laboratory can cause momentary condensation of vapors before they reach the thermometer. A certain amount of superheating of vapors occurs almost invariably under ordinary conditions. Because of these contrary effects a distillation range of 1° actually represents an essentially constant boiling point. With somewhat more refined apparatus and technique a distillation range of 0.1° can be observed for a pure compound.

The temperature reading of a thermometer *in the distilling vapor* represents the boiling point of that particular portion of the distillate. This temperature will be the same as the boiling point of the liquid in the distilling flask only if the distilling vapor and the boiling liquid are identical in composition. Since a pure liquid fulfills this condition, a constant thermometer reading is sometimes used as a criterion of purity of a liquid. It should be noted, however, that certain mixtures (such as azeotropes—page 22) give constant thermometer readings. Occasionally two liquids have such similar boiling points that no appreciable change in thermometer reading will be observed when a mixture of them is distilled.

Ideal Solutions. The pressure and composition of vapor above an ideal mixture of liquids at a given temperature can be calculated if the composition of the mixture and the vapor pressures of the pure components are known. The total pressure is the sum of the partial vapor pressures of all components. The partial pressure of each component is given by Raoult's law.

$$p_A = p_A^\circ X_A \tag{2.1}$$

where p_A (the partial pressure of A) is the vapor pressure of A above the mixture, p_A° is the vapor pressure of pure A, and X_A is the mole fraction of component A in the mixture. Because there is a fixed number of molecules in a mole, Raoult's law states in molecular terms that the vapor pressure of A above a solution is proportional to the mole fraction of the molecules of A in the liquid. Application of Raoult's law to the two-component mixture of carbon tetrachloride and toluene is illustrated graphically in Figure 2.2.

The composition of the vapor, with respect to each component, can be calculated from Dalton's law.

$$Y_A = \frac{p_A}{\text{total vapor pressure}} = \frac{p_A}{p_A + p_B + p_C + \cdots} \qquad (2.2)$$

where Y_A is the mole fraction of component A in the vapor. Dalton's law and Raoult's law together show that for an ideal mixture at any temperature the most volatile component has a greater mole fraction in the vapor than in the solution. In terms of the previously defined symbols, if A is the most volatile component of the mixture, Y_A is greater than X_A.

The boiling point of a mixture is defined as that temperature at which the *total* vapor pressure equals the pressure above (or on) the solution. From Raoult's law (also see Figure 2.1) it is apparent that the total vapor pressure of an ideal mixture is intermediate between the vapor pressures of the pure components. This means that the boiling point also will be intermediate between the boiling points of the pure substances. The general dependence of boiling point on composition of ideal binary mixtures resem-

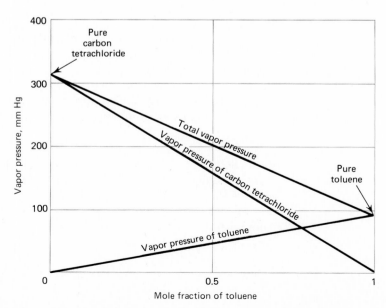

FIGURE 2.2
Graphical Application of Raoult's Law

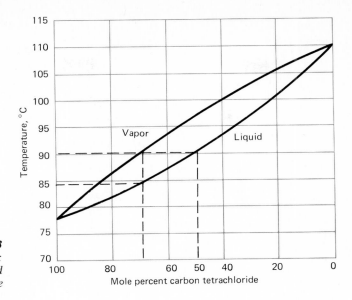

FIGURE 2.3
Boiling Point Diagram:
Carbon Tetrachloride and
Toluene

bles that depicted in Figure 2.3 for the specific system of carbon tetrachloride and toluene. The boiling point of any particular mixture is obtained by erecting a vertical line from the horizontal composition axis until it intersects the *liquid* curve. For example, from Figure 2.3 it will be found that a 50-mole-percent (0.5 mole fraction in carbon tetrachloride) mixture of carbon tetrachloride in toluene boils at 90°. A thermometer placed *in the boiling liquid* would read 90°, but a thermometer placed in the distillation head, as shown in Figure 2.10, would read lower, since it is in contact with the condensing vapor, which is richer in the lower-boiling component.

The composition of the vapor in equilibrium with any particular liquid composition is obtained from Figure 2.3 by projecting a horizontal line from the vertical intersection of the *liquid* curve over to the *vapor* curve and from that intersection back vertically to the composition axis. The vapor above a 50-mole-percent solution of carbon tetrachloride in toluene contains 71 mole percent of carbon tetrachloride. Figure 2.3 demonstrates graphically the previous conclusion that the vapor is richer in the lower-boiling, more volatile component. The reading of a thermometer placed in the distillation head would be the boiling point of the 71-mole-percent mixture, which can be seen from Figure 2.3 to be close to 84°.

The boiling point and vapor composition calculated in this manner apply only to the initial state of a distillation. Because of the higher concentration of carbon tetrachloride in the vapor compared to the liquid remaining in the boiler, the composition of the liquid *gradually* shifts toward pure toluene as the distillation proceeds; the boiling point, reflecting this composition change, climbs gradually also. The actual rate of change of boiling

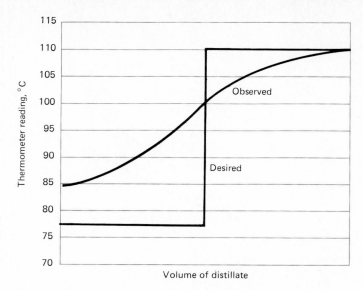

FIGURE 2.4
Distillation Curve for a 50 : 50 Carbon Tetrachloride–Toluene Mixture

point depends on how rapidly the mixture is distilled, but a typical set of observations would resemble those of Figure 2.4, obtained with a 50 : 50 (by volume) carbon tetrachloride–toluene mixture. A 50 : 50 mixture by volume corresponds to 0.52 mole fraction in carbon tetrachloride and can be found from Figure 2.3 to have a boiling point of about 88° and an initial distillation temperature of about 84°. The technique of fractional distillation, to be discussed later in this chapter, is a method for more nearly approaching this perfect separation.

Nonideal Solutions and Azeotropes. Many actual solutions depart widely from Raoult's ideal law. An expression for vapor pressure that encompasses both ideal and nonideal behavior is

$$p_A = f_A X_A \tag{2.3}$$

where p_A is the vapor pressure of A above the mixture, X_A is the mole fraction of component A in the mixture, and f_A is the effective vapor pressure of A. If the solution is ideal, $f_A = p_A^\circ$. If the solution shows negative deviations from Raoult's law, $f_A < p_A^\circ$; that is, component A behaves in solution as though the vapor pressure of pure A were less than it actually is. The analogous description applies for positive deviations from Raoult's law for which $f_A > p_A^\circ$.

The methanol–water system is typical of those that show positive deviation from Raoult's law (Figure 2.5). The boiling point composition curve for methanol–water mixtures shown in Figure 2.6 also reflects the nonideal

Sec. 2.1] PRINCIPLES OF DISTILLATION

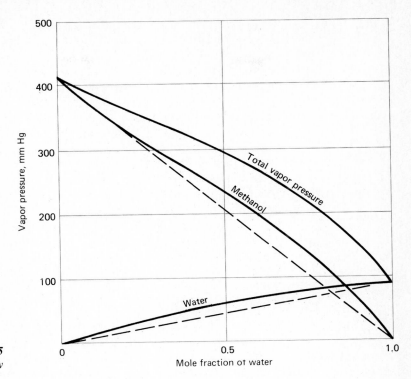

FIGURE 2.5
Deviations from Raoult's Law

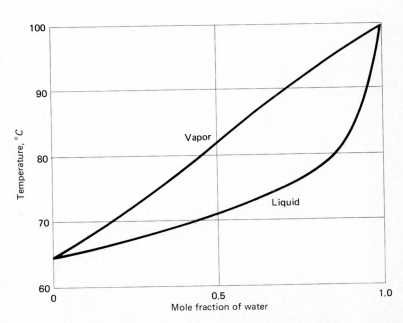

FIGURE 2.6
Boiling Point Diagram:
Methanol–Water

behavior by its distorted shape (compare with Figure 2.3). The origin of the positive deviations for methanol–water mixtures is the disruption of hydrogen bonding between the hydroxyl groups, which leads to enhanced escaping tendencies for the two components. In contrast, there are some binary mixtures in which the two components attract each other particularly strongly and cause the vapor pressure to be lower than ideal (negative deviations from Raoult's law). Other mixtures show positive deviations because the molecules of one component are attracted to each other more strongly than they are to molecules of the other component.

Frequently, the deviations from ideality are so extreme that boiling point–composition diagrams have a maximum or a minimum (Figure 2.7). If a mixture showing this extreme behavior has the composition corresponding to the extreme boiling point (an azeotropic mixture), it will behave like a pure liquid and show a constant boiling point. The components of an azeotropic mixture cannot be separated by ordinary distillation processes because the vapor in equilibrium with the liquid has the same composition as the liquid itself. Table 2.1 gives the composition and boiling point of several examples of binary azeotropic mixtures. Azeotropic mixtures containing three components (ternary systems) are also encoun-

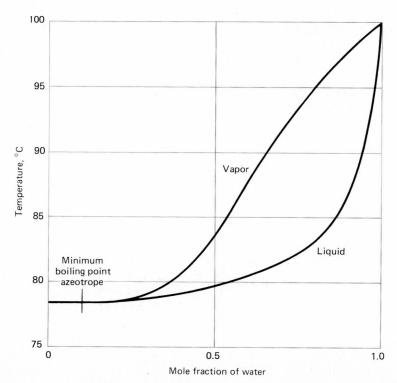

FIGURE 2.7
Boiling Point Diagram: Ethyl Alcohol–Water

TABLE 2.1 Binary Azeotropic Mixtures

Component A		Component B		Azeotropic mixture		
Substance	Boiling point, °C	Substance	Boiling point, °C	Percent of A (weight)	Percent of B (weight)	Boiling point °C
Acetone	56.4	Chloroform	61.2	20	80	64.7 (max.)
Nitric acid	86.0	Water	100.0	68	32	120.5 (max.)
Formic acid	100.7	Water	100.0	77.5	22.5	107.3 (max.)
n-Propyl alcohol	97.2	Water	100.0	71.7	28.3	87.7 (min.)
t-Butyl alcohol	82.5	Water	100.0	88.2	11.8	79.9 (min.)
Ethanol	78.3	Water	100.0	95.6	4.4	78.1 (min.)
Ethanol	78.3	Chloroform	61.2	7	93	59.0 (min.)
Ethanol	78.3	Toluene	110.6	68	32	76.7 (min.)
Acetic acid	118.5	Toluene	110.6	28	72	105.4 (min.)

tered; for example, benzene–water–ethanol or ethanol–water–ethyl acetate give minimum-boiling-point azeotropic mixtures.

The effect of even more extreme deviations from Raoult's law will be considered in connection with steam distillation.

2.2 Fractional Distillation

The common use of the term *fractional distillation* refers to a distillation operation where a *fractionating column* has been inserted between the boiler and the vapor takeoff to the condenser (see Figure 2.12). The effect of this column, when properly operated, is to give in a single distillation a separation equivalent to several successive simple distillations. This represents a considerable saving of time and makes the selection and proper operation of fractionating columns an important subject for chemists.

Fractionating Columns. The easiest approach to understanding the principles by which fractionating columns give their superior separations is to consider first a rather special type of column known as a *bubble plate column*. The essential features of a bubble plate column, illustrated in Figure 2.8, consist of (1) a series of horizontal plates, A, which support a layer of distillate; (2) capped risers, B, through which the distilling vapors ascend; and (3) overflow pipes, C, which return any excess distillate to the next lower plate. At the beginning of a distillation, the vapors coming up from the boiler pass through the first riser and are deflected downward by the cap onto the first plate, where they are condensed. As simple vaporization and condensation continue, the rising vapors are forced to bubble through the liquid on the plate. The liquid level rises to the top of the overflow tube

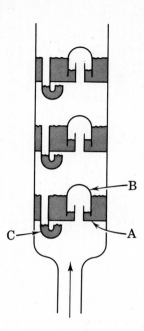

FIGURE 2.8
Bubble Plate Column

and then flows downward to the boiler. The liquid on the first plate corresponds to the first fraction in a simple distillation—it is enriched in the lower-boiling component. It follows that the temperature of the vapor bubbling through the liquid is above the boiling point of the liquid on the plate; through heat exchange the liquid is brought to its boiling point and its vapor rises to the second plate where the same processes are repeated. As the distillation continues, each plate becomes filled with a layer of liquid whose composition is that of the vapor rising from the next lower plate. Under ideal circumstances each plate achieves an increment of separation equivalent to one simple distillation.

The overflow tubes serve a more important function than just acting as returns for excess condensate. Since the vapor leaving any plate is richer in the lower-boiling component than the vapor entering the plate, the higher-boiling materials tend to accumulate on the plate. The overflow returns this higher-boiling material to the lower plate, so that an equilibrium balance of low-boiling to high-boiling components is maintained. In effect, vapor and condensate are passing in opposite directions through the column: the more volatile component ascends the column in the vapor stream, while the less volatile components descend. The counterflow is essential for effective separation in a fractionating column.

The separation process can be understood more clearly by reference to a liquid–vapor composition diagram such as that shown in Figure 2.9 for carbon tetrachloride (bp 77°) and toluene (bp 111°). A liquid mixture containing 50 mole percent carbon tetrachloride (point A on the liquid line) is in equilibrium with vapor containing 71 mole percent carbon tetrachloride (point A' on the vapor curve). If liquid with composition A is partially vaporized and the vapor with composition A' condensed completely on the first bubble plate, the condensate is represented by B (on the liquid line). Repetition of the vaporization–condensation process with liquid B yields a new distillate, C, containing 85% carbon tetrachloride, which condenses on the second bubble plate. Each successive bubble plate achieves an additional increment of separation.

Bubble plate columns have the drawback of requiring large samples for effective operation, and a substantial portion of material is withheld on the plates (*holdup*). To overcome these disadvantages, small-scale laboratory fractionations are usually done with cylindrical columns packed with materials having large surface area (glass beads or helices, small sections of twisted metal, Carborundum chips, and the like). The principles of operation of *packed columns* are quite similar to those of the bubble plate column. The layers of packing material, like the bubble plates, serve as support for films of condensate; vapor passing through the layers is enriched in the lower-boiling component, and the higher-boiling components move downward to lower layers. The scrubbing action of the packing material effects the counterflow of vapor and condensate that is essential for fractionating efficiency.

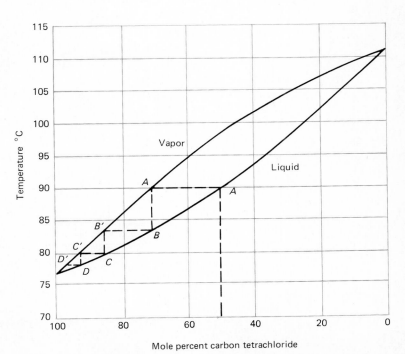

FIGURE 2.9
Boiling Point Diagram:
Carbon
Tetrachloride–Toluene

Relative Efficiency of Fractionating Columns. Since column packings differ widely in efficiency, it is desirable to have a means of comparing their effectiveness for separating mixtures. The enrichment factor, α, which relates the relative volatility of two components of a mixture, is the ratio of their effective volatilities, which, from equations (2.2) and (2.3), can be expressed as the quotient of the ratio of the mole fractions of the components in the vapor to the ratio of their mole fractions in the liquid.

$$\alpha = \frac{f_1}{f_2} = \frac{Y_1/Y_2 \text{ (vapor)}}{X_1/X_2 \text{ (liquid)}} \tag{2.4}$$

A theoretical plate is the unit of separation corresponding to the difference in composition, α, that exists at equilibrium between a liquid mixture and its vapor. This concept may be illustrated by considering a 50:50 mole-percent mixture of carbon tetrachloride and toluene. The vapor in equilibrium with the liquid (bp 90°) contains 71 mole percent carbon tetrachloride and 29 mole percent toluene. This amount of enrichment corresponds to one theoretical plate.

$$\alpha = \frac{71/29}{50/50} = \sim 2.5$$

The length of packed column required to obtain this degree of separation in the mixture is known as the height equivalent to a theoretical plate (usually abbreviated HETP). The smaller the value of the HETP, the more efficient the column is. Although the exact HETP of any given packing depends on operating factors (diameter of the column, density of packing, rate of distillation, etc.), it is useful to have rough estimates of relative values. Table 2.2 records representative values of HETP for several packings as measured under normal working conditions using student apparatus to separate a benzene–toluene mixture. Also shown in Table 2.2 are representative values of the column holdup per plate. Both the HETP and the holdup values will vary with the manner of packing and subsequent treatment of the column.

In addition to packed columns, special columns are available that achieve mixing of the ascending vapor and the descending condensate by

TABLE 2.2
HETP for Common Packing Materials[a]

Packing	HETP, cm	Holdup/plate, g
Carborundum chips	6	1.2
Glass beads	8–9	0.9
Glass helices	4–5	0.6
Metal helices	8–9	0.9
Metal sponge	30	1.6

[a] Obtained with 25-cm packed column using a benzene–toluene mixture.

their special construction. One of the simplest, least expensive, and most widely used is the Vigreux column illustrated in Figure 2.12. Under normal working conditions the Vigreux column has a relatively low efficiency (high HETP of ~10 cm), but its low resistance to vapor flow permits a large throughput (volume of distillate per unit of time) that makes the column well suited to distillation of bulk solvents. Because of its small surface area the column has a low holdup and is sometimes used for preliminary purification of small samples.

The spiral wire column is also widely used. It consists of a wire wound spirally on a glass rod that is held concentrically within an outer glass tube. Spiral wire columns are slightly more efficient than columns packed with glass beads (HETP of ~2 cm) and have about half the holdup of a packed column capable of the same throughput. Its limitation is throughput, which is essentially fixed (~0.5 mL/min maximum), whereas packed columns can be scaled up as needed. Because of their simple construction, spiral wire columns are generally built in the laboratory rather than purchased.

A unique column deserving special mention is the "Auto Annular Still."[1] This unit, resembling a spiral wire column, contains an annular Teflon helix wrapped around a Teflon rod. A motor spins the Teflon helix at high speeds, which mixes the ascending vapors extremely effectively with the descending condensate. It is claimed that the 60-cm column (115 cm with accessories attached) produces more than 150 theoretical plates with a holdup of less than 0.5 mL and a throughput of 15–60 mL/hr. This remarkable unit unfortunately costs several thousand dollars.

Separation Efficiency. The total number of theoretical plates, n, present in a column is equal to the height of the packed portion of the column divided by the HETP of the packing material. The composition of the vapor at the top of the column, (Y_1/Y_2), is related to the composition in the boiler, (X_1/X_2), in the following way.

$$\frac{Y_1}{Y_2} = \alpha^{n+1}\left(\frac{X_1}{X_2}\right) \tag{2.5}$$

The exponent of α is $n + 1$, rather than n, because in vaporizing the mixture in the boiler an additional enrichment factor is introduced. Although this equation has theoretical significance, it is more practical to have an expression for the number of theoretical plates required to separate a given mixture. An approximate expression (equation 2.6) has been derived for fractional distillation of 50:50 mixtures, such that the first 40% of the material distilled will have an average purity of 95% in the lower-boiling

---

[1] This column, introduced by the Nester Faust Mfg. Corp., is now manfactured and distributed by the Perkin–Elmer Corp., Norwalk, CT 06856.

component. Equation (2.6) shows that as the relative volatility, α, approaches unity, the number of theoretical plates required to achieve 95% purity rises steeply.

$$\frac{\text{Number of theoretical plates required}}{\text{to achieve a standard separation}} = \frac{1.53}{\log_{10} \alpha} \quad (2.6)$$

This relationship becomes still more useful (but also more approximate) if one substitutes for log α an expression involving the difference in boiling points of the two components. Equation (2.7) gives results for ideal mixtures being fractionally distilled under perfect conditions.[2] Under practical conditions, discussed in the next section, the number of plates required can double. It is clear that tall high-platage fractionating columns are required for clean separation of materials boiling a few degrees apart. When the required number of theoretical plates is unavailable for practical reasons, it is necessary to collect a smaller portion of the low-boiling distillate.

$$\frac{\text{Number of theoretical plates required}}{\text{for standard separation}} = \frac{120}{T_2 - T_1} \quad (2.7)$$

Reflux Ratio and Holdup. Equations (2.5) and (2.6) were derived for an ideal fractional distillation where there is equilibrium between the rising and descending counterflowing streams of materials. For this equilibrium to be attained, it is essential that vapor reaching the top of the column be condensed and liquid returned to the column (*reflux*). If a large portion of the vapor reaching the top of the column is removed as a distillate (*takeoff*), the equilibrium is seriously disturbed and much lower separation efficiency results. The extreme modes of operation are known as *total reflux* and *total takeoff*. Since the first mode yields no distillate and the second gives distillate of much lower purity than is possible with the column, in practice some intermediate ratio of takeoff to reflux is employed. The best practical compromise is to adjust the reflux ratio so that it equals the number of theoretical plates of the column. Higher rates of collecting distillate (lower reflux ratios) give poorer separations; slower rates are overly time-consuming and do not provide significantly better separations.

Another factor that seriously affects separation efficiency is the total

---

[2] Equations (2.6) and (2.7) differ from those given in the literature by not including the effect of partial equilibration when the reflux ratio is small. The present equations were derived by assuming a constant α and an average boiling point of 100°. They give the theoretical minimum required number of plates without regard to practical factors. Liquid mixtures showing deviations from ideality may require fewer plates (for positive deviations) or more plates (for negative deviations).

amount of liquid and vapor in the column at any instant (holdup). A great drop in separation efficiency occurs if the holdup is more than about 10% of the amount of sample to be distilled.

Under practical conditions of partial takeoff and some holdup, the number of plates required to achieve a given separation can be twice as many as predicted by equation (2.7). If the number of plates available is insufficient, the distillation rate must be slowed to more nearly approach ideal conditions.

2.3 Laboratory Practice

Apparatus for Simple Distillation. A simple distillation apparatus suitable for distillation of samples greater than 5 mL in volume is shown in Figure 2.10. This consists of a round-bottom flask connected by means of a distillation adapter to a water-cooled condenser. A thermometer is held in place in

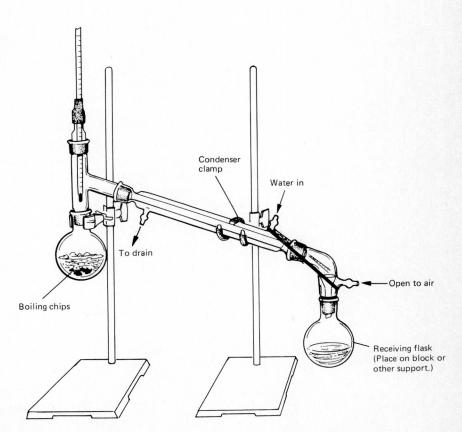

FIGURE 2.10
Apparatus for Simple Distillation

the vertical arm of the distillation adapter by a special rubber connector[3] at a height adjusted so that the top of the mercury bulb is *even* with the bottom of the opening of the side arm. A vacuum adapter is connected to the lower end of the condenser.

The distilled liquid is collected in a clean, dry receiver, commonly a round-bottom flask with its ground-glass joint mated to the lower joint of the distillation adapter. It is permissible to use an Erlenmeyer flask or a graduated cyclinder as a receiver if vapor losses and fire hazards are minimized by inserting the lower end of the adapter well into the mouth of the receiver. *A distilling assembly must have an opening to the atmosphere* to avoid development of a dangerously high pressure within the system when heat is applied. When a mated round-bottom flask is used as the receiver, the side arm on the distillation adapter becomes the opening, and this arm *must not be sealed.*

If a burner is used as a heat source, the distilling flask should rest on a piece of wire gauze (preferably one with a heat-dispersing center),[4] which is supported on an iron ring. The main purpose of the wire gauze is to prevent superheating and decomposition of the liquid or vapor that might result from heating the sides and upper portion of the flask. An alternative, used almost exclusively in advanced work, is to heat the flask in an oil bath or with a heating mantle (see also Section 1.4). Hot oil baths have the advantage of being more even heat sources, but they can cause severe burns if the hot oil spills on the skin. Heating mantles are safer if they are grounded properly and used with the proper electric-current-controlling device.

The size of the distilling flask chosen should be such that the material to be distilled occupies between one-third and two-thirds of the bulb. If the bulb is more than two-thirds filled, there is danger that some of the liquid may splash into the distillate. If the bulb is less than one-third filled, there will be an unnecessarily large loss resulting from the relatively large volume of vapor required to fill the flask. This loss is particularly serious with compounds of high molecular weight. The only exception to the one-third to two-thirds rule is with liquids, such as cyclohexanol, that foam badly on distillation. These liquids require a much larger distillation flask to contain the foam.

Whether a pure compound or a mixture is distilled, a small portion of liquid will always be left in the flask upon cooling. The flask containing the

---

[3] The thermometer can be mated to the vertical arm by means of a short length of soft rubber tubing. However, this practice is hazardous because, if the thermometer is to be held snugly, the tubing must be forced over the much larger adapter arm.

[4] There is more danger of cracking glass apparatus by heating on a *plain* wire gauze with a hot flame than by direct impingement of the flame on the glass, as local hot spots having a temperature of almost 1000° can be developed by the flame of a Bunsen burner directed against a plain metal gauze [see Wooster, *J. Chem. Educ.*, **18**, 196 (1941)]. Gauzes with asbestos centers are suitable, but because of the hazards in manufacturing asbestos, gauzes with ceramic centers are preferred.

material to be distilled should *never* be heated to dryness with a flame since there is a possibility that the flask will crack.

Apparatus for Fractional Distillation. Successful fractional distillation demands a column with an adequate number of plates. As simplified as it is, equation (2.7) is a useful guide to the minimum required number. Estimation of the desired number of plates requires knowledge of the composition and boiling behavior of the mixture to be separated. When this information is lacking, it is desirable to run a preliminary simple distillation and to plot a graph showing the actual relation of distillation temperature to volume of distillate.

It may happen that none of the available fractionating columns has an adequate number of plates. In this case it will be necessary to separate the mixture into a number of fractions of progressively higher boiling point and to refractionate these separately in a systematic way until an acceptable separation of the components is achieved.

In the simpler and less efficient columns (Figure 2.11), the returning liquid is provided merely by atmospheric cooling of the vapor in the upper portion of the column (but the main portion should be insulated to avoid excessive heat loss). In more effective columns, an adequate quantity of refluxing liquid is obtained by placing a partial condenser (cold finger) at the head of the column, which is adjusted to give the desired ratio of distillate to reflux.

The upper portion of a relatively efficient and inexpensive packed column is shown in detail in Figure 2.12. The head of the column is equipped with a reflux condenser and an adjustable stopcock for controlling the reflux ratio. The column is surrounded by a glass jacket on which is wound a spiral heating wire that is heated electrically so as to minimize the heat loss from the column. The temperature of the upper section of the column is indicated by a thermometer taped to the outside of the packed column. To further reduce and even out the heat loss from the column, the heating jacket is surrounded by another glass tube.

Increasing the length of a column or the reflux ratio improves the efficiency of separation, but care must be taken to avoid flooding the column with liquid. Flooding diminishes the contact area between vapor and liquid, and the pressure of ascending vapor may force the liquid upward in the column. To obtain good heat exchange between vapor and liquid, and to prevent flooding, a column should be well insulated. For liquids that distill below 100° a wrapping of glass wool is usually sufficient; for higher-boiling liquids or very long columns, an evacuated or electrically heated jacket may be used.

In a packed column it is essential to leave sufficient free space for the countercurrent flow of liquid and vapor. With packing materials like Car-

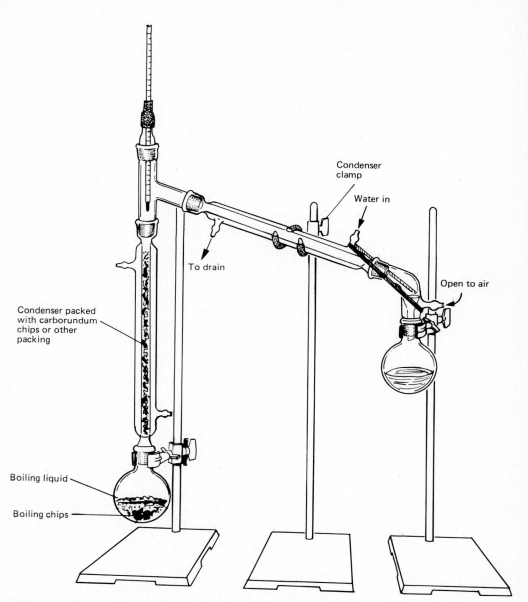

FIGURE 2.11 *Apparatus for Fractional Distillation (with packed column)*

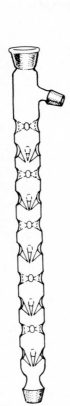
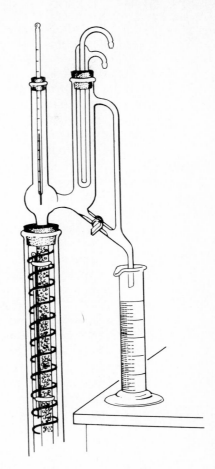

FIGURE 2.12
Special Devices for Fractional Distillation

Vigreux column

Distilling head for total condensation and partial takeoff on column equipped with electrical heating jacket

borundum chips, glass beads, or short lengths of glass tubing the column may be filled simply by pouring in the packing, but with glass or metal spirals the best results are obtained by dropping the spirals into the column singly.

Distillation Procedure. The proper method of carrying out a distillation is to supply just enough heat at the distilling flask so that the liquid distills regularly at a uniform rate. Insufficient supply of heat will arrest the distillation temporarily and permit the bulb of the thermometer to cool below the distilling temperature, resulting in erratic temperature readings. Overheating and unsteady application of heat increase the opportunity for superheating the liquid and result in bumping. Even under the proper conditions of heating it is necessary to introduce one or two tiny boiling chips of porous substance[5] or some other antibump agent[6] into the liquid before heat is applied.

Superheating occurs because the transformation of a liquid into the vapor phase will not take place immediately, even at the boiling point, unless the liquid is in contact with a gaseous phase. Consequently, in a distilling flask, the liquid can vaporize only at the surface unless gas bubbles are introduced into the body of the liquid. Boiling chips consists of porous material containing a large amount of air, which expands on heating and furnishes bubbles that initiate vaporization throughout the liquid. Boiling chips lose their effectiveness after a single use and must be discarded; indeed, fresh boiling chips should be added before resuming a distillation that has been interrupted. It is dangerous to introduce boiling chips into a liquid that is at or near its boiling point, as this will induce sudden and violent bumping.

When the distillation assembly has been completed it is checked for tightness of all connections and for physical stability. The liquid to be distilled is introduced through the neck of the distilling flask with the aid of a funnel to prevent it from contaminating the ground-glass joint. When a condenser is being used, the flow of water through the jacket is started before heat is applied. The water should enter the lower end of the jacket and flow in a direction opposite to that of the organic vapor (countercurrent cooling). The rate of flow through the condenser should not be excessive but adequate to keep the jacket cool; this may be tested from time to time by carefully touching the underside of the adapter through which the distillate is running to see if the distillate is too warm.

The rate of heating must be adjusted (see Section 1.4) so that the liquid boils gently and distills slowly at a uniform rate, generally between 30 and 60 drops (1–2 mL)/min for simple distillation. A slower rate is used for fractional distillation or for vacuum distillation. Heating should be stopped just before the last traces of liquid have been vaporized to avoid decomposition and charring in the flask. The thermometer reading is recorded when the first drops of distillate appear at the end of the side arm or on the walls of the condenser; this is called the "initial boiling point."[7] Thereafter the

---

[5] Tiny fragments of porous unglazed clay plate or brick are frequently used for this purpose, and it is convenient to keep a supply of these on hand in a small sample tube. The commercial material usually consists mainly of rather large and heavy pieces that must be broken up into small fragments before use. Carborundum chips, No. 12 mesh, are suitable also. They are available from Carborundum Electro Metals Co., Nigara Falls, NY, 14302.

[6] A convenient substitute is a Peerless wood applicator, which may be purchased in any drugstore. The effective surface of these wooden splints is greatly increased by breaking them and inserting the broken end into the liquid.

[7] A short time lag is necessary to permit the thermometer to warm up to the temperature of the distilling vapor. An incorrect value will be obtained if the temperature is read when the first drops of liquid appear on the thermometer bulb. During this lag the thermometer is coming into equilibrium with the distilling vapor, and the mercury column is rising rapidly.

temperature and the volume of the distillate are recorded at frequent intervals. If the purpose of the distillation is to determine the composition of the liquid, many temperature–volume readings are required, and it convenient to collect the distillate directly in a graduated cylinder. The results should be recorded in a tabular form.

It is useful to plot a temperature–volume curve, from which the presence and amount of low-boiling inpurities, the approximate distilling range of constant-boiling components of a mixture, and so on, can be determined. When a substance containing small amounts of impurities is distilled, the first portion of distillate (called the forerun, or low-boiling fraction) will contain the more volatile impurity and a certain amount of the main liquid that is carried with it. As the temperature continues to rise, the bulk of the principal liquid will distill over a short temperature range, usually 2–3° (called the principal fraction, or main fraction). After this fraction has distilled, the boiling point will rise, owing to the presence of the less volatile impurity. The next fraction (called the afterrun or high-boiling fraction) will consist of a mixture of the principal liquid and the less volatile impurity. The residual liquid in the distilling flask will contain the less volatile impurity along with some of the principal liquid, which it holds back from distilling. However, even a pure substance will always leave a small amount of residual liquid. (Why?)

When the distillation is being carried out to purify a liquid, it is better to use tared[8] flasks to collect the different fractions. If the distillation behavior is known or can be estimated (as when a liquid of known boiling point is being purified), it is a simple matter to use three receivers and to collect the forerun, the main fraction, and the afterrun over the proper temperature ranges. When a liquid with unknown properties is being purified and a sufficient sample is available, it is a good strategy to determine first the temperature–volume distillation curve. If the losses of two distillations cannot be tolerated, it is necessary to deduce the boiling behavior of the sample as the distillation proceeds. This requires close attention to the thermometer readings; it is desirable that several extra tared flasks be available in case the collection of the main fraction is begun or ended prematurely.

Correction of Boiling Temperatures. The temperature readings registered directly on an ordinary thermometer in the course of laboratory distillations (or determinations of melting points) are subject to several sources of error. Two rather important contributions are errors resulting from exposure of a portion of the mercury column to atmospheric cooling and inaccurate or incorrect graduations of the thermometer scale.

---

[8] A tared flask is a vessel that is weighed when it is clean and dry. The amount of liquid distilled is then easily calculated by subtracting the weight of the tared flask from the total weight of liquid plus flask.

When a typical thermometer with a long scale (250–300 mm) is used, the true boiling point (or melting point) is not registered because the mercury column is not entirely at the temperature of the mercury in the bulb of the thermometer. The portion of the mercury column that extends above the stopper of the distillation adapter (or the surface of a melting-point bath) is cooled by the surrounding atmosphere, and the registered temperature is therefore below the true temperature of the vapor in the distilling flask. For temperatures below 100° this cooling effect does not cause any considerable error, but for high temperatures the observed reading may be several degrees below the true temperature. This error can be corrected by adding a *stem correction* calculated by the formula

$$\text{Stem correction (deg)} = 0.000154(t - t')N$$

where 0.000154 is the coefficient of apparent expansion of mercury in glass, N is the number of degrees on the stem of the thermometer from the lower exposed level to the temperature read, t is the temperature read, and t' is the average temperature of the exposed mercury column. In practice, this correction is subject to an error, since t' is not accurately known, but it may be taken roughly to be one-half of the difference between room temperature and the observed temperature.

Some thermometers have graduated scales that already include a correction for an assumed 3-in. (76-mm) immersion of the stem, and temperature readings taken with them should not be corrected. Such partial immersion thermometers are designated by having an engraved line circling the stem 76 mm above the bottom of the mercury bulb.

Many important errors of temperature readings in ordinary laboratory work may be due to incorrect graduation and calibration of the thermometer scale. To determine whether or not a thermometer registers correctly you may test it by verification at several temperatures against the boiling points of pure liquids or the melting points of pure solids. (Table 2.3) or by comparison with previously standardized thermometers.

In a simple distillation, the pressure on a liquid is the atmospheric

TABLE 2.3
Reference Temperatures for Calibration

Liquid	Boiling point, °C (at 760 mm)[a]	Solid	Melting point, °C
Acetone	56.1	Water–ice	0.0
Water	100.0	Benzoic acid	121.7
Aniline	184.4	Benzilic acid	150
Nitrobenzene	210.9	Hippuric acid	187.5
2-Bromonaphthalene	281.1	3,5-Dinitrobenzoic acid	204
Benzophenone	305.9	*p*-Nitrobenzoic acid	241

[a] The boiling points of azeotropes also may be used for reference temperatures.

TABLE 2.4
Effect of Pressure on Boiling Points

Pressure, mm	Boiling point, °C			
	Ethanol	Benzene	Water	Aniline
780	79.0	81.1	100.73	185.6
770	78.6	80.6	100.37	184.9
760	78.32	80.2	100.00	184.4
750	78.0	79.8	99.63	183.8
740	77.6	79.4	99.26	183.3
730	77.3	79.0	98.88	182.5
100	34.3	25.8	51.58	121.0
20	7.1	−5.6	22.14	81.9

pressure. For ordinary work, the variation in boiling point due to small deviations in pressure from 1 atm (760 mm) may be neglected, but for accurate work or when you are working at higher elevations, it is necessary to record the barometric pressure during distillation. Examples of the effect of pressure changes are shown in Table 2.4.

The boiling point of a reference liquid must be corrected if the atmospheric pressure during standardization is other than 760 mm. For water and several other liquids, the changes of boiling point at pressures near 760 mm are given in Table 2.4. The boiling point at pressures in the region of 760 mm can be calculated with sufficient accuracy for most purposes by the rule of Crafts, in the following convenient form.

$$\text{Bp at } p \text{ mm} = \text{bp at 760 mm} - \frac{(273 + \text{bp at 760 mm})(760 - p)}{10,000}$$

No correction for variations from 760 mm is needed when standardizations are made by means of melting points, since the effect of small pressure changes on melting points is negligible.

It is best to calibrate the thermometer using the same conditions under which the thermometer is to be employed. Thus, a thermometer normally should be calibrated for a fixed partial immersion of the stem: for example, a thermometer to be used for distillation should be calibrated for 3-in. (76-mm) immersion of the stem; one to be used for melting-point determinations should be calibrated for 1-in. (25-mm) immersion of the stem.

2.4 Representative Simple and Fractional Distillations

The purpose of this section is to provide sufficient practice in purification of liquids by distillation so that this operation can subsequently be carried out skillfully and without reference to detailed directions. Usually only one or two of these procedures will be assigned.

Simple Distillation. Arrange a distillation assembly similar to the one shown in Figure 2.10. Use a 50-mL boiling flask and follow the correct methods for supporting the apparatus and lubricating the joints as described in the text (Section 2.3).

(A) Distillation of a pure compound. In the dry 50-mL boiling flask introduce 25 mL of pure, dry methanol (*caution—flammable liquid*) by means of a clean, dry funnel. Add one or two tiny boiling chips, attach the boiling flask, and make certain that all connections are tight. Arrange a graduated cylinder to serve as receiver. Heat the flask gently until the liquid begins to boil. Adjust the heating rate until the ring of vapor condensation moves up the wall of the flask and past the thermometer into the condenser. Record the temperature when the first drops of distillate collect in the condenser. Continue to distill the liquid slowly (not over 2 mL/min) and record the distilling temperature at regular intervals during the distillation—when the total distillate amounts to 5, 10, 15, and 20 mL, and so on. Discontinue the distillation (and turn off the heat source) when all but 2–3 mL of the liquid has distilled. Record the temperature range from the beginning to the end of the distillation; this is the observed boiling point. If the boiling point differs from the literature value, record the correction in your laboratory notebook for future reference.

Transfer the used methanol to a bottle provided for this purpose. From your data, draw a rough distillation graph for pure methanol, plotting distilling temperatures on the vertical axis against total volume of distillate on the horizontal axis.

(B) Distillation of a mixture. By means of a clean, dry funnel, introduce 25 mL of a mixture of methanol and water into the distilling flask, add a few tiny boiling chips, and distill the mixture slowly. Follow the same procedure used for distilling pure methanol. Draw a rough distillation graph and compare it with that observed for pure methanol. From the graph, estimate the composition of the liquid and record the analysis in your notebook.

Transfer the distillate to a bottle for this purpose (labeled "Recovered Methanol–Water Mixture").

(C) Purification of an unknown liquid. From your instructor obtain a 25-mL sample of an impure unknown. Carry out a preliminary distillation to determine the distillation behavior of the mixture and its approximate composition. Redistill the liquid and collect in a tared receiver the main fraction boiling over a 4–5° range. Record the boiling range and weight of the main fraction.

Fractional Distillation. Arrange an assembly for fractional distillation as shown in Figure 2.11. A condenser makes an effective thermally insulated distillation column. A small wad of stainless steel sponge poked into the lower end of the column will retain the packing without interfering with the flow of vapor or liquid reflux.

The laboratory instructor will indicate what kind of packing is to be used and issue any special instructions for placing it in the column. The column should have about five plates if a "standard separation" is to be achieved. Prepare five clean, dry receivers (50- to 150-mL Erlenmeyer flasks), provide each with a clean tightly fitting cork, and label them A, B, C, D, and E (residue). The boiling flask should be selected so that it is approximately 60% full at the beginning of the distillation.

Before starting the distillation, check your apparatus carefully and have it approved by the laboratory instructor. Be careful to carry out a fractional distillation methodically, since haste will lead to sharply lowered separation efficiency.

(D) Cyclohexane and *n*-heptane.[9] In the distilling flask place 120 mL of a mixture of cyclohexane and *n*-heptane, 1:1 by volume, add two boiling chips, and fit the flask securely to the column.

If a burner is used as a heat source, heat the flask on a wire gauze with a heat-dispersing center, using a small flame that impinges directly below the flask. As soon as the mixture starts to boil, regulate the heat source with particular care so that the liquid distills slowly and regularly at a rate of about 2 mL (60 drops)/min. In the first distillation, collect in flask A the fraction (if any) that distills between 81 and 84° (the 81–84° fraction); in flask B, the 84–88° fraction; in flask C, the 88–92° fraction; and in flask D the 92–96° fraction.[10] After fraction D has distilled, remove the heat source, cool the flask, allow the column to drain, disconnect the flask, and pour the residue into E. Measure the volume of each fraction and record the results in tabular form (Table 2.5).[11]

If the separation efficiency of the column was not adequate, it will be necessary to redistill the different fractions. In the subsequent distillations proceed in the following way: Pour the contents of flask A into the round-

---

[9] The composition of the distillate may be determined by gas chromatography using a 5-ft 5% SE 30/Chromosorb W column at 25°. With these data, the number of theoretical plates can be calculated from equation (2.4), and an approximate average value for α of 1.9.

[10] In the first distillation there may be no distillate in the range 76–81°, so that flask A will remain empty and flask B will be used. In the subsequent distillations, likewise, there may be little or no distillate in the intermediate ranges B, C, or D. The results vary widely depending on the type of column and the care used in operation.

[11] If you desire to record the weight of each fraction instead of the volume, it is convenient to weigh each receiver empty, with its cork, and record this weight (called the tare) on the label of the receiver. It is good practice to record the tare of each receiver also in the notebook; when the receivers with distillate are weighed after a distillation, the gross weight is recorded, the tare subtracted, and the net weight of the fraction entered in the tabular form. If all of the weights are recorded in the notebook, you can check the figures at a later date for arithmetic errors if a discrepancy shows up.

TABLE 2.5
Fractional Distillation of Cyclohexane and n-Heptane Mixture

Fraction	Temperature range, °C	Volume, mL		
		1st dist'n	2nd dist'n	3rd dist'n
A	81–84			
B	84–88			
C	88–92			
D	92–96			
E	Residue			
Total volume of fractions				

bottom flask, add one or two tiny boiling chips,[12] and redistill, collecting the 81–84° distillate in the same flask A. When the thermometer reaches 84°, stop the distillation and add the contents of flask B. Continue the distillation, and collect the 81–84° fraction in flask A, and the 84–88° fraction in flask B. When the thermometer reaches 88°, stop the distillation and add the contents of flask C. Continue the distillation and collect the 81–84° fraction in flask A, the 84–88° fraction in flask B, and the 88–92° fraction in flask C. When the thermometer reaches 92°, stop the distillation and add the contents of flask D. Continue the distillation and collect the fractions A, B, C, and D. When the thermometer reaches 96°, stop the distillation and add the contents of flask E. Continue the distillation and collect the fractions A, B, C, and D. After fraction D has distilled, extinguish the flame, cool the flask, allow the column to drain, disconnect the flask, and pour the residue into E. Measure the volume of each fraction, and record the results in tabular form.

If B, C, and D at this stage contain a total of more than 15–20 mL of liquid, carry out a third distillation in the same manner. If necessary, carry out a fourth or fifth distillation, so that the fraction A will contain almost all of the cyclohexane and the residue E almost all of the *n*-heptane. To obtain almost pure *n*-heptane, E may be redistilled from a small distilling flask without a column.

Draw rough distillation graphs for each successive distillation, plotting the midpoint of the temperature range of the fractions against total volume of distillate.

In a research laboratory, it is customary to follow the progress of a distillation by some convenient analytical procedure. Gas chromatography (Section 7.4) is used commonly.

➤ CAUTION Extinguish or remove any flame when transferring fresh fractions into the round-bottom flask, since both cyclohexane and *n*-heptane have high vapor pressures and are flammable. If the lower end of the condenser is fitted with an adapter, there is

---

[12] It is desirable to add a tiny fresh boiling chip each time the distillation is stopped and a new fraction introduced. At the end of the distillation series, the residual liquid in the still is poured off and the accumulated used chips are discarded.

usually less loss of material by evaporation and less danger of fire. A loose plug of cotton between the adapter and receiver will also diminish the evaporation losses. Care must be taken to prevent the cotton from coming in contact with the distillate and from plugging the opening tightly!

(E) Acetic acid and water. For the separation of glacial acetic acid and water (100 : 31.5 by volume, 1 : 1 mole ratio), the following temperature ranges are satisfactory for the fractions: A, 100–102°; B, 102–107°; C, 107–112°; D, 112–117°; and E, residue. The progress of the separation of this mixture may be followed conveniently by titrating a 1-mL portion of each fraction against a standardized aqueous sodium hydroxide with phenolphthalein indicator to determine the acetic acid content. The acetic acid content of the original mixture should be determined in the same way before the material is fractionated. If a column having a large number of plates is used, it will be desirable to use larger portions of the early fractions.

(F) Methanol and water. For the separation of a mixture of methanol and water, the following temperature ranges are satisfactory for the fractions: A, 64–70°; B, 70–80; C, 80–90°; D, 90–95°; and E, residue. A simple way to verify the quality of separation is to determine the specific gravity of each fraction.

Questions

1. (a) Define accurately the term boiling point.
   (b) What effect does a reduction of the external pressure have upon the boiling point?
2. What effect on the temperature of a boiling liquid and on its distillation temperature is produced by each of the following?
   (a) a soluble nonvolatile impurity
   (b) an insoluble admixed substance such as sand or fragments of wood or cork
3. Why should a distilling flask at the beginning of a distillation be filled to not more than two-thirds of its capacity *and* filled to not less than one-third of its capacity?
4. Calculate the weight of vapor of each substance required to fill a 50-mL flask at the boiling point under normal atmospheric pressure.
   (a) cyclohexane, C_6H_{12}
   (b) toluene, C_7H_8
   (c) carbon tetrachloride, CCl_4
5. Why is the apparatus shown in Figure 2.10 not suitable for distillation of samples with a volume of 5 mL or less?
6. Why is it dangerous to heat an organic compound in a distilling assembly that is closed tightly at every joint and has no vent or opening to the atmosphere?
7. Calculate the stem correction for observed temperature readings of 125, 175, and 250°; assume that the thermometer scale is exposed above the 25° mark and the average temperature of the exposed portion is half the difference between the observed temperature and room temperature (20°).

8. The composition of the vapor above methanol–water mixtures has been measured and is given by the following formula in which X is the mole fraction of methanol in the liquid and Y is the mole fraction in the vapor.

$$Y(CH_3OH) = 6.05X - 21.37X^2 + 39.05X^3 - 34.02X^4 + 11.29X^5$$

What are the mole fractions of methanol in the vapors above mixtures with X equal to 0.25, 0.50, and 0.75?

9. Why is it necessary to have liquid flowing back through the fractionating column to obtain efficient fractionation?

10. (a) What is an azeotropic mixture?
    (b) Why cannot its components be separated by fractional distillation?

11. What physical constants may be used to test the purity of the samples of purified material obtained after a fractional distillation?

12. (a) What is meant by the temperature gradient of a column?
    (b) Why is it desirable to maintain a uniform temperature gradient and how is this achieved?

13. It is common to use an oil bath to heat the boiling flask during a fractional distillation. What is the advantage of this indirect heating?

3 Vacuum Distillation

Since the boiling temperature of a liquid is decreased by diminishing the pressure on its surface, it is possible to distill at a lower temperature by using a closed system inside which the pressure has been reduced. This procedure is useful for purifying liquids (or low-melting solids) that are decomposed at elevated temperatures. For example, glycerol boils with some decomposition at 290° under 760-mm pressure but may be distilled without decomposition under 12-mm pressure, where its boiling point is 180°. A possible disadvantage of fractional distillation under reduced pressure is the reduction in separation efficiency of most fractionating columns.

In planning a vacuum distillation, three aspects must be considered: the pressure needed to achieve the desired boiling point, the type of vacuum pump needed to lower the pressure to the required level, and the associated glassware, pressure measuring, and heating devices.

3.1 Effect of Pressure on Boiling Point

Estimation of Boiling Point. One useful relationship between pressure and boiling point is given in equation (3.1), where P is the pressure over the liquid and T is the boiling point at this pressure.[1] In this equation, both boiling temperatures are expressed in degrees Kelvin (K = °C + 273). The

---

[1] In equation (3.1) the value of the constant has been changed from its usual theoretical value of 4.81 to 5.46 to correlate a wider range of data.

equation is fairly precise for most organic liquids, but is in error for substances possessing unusually large attractions between molecules (water, alcohols, acids). More precise relationships have been developed, but the extra work required to use them is not justified for preparative organic chemistry.

$$\log_{10}\left(\frac{760}{P}\right) = 5.46\left(\frac{\text{normal boiling point}}{T} - 1\right) \quad (3.1)$$

An example of the use of equation (3.1) is outlined here for nitrobenzene.

1. Normal boiling point of nitrobenzene = 211°C = 484 K.
2. If the desired boiling point is 100°C = 373 K, the equation becomes

$$\log_{10}\left(\frac{760}{P}\right) = 5.46\left(\frac{484}{373} - 1\right) = 1.62$$

3. The expression is solved for P.

$$P_{\text{predicted}} = 18.0 \text{ mm}$$

4. At a pressure of 18.0 mm, it is observed that nitrobenzene boils at 98°C instead of the desired 100°C.

Another approach to estimating the boiling points at reduced pressure is to use a monograph, such as the one shown in Figure 3.1. To estimate the boiling point at some reduced pressure of a hydrocarbon for which the boiling point at 760 mm is known, place a straightedge (preferably transparent) on the nomograph connecting the known boiling point on scale B and the reduced pressure on line C for hydrocarbons. The boiling point of the compound at the reduced pressure is read from the intersection of the straightedge and the left-hand A scale. If the compound contains a carboxylic acid functional group, line C for carboxylic acids is used instead. The boiling points of polar molecules and less strongly hydrogen-bonded molecules can be estimated as the average of the boiling points obtained by using both the hydrocarbon and carboxylic acid lines. Note that, for compounds boiling near 100°C, both lines give about the same result; it is only for high-boiling liquids that different predicted boiling points are obtained.

3.2 Vacuum Pumps

The pump used to reduce the pressure in the system is selected according to the range of pressure required. An aspirator (water pump) is used for pressures above about 25 mm, a rotary oil pump is usually used for the range of

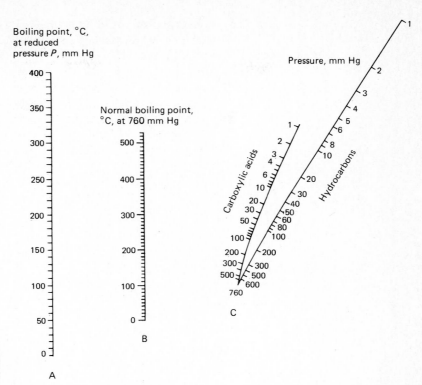

FIGURE 3.1
Nomograph for Boiling Points at Reduced Pressures

0.01–25 mm, and a diffusion pump is used for pressures below about 0.01 mm. Portable rotary vane pumps are available that are suitable for the pressure range of 760–775 mm. The user should be aware of the characteristic features of each type of pump.

An aspirator in good condition can produce a vacuum almost down to the vapor pressure of the water flowing through it. With room-temperature water, a pressure of about 25 mm can be produced, but during the winter, if the water is quite chilled, pressures of 10 mm or less may be obtained. It is not feasible to vary the water flow through the aspirator to regulate the pressure. Instead, turn the water flow fully on and control the pressure by adjusting a valve that leaks air into the system. An inexpensive valve can be obtained by using the valve at the base of an adjustable Bunsen burner. A hose is connected from the apparatus to the gas inlet tube of the burner, and the reverse flow of air into the apparatus is regulated at the burner base.

A troublesome characteristic of aspirators is their tendency to allow water to flow back into the system if the water pressure drops momentarily. Modern aspirators come with internal check valves to prevent this backflow, but they should not be trusted; it is wise to include a safety trap between the aspirator and the rest of the system. The safety trap also provides a convenient means of interconnecting the pump, the air leak, and the manometer used to measure the internal pressure. A practical arrangement is shown in Figure 3.2.

The pump oil in a rotary oil pump has such a low vapor pressure that contamination of the distillate is improbable; however, one must guard against contamination of oil by any uncondensed materials from the distillation. Oil pumps work by compressing small samples drawn from the vapor inside the distillation apparatus to a pressure above the atmospheric pressure and then releasing them from the pump. This pumping operation requires good seals between two reciprocating vanes and a rotating eccentric piston. If the oil becomes contaminated with acidic materials, the movable vanes will corrode and the pumping capacity of the pump will be sharply diminished. Other nonacidic organic materials may polymerize in

FIGURE 3.2 Assembly for Distillation Under Diminished Pressure

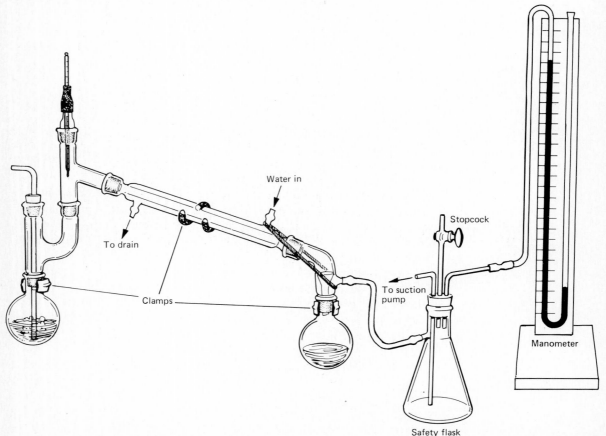

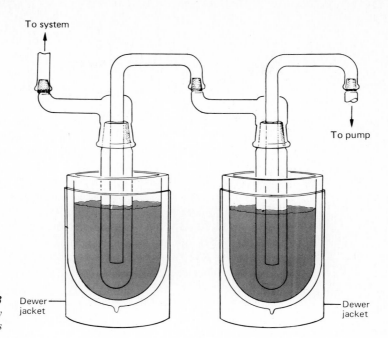

FIGURE 3.3
Trap Arrangement for Rotary Oil Pumps

the pump to give sludges that also will wear the movable vanes with a corresponding reduction in pump effectiveness. It is important that the pump be protected by at least one vapor trap that is maintained at the dry ice sublimation temperature ($-78°$) or lower. If the mixture being distilled evolves gases that would sweep vapors into the pump, much more efficient trapping devices are required. Figure 3.3 displays one widely used trap arrangement for rotary oil pumps.

An alternative to the rotary oil pump, suitable for the higher pressure range of 760–775 mm, is the much less expensive pressure–vacuum pump, such as the Gast pump. This unit also uses rotating vanes to develop a vacuum, but it is built to lower tolerances and requires less care in its operation. The vacuum (or pressure) is adjusted easily with a needle valve built into the pump. Because the motor is connected directly to the pump, the hazard of moving belts is eliminated. The most significant drawback of the pump, beside its limited range, is its noisiness.

Like the water aspirator, diffusion pumps work on the Bernoulli principle,[2] except that in place of a rapidly flowing stream of water they use a jet of mercury or oil vapor. Diffusion pumps produce pressures in the range of 10^{-2}–10^{-6} mm. They require heaters to vaporize the mercury or oil and

---

[2] When liquids or gases flow through a pipe of variable cross section, the pressure is smallest where the cross section is least and the velocity is greatest.

cooling devices for recondensing the vapor after it has passed through the jet. Unlike aspirators and mechanical pumps, diffusion pumps require an additional forepump ("backing pump") that reduces the pressure in the pump to a critical level (usually about 0.1–0.01 mm).

Manometers. There are many styles of manometers for measuring the pressure in the system, each designed for maximum precision over a small range of pressures. Two general-purpose manometers that together cover a sufficient range with adequate precision for preparative work are the tilting McLeod gauge (10^{-2}–10 mm) and closed-end U-tube mercury manometer (5–200 mm), which are shown in Figure 3.4.

The McLeod gauge is operated by tilting the movable section of the gauge toward the upright position until the higher of the two mercury columns is level with the top of the bore of the lower capillary column. If the gauge has been calibrated properly, the pressure can be read directly from the height of the lower capillary column against the gauge markings.

With the open-tube manometer (Figure 3.2) it is necessary to subtract the net height of the mercury column in the manometer from the barometric pressure.

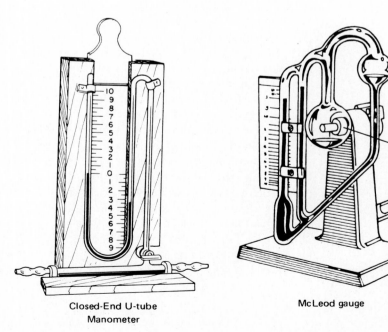

FIGURE 3.4
Manometers

Closed-End U-tube Manometer

McLeod gauge

3.3 Laboratory Practice

An apparatus for vacuum distillation using standard taper glassware is depicted in Figure 3.2. If corks or rubber stoppers are being used, the round-bottom flask and the triply jointed Claisen adapter are replaced by a special one-piece Claisen distillation flask; the condenser and vacuum adapter are replaced by a side-arm distillation flask. For samples larger than about 10 mL the ground-jointed apparatus is easier to use; however, for smaller samples the one-piece Claisen flask is recommended to avoid excessive mechanical losses.

The principal purpose of the Claisen adapter (or flask) is to diminish the chance of contamination of the distillate from frothing or violent bumping. Both of these conditions are more troublesome in vacuum distillation than in ordinary distillation. The size of the boiler should be such that it is not quite half filled at the start of the distillation. The flask is usually heated in an oil bath or sand bath to insure regular heating. The bath temperature is usually 15–25° higher than that of the distilling liquid.

Boiling chips are frequently ineffective in vacuum distillation because of their limited air supply. Somewhat better are long wooden splints of the kind that can be purchased in any drugstore. These can be broken off to a size that will fit in the flask comfortably. The best procedure, although it is somewhat inconvenient, is to introduce a fine stream of air bubbles through a thin flexible capillary tube.[3] If the substance is easily oxidized, argon or nitrogen can be substituted. A different approach that works quite well is to stir the boiling mixture rapidly with a magnetic stirrer.

With vacuum distillations using ground-jointed apparatus, it is imperative that the joints be properly lubricated to prevent leaks during the distillation and simplify separation of the joints afterwards. Remember that, if a distillation is carried out at a pressure as high as 100 mm, there is still almost 1 atm ($660/760 = 0.9$ atm) pressing in on each joint and the walls of the flasks.

With distillations using corks or rubber stoppers, ordinary side-arm distilling flasks are used for receivers. Several of these, sufficient to receive all the fractions, are prepared before the distillation is begun. The necks of all the receiving flasks must be about the same dimensions so that they will all fit tightly on the stopper of the side tube of the Claisen flask. The delivery tube of the Claisen flask should reach just into the bulb of the

---

[3] The capillary is prepared by drawing out a piece of ordinary glass tubing, 7–8 mm in diameter, to capillary dimensions and drawing out this first capillary, in a small luminous flame, to an extremely fine and flexible capillary thread. The capillary thread is tested by blowing into the tube while the thread is held under ethanol or ether; it should emit a fine stream of very minute bubbles. The top of the tube, which bears the capillary, should be bent at right angles to facilitate adjustment of the depth of the capillary tube so that it will reach exactly to the bottom of the distilling flask.

receiver. To simplify the apparatus the ordinary condenser is omitted; instead the receiver is supported above a large funnel and is cooled directly by a jet of cold water. This type of cooling is satisfactory for liquids that distill above 50°; for liquids that distill below this temperature, a more elaborate cooling device is necessary.

In advanced work the single receiver is replaced by a device consisting of a flask with two or more arms, to each of which is attached a receiving flask. By rotation of this flask the different receivers in turn can be brought in line with the drops of condensate without having to break any of the vacuum seals.

The following points should be observed in carrying out a vacuum distillation.

1. Never use Erlenmeyer flasks as receivers. Even with the small 50-mL Erlenmeyer flask the force acting on the flat bottom is about 50 lb. Remember that the force at any point is proportional to the difference between the internal and external pressure. A distillation at 100 mm places nearly as much stress on the apparatus as one at 10^{-6} mm.

2. Test the completely assembled apparatus before placing the liquid in the boiler flask, to detect leaks and to make certain that all of the parts of the apparatus will withstand the external pressure. *Use safety goggles to protect your eyes.*

3. When using a water aspirator, turn on the water to the full pressure, otherwise water may be sucked back into the safety flask.

4. To release the vacuum in the distilling flask, open the stopcock on the safety flask and gradually allow the pressure to reach atmospheric pressure in the apparatus *before* shutting off the aspirator.

5. When using a rotary pump, be certain it is adequately guarded with traps and that they are filled with coolant. A slush of dry ice and isopropanol ($-78°$) is reasonably effective and inexpensive. In some research laboratories liquid nitrogen ($-196°$) is used because of the greater protection it affords the pump.

6. When changing receivers, remove the flame from beneath the bath and allow the distilling flask to cool slightly before releasing the vacuum (by lowering the bath or raising the flask from the bath). After the receiver has been changed the system should be evacuated again before heating is resumed.

7. If a closed-tube manometer is used, close the stopcock of the manometer before releasing the vacuum. If this is not done, the abrupt surge of the mercury column may break the glass tube.

3.4 Representative Vacuum Distillations

(A) Purification of Benzaldehyde. Purify a 60-g (58-mL) sample of technical benzaldehyde[4] in the following way. Wash with two 20-mL portions of sodium carbonate solution (10%), then with water, and dry over anhydrous magnesium sulfate. It is advantageous to add a few small crystals of hydroquinone or some other antioxidant during the drying operation.

Arrange an assembly as shown in Figure 3.2 with a boiler flask of appropriate size, and decant the benzaldehyde through a fluted filter into the boiler. Distill under diminished pressure (preferably below 30 mm) in the manner previously described. Be certain to wear your safety *goggles*. Place a few small crystals of the antioxidant in the receiving flask in which the purified benzaldehyde is to be collected. To determine the boiling point of benzaldehyde under the particular pressure in your apparatus, use the nomograph in Figure 3.1; the boiling point at 760 mm is 180°.

Autooxidation. In common with many oxidizable substances, benzaldehyde is capable of combining directly with oxygen of the air (autooxidation) and is converted eventually to benzoic acid.

Autooxidation is extremely sensitive to the effect of catalysts, which are considered to act on an unstable intermediate complex of "peroxide" character that is formed in the initial step of oxidation. Catalysts that accelerate autooxidation are called *prooxidants*; those that retard or inhibit autooxidation are called *antioxidants*. The latter find important technical application for the preservation of organic materials; for example, the deterioration of rubber is greatly retarded by the incorporation of antioxidants. Likewise, the autooxidation of benzaldehyde can be effectively inhibited by the addition of a trace (less than 0.1% is sufficient) of hydroquinone or some other antioxidant.

(B) Purification of Ethyl Acetoacetate. Place a 60-g (60-mL) sample of technical ethyl acetoacetate[5] in a vacuum distillation assembly of suitable size and distill under diminished pressure (preferably below 30 mm) in the manner described earlier. Be certain to wear your safety *goggles*.

During the distillation of the low-boiling fraction containing ethyl acetate, the high vapor pressure of the latter may increase the pressure in

---

[4] Technical benzaldehyde usually contains a small amount of benzoic acid.

[5] Technical ethyl acetoacetate may contain a little ethyl acetate, acetic acid, and water. Since ethyl acetoacetate decomposes to some extent on heating to its boiling point at atmospheric pressure with the formation of dehydroacetic acid, it is purified by distillation under diminished pressure. Small quantities may also be purified by rapid distillation at atmospheric pressure.

the system above 30 mm. If this occurs, the distillation of the first fraction may be carried out at a higher pressure, but the pressure should be maintained below 30 mm for the distillation of the remaining fractions. To obtain the correct boiling point of ethyl acetoacetate under the particular pressure in your apparatus, use the nomograph of Figure 3.1; the boiling point at 760 mm is 180°.

Collect the purified ethyl acetoacetate over an interval of 5–6°, determined from the nomograph, and calculate the percentage recovery from the crude product.

Questions

1. Estimate from the nomograph in Figure 3.1 the boiling point of each compound at 1 mm pressure.
   (a) pentanoic acid (bp 184°)
   (b) ethylbenzene (bp 136°)
   (c) *o*-chlorotoluene (bp 159°)
2. Estimate the pressure under which each compound could be vacuum distilled at 80°.
   (a) pentanoic acid
   (b) ethylbenzene
   (c) bromobenzene (bp 156°)
3. Why is bumping more troublesome in vacuum distillation than in ordinary distillation?
4. What precautions must be observed in using an aspirator for vacuum distillation?

4 Steam Distillation

Steam distillation consists of distilling a mixture of water and an immiscible or partly immiscible substance.[1] The practical advantage of steam distillation is that the mixture usually distills at a temperature below the boiling point of the lower-boiling component. Consequently, it is possible to effect steam distillation of a high-boiling organic compound at a temperature much below its boiling point (in fact, below 100°) without resorting to vacuum distillation. Steam distillation is useful also in separating mixtures when one component has an appreciable vapor pressure (at least 5 mm) in the vicinity of 100° and the other has a negligibly low vapor pressure. The process of steam distillation is employed widely in the laboratory and in industry; for example, for the isolation of α-pinene, aniline, nitrobenzene, and many natural essences and flavoring oils.

4.1 Principles of Steam Distillation

Mixtures of two immiscible substances behave quite differently from homogeneous solutions, and the description of their behavior requires a different physical law. The basis of the new law can be grasped by considering the consequence of increasingly positive deviations from Raoult's law (see Section 2.1). One symptom of small positive deviations is a skewed boiling-

---

[1] Occasionally the principle of steam distillation is extended to mixtures of two immiscible organic compounds such as ethylene glycol and hydrocarbons and is then called codistillation.

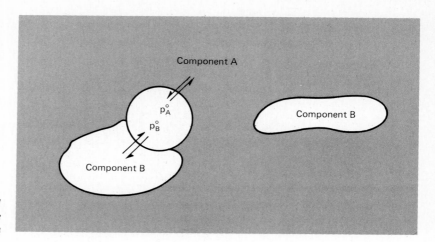

FIGURE 4.1
Vapor Pressure Inside Bubble During Steam Distillation

point composition diagram, as is found with methanol–water solutions (see Figure 2.6). Greater positive deviations, as occur in ethanol–water solutions, lead to maxima in the total vapor pressure curve and to low-boiling azeotropes (Figure 2.7). With still greater positive deviations the two components can separate into two immiscible layers. In the limit of very large positive deviations from ideal solution behavior, the two components are completely insoluble and each component vaporizes independently of the other to give a constant total vapor pressure that is the sum of the individual vapor pressures. The physical basis for this independent behavior of the two components is depicted in Figure 4.1. In this diagram, component B is represented as two globules suspended in component A; an incipient bubble in contact with both components is shown in the center of the diagram. The vapor pressure of component A inside the bubble is p_A° (the vapor pressure of pure A), just as it would be if no B were present. In the same fashion, the vapor pressure of B inside the bubble is p_B°. Most water-insoluble organic compounds approximate this extreme behavior so that steam distillation calculations are normally based on the simple law: *the total vapor pressure equals the sum of the vapor pressures of the two components.*

4.2 Distillation Temperature and Composition of Distillate

As with ordinary distillations, the boiling point is the temperature at which the total vapor pressure equals the atmospheric pressure. If the vapor pressures of the two components are known at several temperatures, the distillation temperature is readily found by plotting the vapor pressure curves of the individual components and making a third curve showing the sum of the vapor pressures at the various temperatures (Figure 4.2). The steam distillation temperatures will be the point at which the sum equals the atmospheric pressure.

Knowing the distillation temperature of the mixture and the vapor pressures of the pure components at that temperature, one can calculate the composition of the distillate by means of Dalton's law of partial pressures.

According to Dalton's law, the total pressure (P) in any mixture of gases is equal to the sum of the partial pressures of the individual gaseous components (p_A, p_B, etc). The proportion by *volume* of the two components in the distilling vapor will consequently be equal to the ratio of the partial pressures at that temperature; *the molar proportion* of the two components (n_A and n_B) in steam distillation will be given by the relationship $n_A/n_B = p_A/p_B$, where $p_A + p_B$ equals the atmospheric pressure. The weight proportion of the components is obtained by introducing the molecular weights (M_A and M_B).

$$\frac{\text{Weight of A}}{\text{Weight of B}} = \frac{p_A \times M_A}{p_B \times M_B}$$

Example. Consider a specific case, such as the steam distillation of bromobenzene and water. Since the sum of the individual vapor pressures (see Figure 4.2) attains 760 mm at 95.2°, the mixture will distill at this temperature. At 95.2° the vapor pressures are bromobenzene, 120 mm; water, 640 mm. According to Dalton's law, the vapor at 95.2° will be composed of molecules of bromobenzene and of water in the proportion of

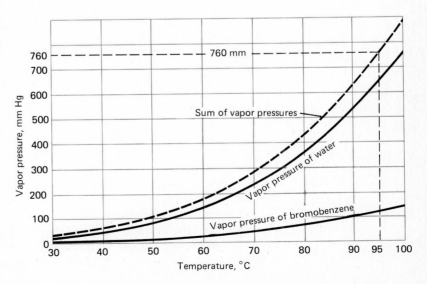

FIGURE 4.2
Vapor Pressure Curves for Water and Bromobenzene

120 : 640. The proportion by weight of the components can be obtained by introducing their molecular weights.

$$\frac{\text{Weight of bromobenzene}}{\text{Weight of water}} = \frac{120 \times 157}{640 \times 18} = \frac{1.63}{1.00}$$

The weight composition of the distillate will therefore be 62% bromobenzene and 38% water.

$$\text{Bromobenzene} = \left(\frac{1.63}{1.00 + 1.63} \times 100\right) = 62\%$$

$$\text{Water} = \left(\frac{1.00}{1.00 + 1.63} \times 100\right) = 38\%$$

This calculation gives the minimum amount of water in the distillate. In practice, an excess of water or steam is used in the distilling flask to sweep out the vapor mixture and to compensate for imperfect mixing.

It can be seen from calculations of the type illustrated above with bromobenzene that there are several requirements for the practical use of steam distillation in the laboratory: the substance to be steam distilled must be insoluble, or only sparingly soluble, in water; it must not be decomposed by prolonged contact with boiling water or steam; and it must have an appreciable vapor pressure (preferably, at least 5 mm) in the neighborhood of 100°. That water has a very low molecular weight (18) compared with those of typical organic compounds is a favorable circumstance for steam distillation, since this permits a substance to be steam distilled at a practical rate even though its vapor pressure is relatively small near 100°.

4.3 Laboratory Practice

A simple assembly for small-scale steam distillations is shown in Figure 4.3. A large round-bottom flask, which serves as the boiler, is fitted with a Claisen adapter. The central arm of the adapter is stoppered; the side arm is attached to a distillation adapter equipped with a thermometer and connected to a condenser. A vacuum adapter is attached to the lower end of the condenser and held in place by a rubber band around the condenser water inlet and the side arm of the vacuum adapter.

The flask should be less than half filled with the mixture to be steam distilled and about twice the volume of water anticipated to be needed for

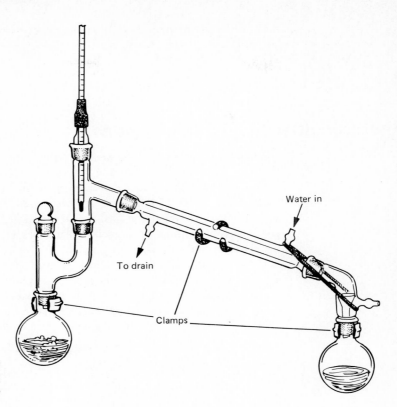

FIGURE 4.3
Small-Scale Steam Distillation Assembly

the distillation.[2] If a large enough flask is not available, the stopper in the Claisen adapter can be replaced by a dropping funnel, so that additional water can be added as the distillation proceeds. In carrying out a steam distillation with internal generation of steam, it is essential that the mixture be boiled vigorously, since success depends on thorough mixing of the water-insoluble compound with the boiling water. The purpose of the Claisen adapter is to prevent fine spray from being carried over mechanically into the distillate. If the mixture tends to froth, the Claisen adapter will help trap it, or at least give some warning that unvolatilized material is about to be carried over and allow the operator to lower the heat.

Many substances that are solids at room temperature may be steam distilled. With such materials, which may solidify in the condenser and form a mass that obstructs the tube, it is *essential* to watch the condenser tube carefully. If a mass of crystals forms, the flow of water through the jacket is stopped *temporarily*, and the water is allowed to drain out of the jacket. The

---

[2] If the boiling points of the components are known their vapor pressures at 100° can be estimated by the nomograph in Figure 3.7; the volume of water can be estimated from these data using the method of the preceding section.

heat from the hot distillate will cause the crystalline mass to melt and flow out into the receiver. When the tube is clear, the flow of water through the jacket is started again.

4.4 Representative Steam Distillations

(A) Steam Distillation of Turpentine.[3] Arrange an apparatus for steam distillation as shown in Figure 4.3 using a 250-mL distilling flask and a 100-mL graduated cylinder as the receiver. In the flask, place 30 mL (26 g) of turpentine (bp 156–165° at 760 mm) and 75 mL of water. Add two boiling chips and heat the flask on a wire gauze with a flame adjusted to give vigorous boiling. It is essential for the success of this experiment that the mixture boil rapidly with good mixing of the two phases. Discard the first 10 mL of distillate, and collect the next 30 mL. Record the volumes of the water and turpentine layers in this distillate. Compare the ratio of the volumes actually found with the ratio calculated from the ideal steam distillation law using the tabulated vapor pressures and densities. Compare the observed distillation temperature with the calculated value.

(B) Separation of a Mixture by Steam Distillation. Arrange an apparatus for steam distillation as shown in Figure 4.3 using a 500-mL round-bottom flask as the boiler and a 250-mL Erlenmeyer flask as a receiver. Make sure that the apparatus is supported firmly and that all stoppers and connections are tight.

Weigh out 5-g samples of *p*-dichlorobenzene and of salicylic acid, and determine the melting point of each sample.

In a clean porcelain mortar, thoroughly mix together the samples of *p*-dichlorobenzene and salicylic acid and determine the melting point of the mixture. Transfer the mixture to the steam distillation flask, add about 200 mL of water, and heat the mixture until it boils vigorously. Continue to distill with steam until a test portion of the distillate shows that no more water-insoluble material is passing over. When the distillation is finished, save the distillate and the residue in the round-bottom flask for further examination.

▶ *CAUTION* Since the material that distills with steam may solidify in the condenser, watch carefully to avoid the formation of a crystalline mass that may completely obstruct

---

[3] This experiment illustrates the level of agreement to be expected between predictions from the ideal steam distillation law and an actual distillation. Commercial turpentine is largely α-pinene and can be treated as such for the purposes of this experiment. The vapor pressures of water and α-pinene, $C_{10}H_{16}$, are given in the Appendix. The density of α-pinene at 25° is 0.86.

the condenser tube. If a large crystalline mass collects in the tube, stop the flow of water through the condenser *momentarily*, and drain the water from the condenser jacket. The heat from the vapors will then melt the crystals, and the obstruction will be removed. As soon as this occurs, start the water again through the condenser jacket.

Before the residue in the flask cools, filter the solution through a fluted filter (see Figure 5.5) and collect the filtrate in a clean beaker. Cool the solution rapidly to room temperature or below, and collect the crystals with suction in a Büchner funnel. Allow the crystals to dry and determine their melting point. What is this substance that did not distill with steam?

Separate the solid material in the distillate by filtering with suction, and press it as dry as possible with a clean cork or spatula. Allow the crystals to dry completely and determine their melting point. What is the substance that distilled with steam?

Questions

1. What properties must a substance have for steam distillation to be practical?
2. What are the advantages and disadvantages of steam distillation as a method of purification?
3. At 100° the vapor pressure of limonene, $C_{10}H_{16}$, is 70 mm. Estimate the amount of steam required to steam distill 13.6 g (0.1 mole) of limonene.
4. If a mixture of bromobenzene and water were subjected to steam distillation at 100 mm pressure (see the Appendix for vapor pressure data), what would be the temperature of distillation and the weight composition of the distillate? Compare the results with the composition at 760 mm, in Figure 4.2.
5. (a) Calculate the distillation temperature and the theoretical weight composition of the distillate for the steam distillation of *p*-dibromobenzene at 760 mm.
   (b) How would a mixture of bromobenzene and *p*-dibromobenzene behave when subjected to steam distillation?

5 Melting Points, Crystallization, and Sublimation

This chapter is concerned with solids—their melting behavior as a criterion of purity and two common methods of purification, crystallization and sublimation.

5.1 Melting Points

The Theory of Melting Point Depression. A pure crystalline substance usually possesses a sharp melting point; that is, it melts completely over a very short temperature range (in practice, not more than 0.5–1.0°, provided good technique is used). The presence of even small amounts of impurities soluble in the *molten* compound will usually produce a marked depression of the temperature at which fusion begins as well as the temperature at which the last crystal disappears, resulting in a large increase in the melting point range. The lowering of the final temperature at which the last crystal disappears is called melting-point depression.

The physical basis of melting point depression by impurities is a consequence of the different effects of impurities on the vapor pressure of solutions and of solid mixtures. As illustrated in Figure 5.1, the melting point of a pure substance is the temperature at which the solid phase is in equilibrium with the liquid phase, that is, the temperatures at which the solid and liquid have the same vapor pressure.

Addition of an impurity to the molten substance will reduce its mole fraction and lower the vapor pressure of the substance above its melt. The

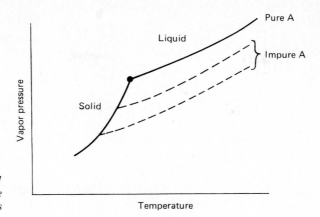

FIGURE 5.1
Vapor Pressure of Substance A in Solid and Liquid Phases

vapor pressure curves at different temperatures for two concentrations of impurities are shown as dashed lines in Figure 5.1.

The situation is entirely different in the solid phase. Here, the original substance and the impurities usually form a *heterogeneous* mixture of crystals of each substance. The crystals are so intimately mixed that it is impractical to try to separate them, but at a molecular level, they behave as though they were independent of each other. As a consequence, the vapor pressure of the substance is essentially unaltered by the presence of impurities, which may be thousands of angstroms (Å) away. This behavior is shown in Figure 5.1 as a vapor pressure curve for the solid that is independent of the presence of impurities. Since the temperature at which fusion ends is the temperature at which the solid and melt have the same vapor pressure (the point of intersection of the solid and liquid curves), the final temperature will be depressed. The greater the amount of impurity (at least up to a point, to be described later), the larger the melting-point depression is.

Molecular Weight Determination. An instructive application of the phenomenon of melting point depression is the Rast method for determining molecular weights. Many substances, frequently those that have approximately spherical shapes, show unusually large depressions of melting points when impure. For example, the melting point of camphor is depressed by 37–40° when the concentration of impurity is 1 mole/1000 g of camphor. In the Rast molecular weight method a weighed sample of the unknown is intimately mixed with a weighed sample of camphor and the final melting point of the mixture determined. From the observed lowering of the camphors' melting point, the molecular weight of the unknown is calculated from the following expression.

$$\text{Molecular weight} = 38.5 \times 1000 \times \frac{\text{Weight of unknown}}{\text{weight of camphor}} \times \Delta T$$

In this expression, the figure 38.5 is called the melting-point constant and is characteristic of camphor.

A useful variation of the Rast method is to determine the freezing point of cyclohexanol solutions of the unknown. Cyclohexanol (mp 24.7°) has a melting-point depression constant of 42.5.

Chemical Identification. The melting points of mixtures (colloquially referred to as *mixed melting points* are used frequently to establish the identity of two samples.

A typical illustration is afforded by the situation in which a substance to be identified, designated by X, is suspected of being identical with one or the other of two known substances, A or B, and where the three pure compounds, A and B and X, have approximately the same melting point. Mixtures of about equal amounts of A and X, and B and X, are prepared and the melting points of these mixtures are determined.[1] If A and X are identical, the melting point of a mixture of A and X in any proportions will always be the same as that of A or X alone, apart from any slight differences resulting from different degrees of purity of the samples of A and X. If A and X are different from one another, the melting point of the mixture of A and X will usually, but not invariably, be lower than that of either A or X. Similar reasoning is applied to the mixture of B and X. If the melting points of the mixtures A and X, and of B and X, are both below the original melting points of the pure substances, one can safely conclude that X is not identical with either A or B. There are some mixtures that are exceptions to the general rule that the melting point of a pure compound will be lowered by the presence of a soluble impurity or by admixture with a different substance (see the next paragraph). Consequently, if the melting point of X is not depressed by mixing with A (or with B), one can conclude that X is *probably* identical with A (or B), but one cannot assert this with absolute

---

[1] In practice it is highly desirable to determine the melting points of the mixtures and of a control sample of pure X in a single operation, so that the rate of heating will be the same for the three samples. To avoid confusion the individual tubes are given some means of identification (by making one, two, and three small file scratches at the upper end, or cutting the tubes to three different lengths, etc.) and arranged in a definite manner on the thermometer. The symbols and arrangement should be jotted down in the notebook: for example, "X (medium tube, center"; "A + X (long tube, at left)"; "B + X (short tube, at right)"; and so on. It is difficult to observe carefully the melting point behavior of more than three samples and to avoid mistakes in the identity of the melting point tubes. In careful work, a small pocket magnifying glass is used to observe closely the behavior of the samples on melting and to read the thermometer more accurately. Differences in the behavior of the samples are more significant than the actual temperature readings.

Sec. 5.1] MELTING POINTS

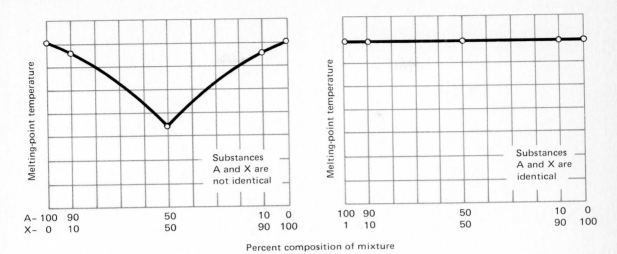

FIGURE 5.2 *Melting Point Behavior of Mixtures*

certainty. Comparisons of other physical constants of the two materials are necessary to establish an identity beyond any reasonable doubt.

In practice, it is often advisable to prepare a number of mixtures that contain varying proportions of the substances to be identified and the known substance; for example, a mixture containing 10% of A and 90% of X, another containing 50% of each, and a third mixture containing 90% of A and 10% of X. The results of the melting point determinations of these mixtures may then be plotted and rough melting point curve may be drawn. For this purpose, the temperature plotted is that at which the mixture liquefies completely. The nature of the melting-point curves thus obtained is shown in Figure 5.2.

Melting Point Apparatus. One common method for determining melting points is to use the Thiele apparatus illustrated in Figure 5.3. The sample is contained in a melting point capillary tube. The thermometer is inserted into a 20- to 25-mm cork containing a properly bored hole of the correct size and supported by means of a buret clamp. The thermometer position is adjusted so that it is well centered vertically, with the upper end of the mercury bulb about 1 cm below the side arm of the Thiele tube. One end of the capillary tube is sealed by touching it to the edge of a small hot flame; no attempt is made to fire-polish the other end, as this is unnecessary and is extremely difficult to accomplish without constricting or sealing the opening.

A small amount of the material to be examined (0.1 g is ample) is pulverized finely by crushing with a clean spatula or knife blade on a piece of smooth hard paper or a watch glass. The crushed material is collected

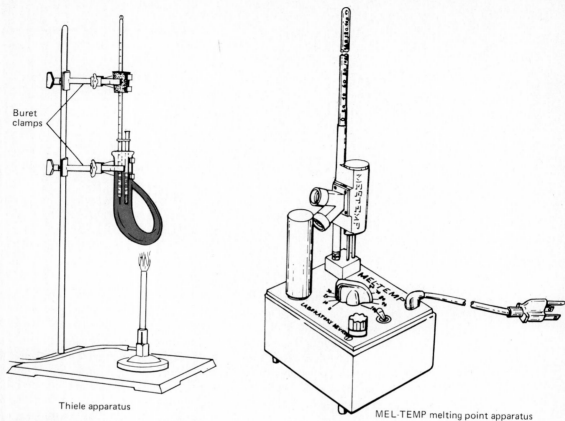

FIGURE 5.3 *Melting Point Apparatus*

into a small mound, and the open end of the melting point tube is thrust into it. The solid is shaken down into the tube by drawing a triangular file gently along the upper portion of the tube and then tapping the lower end on the desk top; or it may be forced down by dropping the tube (sealed end downward) through a length of ordinary glass tubing onto the desk top. Further increments of the sample are introduced in the same way, until the material forms a compact column 3–5 mm high at the bottom of the tube after repeated tapping. It is essential that the material be pulverized finely and packed tightly.

Although the capillary tube will usually adhere to the thermometer by capillary action of a thin film of bath liquid[2] it is advisable to attach it

---

[2] Liquids suitable for bath temperatures up to about 250° are medicinal paraffin oil (Nujol, etc.), anhydrous glycerol, di-*n*-butyl phthalate, medicinal castor oil, and Dow–Corning Silicon fluid #500 (mixtures of organosilicon compounds). Commercial tristearin or hydrogenated fatty oils (Crisco, Spry, etc.) are quite satisfactory, although the bath material forms a soft greasy solid at room temperature. Concentrated sulfuric acid is unsafe.

more firmly by means of a thin slice of rubber tubing or a small rubber band. The tube should be adjusted so that the sample is placed just alongside the mercury bulb of the thermometer, and the rubber fastening should be above the level of the bath liquid (to avoid softening of the rubber and discoloration of the bath).

The Thiele tube is heated at its lowest point. The resulting convection currents will circulate the oil in a counterclockwise direction if the apparatus is set up as shown in Figure 5.3. The tube may be heated at a fairly rapid rate until the bath temperature approaches within about 15–20° of the melting point (roughly determined in a preliminary trial, if necessary). Heating is then continued with a very small flame, adjusted so that the temperature rises slowly and regularly, at a rate of about 2° min. The observed melting point is reported as the temperature range beginning with the thermometer reading when the substance starts to liquefy and ending with the reading when the melt becomes clear. The temperature readings and any other observations are recorded at once in the notebook.

The thermometer used for melting points should be one that has been standardized by one of the methods described earlier (Section 2.3, Correction of Boiling Temperature). The observed readings are subject also to a correction for the exposure of the mercury column to atmospheric cooling. Often the stem correction is not applied; one may indicate whether or not this has been done by a notation such as, "mp 172–173° (uncor)" or "mp 172–173° (cor)." Melting point tubes are discarded after a single use in a "waste glass" container.

In research laboratories the simple Thiele apparatus is modified to include a small electrical stirrer and is heated by a coil of resistance wire. Another form uses a beaker, in place of the Thiele tube, with an electrical stirrer and heater. Usually these research devices equipped with some optical magnifying device for easier viewing of the sample. The research instruments are quite expensive, and, although they are more convenient, they do not give a more precise melting point than the simple apparatus illustrated in Figure 5.3.

One of the problems with the Thiele apparatus and its more refined variations is that at temperatures above 220–250° the usual bath liquids begin to smoke and decompose. For melting points in the higher temperature ranges, a metal block with vertical holes for the thermometer and capillary tube and a small window for observation of the sample may be used. A very convenient commercial version of such a device is the MEL-TEMP unit[3] (Figure 5.3). It is electrically heated, allows up to three samples to be melted simultaneously, and contains a built-in light and optical viewer. Heated blocks depend on thermal conduction rather than convection for heat transfer from the heat source. For this reason, heating

---

[3] The MEL-TEMP unit is available from Laboratory Devices, P.O. Box 68, Cambridge, MA 02139.

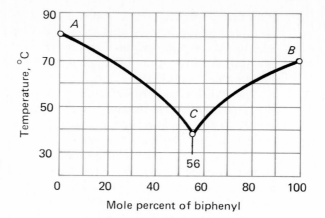

FIGURE 5.4
Typical Melting Point–Composition Diagram for a Mixture

blocks are less responsive to changes in the heat level, and care must be taken not to raise their temperature too quickly and shoot past the melting point range.

Formation of Eutectic Mixtures. From the previous discussion one might assume that as the amount of impurity increased, the magnitude of the melting point depression would increase indefinitely. Actually, if the impurity concentration becomes very large, it changes roles and becomes the main substance. If one examines the equilibrium melting temperatures of the entire range of concentrations of two substances, the behavior illustrated in Figure 5.4 is typically observed. This is the curve for mixtures of naphthalene and biphenyl. The curve *A–C* shows the lowering of the final melting point of naphthalene (*A*, mp 80°) by addition of biphenyl, and *B–C* the lowering of the melting point of biphenyl (*B*, mp 69°) by naphthalene. The intersection *C* is called the eutectic point and is the temperature (39.4°) at which both crystalline solids *A* and *B* can exist in equilibrium with a melt of fixed composition (44 mole percent naphthalene + 56 mole percent biphenyl).

5.2 Crystallization

In a typical laboratory preparation, a crystalline solid separating from a reaction mixture is contaminated with some impurities. Purification is accomplished by crystallization from an appropriate solvent. The procedure consists of the following steps: (1) dissolving the substance in the solvent at an elevated temperature; (2) filtering the hot solution to remove insoluble impurities; (3) allowing the hot solution to cool and deposit crystals of the substance; (4) separating the crystals from the supernatant solution (the mother liquor); (5) washing the crystals to remove adhering mother liquor; (6) drying the crystals to remove the last traces of solvent.

The selection of a suitable solvent for crystallization is of great practical importance. A good solvent for crystallization is one that will dissolve a moderate quantity of the substance at an elevated temperature but only a small quantity at low temperatures. The solvent should dissolve the impurities readily, even at low temperatures, and should be easy to remove from the crystals of the purified substance. It is essential that the solvent not react in any way with the substance to be purified. Other factors such as flammability and cost should also be taken into consideration. Table 5.1 lists some of the more common solvents used for crystallization.

In selecting a solvent for the purification of a given substance, one should consider its effectiveness for removal of the particular impurities that are likely to be present. The following general categories of impurities may be encountered.

Mechanical impurities (dust, grit, particles of paper, cork, etc.) are readily removed by filtering the hot solution, since they are insoluble in all of the common solvents. Inorganic salts may often be separated in this way by using an organic solvent in which they are insoluble; an alternative method is to wash the crystals before recrystalization with a solvent such as water, in which the inorganic salts are soluble and the organic compound is insoluble.

Traces of coloring matter and resinous impurities may often be removed by warming the solution with a small amount of decolorizing carbon or other absorbent (Norit, Darco, Nuchar, etc.) before filtering the hot solution. The action of decolorizing agents varies widely, and effectiveness in removing a particular impurity may differ markedly from one solvent to another. An excessive amount of decolorizing agent must be avoided, since it will adsorb the compound that is being purified, thus reducing the amount of pure compound isolated.

TABLE 5.1
Common Solvents for Crystallization

Solvent	Bp, °C	Solvent	Bp, °C
Water	100	Methylene chloride	40
Methanol[a]	64	Carbon tetrachloride[b]	77
95% Ethanol[a]	78	Cyclohexane[a]	81
2-Propanol[a]	82	Benzene[a,b]	80
Diethyl ether[a]	34	Petroleum ether[a]	60–90
Acetone[a]	56	Ligroin[a]	90–150
Ethyl acetate[a]	78	Toluene[a]	111
Acetic acid (glacial)[a]	118	Xylene[a]	139

[a] Flammable liquid—requires caution against fire hazards.
[b] Prolonged exposure may induce cancer.

Impurities more soluble in the solvent are readily removed by crystallization, since they will be retained in the mother liquor. Likewise, impurities having about the same solubility as the substance being purified, when present in moderately small amounts, are readily eliminated in the mother liquor.

Impurities less soluble in the solvent are very difficult to remove if they are present in considerable amount, since the hot solvent will dissolve an appreciable amount of the impurity and, on cooling, the impurity will crystallize out and contaminate the product. It is for this reason that one tries to select a solvent that will readily, dissolve the impurities, even at room temperature.

In many instances information is available concerning the solvents suitable for crystallizing the substance that is to be purified. If such information is lacking, the common solvents may be tested experimentally, on a small scale, to select a satisfactory one.[4] This is done conveniently by using a series of small test tubes and placing in each tube a small quantity of the substance to be purified and a small quantity of the solvent to be tested. The action of the solvent is tested in cold and at the boiling point, and one notes whether well-formed crystals are produced abundantly on allowing the hot solution to cool. If two or more solvents appear to be promising, each may be tested more thoroughly with larger, weighed samples of material, to determine the loss of weight in the crystallization process and to compare the purity of the recrystallized samples.

To secure a satisfactory recovery of the purified material it is essential to avoid using an unnecessarily large amount of solvent. The quantity of the substance lost through retention in the mother liquor will be minimized if the sample is dissolved in the smallest possible amount of the hot solvent. In practice it is desirable to employ *slightly more than the minimum quantity of hot solvent* required to dissolve the sample (2–5% more), so that the hot solution will be not quite saturated with the solute. This aids in avoiding separation of crystals as a result of slight cooling during filtration of the hot solution, which may clog the filter and funnel. With a substance that melts at a temperature below the boiling point of the solvent, enough of the solvent should be used to allow the solution to cool to a temperature below the melting point of the dissolved substance before the latter separates out; otherwise, the material will separate at first in oily droplets instead of well-formed crystals.

During the preparation of the solution in the hot solvent, the liquid should be stirred or shaken to aid the solution process, as many organic substances dissolve quite slowly. It is advantageous to crush any large crystals or lumps of the sample beforehand. One should avoid using an excessive amount of solvent through haste or attempting to bring insoluble foreign matter into solution. It is better to err on the side of an insufficient

---

[4] For hints on general solubility relationships, see Section 6.3.

amount of solvent; the undissolved residue remaining after filtration or decantation of the hot solution may then be tested with a fresh portion of solvent to see if more of it will dissolve.

Frequently a mixture of two solvents (solvent-pair) is more satisfactory than a single solvent. Such solvent-pairs are made up of two mutually soluble liquids, one of which dissolves the substance readily and another that dissolves it very sparingly. Examples of solvent-pairs are ethanol and water, glacial acetic acid and water, 2-propanol and petroleum ether, and cyclohexane and ethyl acetate. A typical procedure in using a solvent-pair, such as ethanol and water, is to dissolve the substance in the better solvent at the boiling point and add the poorer solvent dropwise, with shaking or stirring, until a faint turbidity persists in the hot solution. A few drops of the better solvent are then added, slightly more than the minimum amount required to form a clear solution, so that the hot solvent mixture will not be quite saturated with the solute. The hot solution is subsequently treated in the usual way (clarified with carbon, filtered, etc.).

Filtration of the hot solution to remove insoluble impurities, decolorizing carbon, and so on, must be carried out rapidly and efficiently to avoid crystallization of the dissolved substance in the filter and funnel. For this purpose, a large fluted filter paper (Figure 5.5) may be used to obtain rapid filtration, and a large funnel with a very short wide stem (or a stemless funnel) is employed to avoid clogging due to separation of crystals in

FIGURE 5.5 Preparation of a Fluted Filter Paper

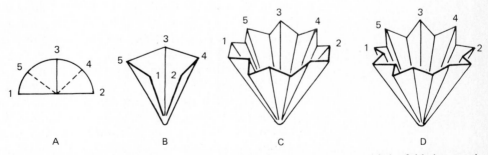

Select a filter paper and glass funnel of such dimensions that the top of the folded paper is 5–10 mm *below* the upper rim of the funnel (120-mm-diam. paper for a 65-mm funnel, 185-mm paper for a 100- to 120-mm funnel, etc.). *A*: Fold the paper in half, and again into quarters (fold *3*). *B*: Bring each edge *in to* the center fold and crease again, producing new folds at *4* and *5*. Do not crease the folds tightly at the center—this would weaken the central portion so that it might break during filtration. Grasp the folded filter cone (*B*) in the left hand, and make a new fold in each segment—between *2* and *4*, between *4* and *3*, and so on—*in a direction opposite* to the first series. The result is a bellows or fan arrangement. *C*: Open the filter paper completely and note the two places, next to folds *1* and *2*, where the paper would lie flat against the side of the funnel. *D*: Fold each of these sections in half, forming two smaller flutings, only half as deep as the others. Strengthen all of the flutings by creasing lightly a second time, and the paper is ready for use.

the stem. Just before the filtration is started the funnel should be preheated by placing it in the mouth of the flask in which the hot solution is being prepared and allowing the hot vapors to warm the funnel.

It is usually desirable to produce small, uniform crystals of the purified material. To accomplish this, reheat the hot filtered solution to redissolve any crystals that have deposited during filtration; crystals that have formed in the funnel may be recovered by placing the funnel in the mouth of the flask, where condensed vapor of the solvent will redissolve the crystals. The clear solution is covered and set aside to cool undisturbed. With a substance that contains only traces of impurity and crystallizes readily, it may be advantageous to cool the hot solution rapidly with vigorous stirring, so that small, uniform crystals are obtained. Since organic substances differ greatly in their rates of crystallization and materials of varying degrees of impurity will be encountered, it is necessary to adapt the crystallization procedure to the specific material at hand.

Not infrequently, a crystalline organic compound is encountered that exhibits a marked tendency to separate from solution, even at temperatures well below the melting point of the pure substance, in the form of an oily liquid that cannot easily be induced to crystallize. This situation arises especially with low-melting solids and with mixtures of closely related compounds. In such cases satisfactory crystallization may be obtained by allowing the solution to cool slowly and to remain undisturbed for a considerable period of time. Often it is advantageous to inoculate the solution with a few *tiny* crystals of the desired product (seed crystals), which will serve as crystal nuclei, and then to allow the solution to remain undisturbed. The necessary seed crystals may be secured by inducing crystallization in a test portion by vigorous scratching with a glass rod or by reserving a small portion of the original sample of material. If a solution is already supersaturated, cooling to a very low temperature may not hasten crystallization, since the rate of crystallization is apt to be reduced by lowering the temperature.

It may be necessary to prepare a dilute solution and allow the solvent to evaporate slowly to secure crystals, but this practice should be used only as a last resort, since it is difficult to obtain a pure product in such a manner. (Why?)

For collecting the purified crystals and separating them from the mother liquor, filtration is carried out by means of a Büchner funnel and a suction filtering flask of heavy glass with a side tube for attachment to a water pump to furnish suction (Figure 5.6). All parts of the Büchner funnel, including the inner inaccessible portion, should be perfectly clean and all holes in the perforated plate should be open (clean with a pin if necessary). The filter paper for the Büchner funnel should be selected (or trimmed if necessary) so that it lies flat on the perforated plate and does not fold up against the side. It is desirable to moisten the filter paper with the solvent

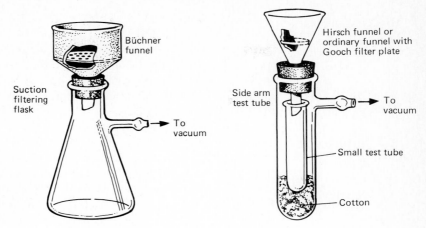

FIGURE 5.6
Apparatus for Suction Filtration

used in the crystallization process and apply suction before the filtration is started. The suspension of crystals is then poured onto the filter in such a way that a layer of uniform thickness is collected. Often it is convenient to use some of the filtrate to wash out crystals that adhere to the walls of the flask in which crystallization was carried out.

An objective of paramount importance in collecting a purified product is the complete *separation of the crystals from the mother liquor containing the dissolved impurities.* When carried out properly, suction filtration is very effective for this purpose. The crystals are first pressed down firmly on the filter with a clean cork or spatula (or a flat glass stopper) and sucked as dry as possible. Before the crystals are washed, the suction tube is disconnected and the crystalline cake is loosened *carefully* (avoid tearing the filter paper) with a spatula. The fresh cold solvent for washing is added in small portions and the material is stirred into a smooth paste. When this has been accomplished, suction is applied again and the crystals are pressed down firmly as before to remove the wash liquid as completely as possible. Washing the crystals with two or three small portions of solvent is more effective than a single washing with the same amount of solvent. It is particularly important in the washing operations to stop the suction and break up the crystalline cake, so that the whole mass comes into contact with the fresh solvent.

After the washed crystals have been pressed firmly and sucked as dry as possible, the crystalline cake is removed and spread out on a larger filter paper to permit evaporation of remaining traces of solvent. The final stages of drying may be done in a desiccator over an appropriate solid drying agent or in a drying oven regulated to a temperature well below the melting point of the substance. All traces of the solvent should be removed before the melting point of the purified substance is taken because the presence of solvent may lower the melting point appreciably.

5.3 Sublimation

The sublimation of a solid substance is an unusual type of distillation process in which the solid undergoes direct vaporization and the vapor is condensed directly to the solid state, *without the intermediate appearance* of a liquid state. For sublimation to occur, it is necessary that the solid have a relatively high vapor pressure at a temperature below its melting point. As relatively few organic compounds fulfill this condition, this method is not used frequently in ordinary laboratory work. When applicable, sublimation often affords a method of obtaining a product of high purity; it is particularly valuable for the isolation of a volatile solid from colored gums and tars and is best adapted to use with small amounts of material.

A simple apparatus suitable for sublimation of 1- to 2-g samples at ordinary or reduced pressures is diagrammed in Figure 5.7. The sample is contained in a 250-mL heavy-walled filtering flask that rests in a sand bath heated by a micro-burner. The flask is fitted with a rubber stopper that has been drilled to accept a 16 × 150-mm test tube, which, when filled with ice, serves as a cold-finger to condense the rising vapors. If the sample has a high enough vapor pressure to be sublimed at atmospheric pressure, the side-arm of the filtering flask is connected to a drying tube (Section 8.1) to prevent atmospheric moisture from condensing on the test tube. If the sample has too low a vapor pressure, the drying tube is omitted and the side-arm is connected directly to the vacuum system. The reasons for heating with a sand bath are two-fold. First, if a flame were applied directly to the bottom of the flask, the sample could easily be overheated and char. Second, the filtering flask has thick walls to prevent it from imploding when a vacuum is applied, and thick-walled vessels, even Pyrex ones, can crack if a flame is applied directly.

Examples of substances that can be purified by sublimation at atmo-

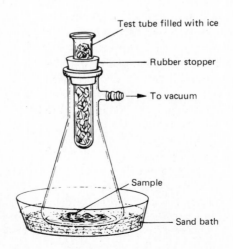

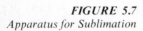

FIGURE 5.7
Apparatus for Sublimation

spheric pressure are camphor, hexachloroethane, anthraquinone, and many other quinones.

5.4 Representative Procedures

(A) Melting Points of Mixtures. Prepare three different mixtures of benzoic acid and 2-naphthol having roughly the following proportions of the two components: (1) about 90% benzoic acid and 10% 2-naphthol; (2) about 50% of each; (3) about 10% benzoic acid and 90% 2-naphthol. For the purpose of this experiment it is permissible to judge the proportions very roughly by the relative lengths of long thin piles of the powdered crystals (e.g., 9 parts benzoic acid and 1 part 2-naphthol for mixture No. 1, etc.). The components must be powdered finely and each sample mixed *very thoroughly*, on a piece of smooth hard paper or a watch glass, by means of a clean spatula or knife blade.

Introduce the mixture into melting point tubes marked appropriately for identification (see footnote 1, Section 5.1) and determine the melting points of the mixtures by the following procedure.

Apply heat at a moderate rate until the bath liquid is within 15–20° of the melting point (in this instance to about 100°; when necessary, make a rough preliminary determination). Continue the heating so that the temperature rises slowly and at a uniform rate (about 2°/min). Observe carefully the samples in the melting point tubes and the thermometer reading. Record as the observed melting point the range between the thermometer reading when the sample starts to liquefy and that when the melt is clear. Note also whether the sample undergoes preliminary fusing together (sintering) or discoloration, melts sharply or slowly over a wide range, and so on.

After the samples have melted, extinguish the flame and allow the bath to cool. Melting point tubes are discarded into the waste crock (not into the sink) after a single use. Record the observed melting points directly in your notebook. From your results, draw a rough melting point curve for mixtures of benzoic acid and 2-naphthol, plotting compositions on the x-axis and melting points on the y-axis. For this purpose, the temperature taken as ordinate is that at which the mixture liquefies completely.

(B) Crystallization from Water. Weigh out 2 g of impure benzoic acid for recrystallization,[5] and transfer it to a 125-mL Erlenmeyer flask. Add about 40 mL of water, and bring the mixture to the boiling point by heating it on a wire gauze (preferably one with a heat-dispersing center). Add successive

---

[5] The impurities may include soluble and insoluble materials as well as colored substances.

small portions of hot water, while stirring the mixture and boiling gently, until the benzoic acid has dissolved completely; then add an additional 10–15 mL of the hot solvent. The objective is to dissolve the solid in slightly more than the minimum amount of hot solvent. Do not use too much solvent by attempting to dissolve resins, mechanical impurities, and the like.

Remove the hot solution from the heat source and allow it to cool slightly. To the hot solution add gradually, with care to avoid excessive foaming, about 0.2 g ($\frac{1}{10}$ teaspoon) of decolorizing carbon (Norit, Darco, etc.), and boil for a short while longer to aid in removing small amounts of colored impurities. Meanwhile prepare a funnel, a fluted filter (see Figure 5.5), and a clean flask to receive the filtrate. The funnel should be at least 120 mm in diameter and have a very short stem (15–30 mm).

Heat the funnel by inverting it on a steam bath, dry with a clean towel, place the fluted filter in the funnel, and arrange the receiving flask to collect the hot filtrate. Without allowing the funnel or the solution to cool, pour the solution into the filter. If the solution cannot be poured into the filter in a single portion, replace it on the wire gauze and continue to heat it to prevent cooling. As soon as all of the solution has been filtered, cover the mouth of the flask containing the hot filtrate with a watch glass and allow it to cool and stand undisturbed. Do not close the mouth of the flask tightly with a stopper. (Why?)

If crystals have separated in the hot filtrate during filtration, the filtrate should be heated to redissolve them. Crystals that have formed in the funnel may be recovered by placing the funnel in the mouth of the receiver and allowing condensing vapor from the hot solution to redissolve them.

When the product has separated completely, filter the crystals with suction on a properly fitted filter paper in a clean Büchner funnel. Wash the crystals twice with a little cold water, then press them as dry as possible on the funnel with a clean spatula or cork. Spread the crystals on a large clean paper, *allow them to dry completely*, and weigh them. Determine the melting point (see part (A)); if the product is not sufficiently pure, as shown by a melting range greater than 1–2°, recrystallize the material in a similar way. Calculate the precentage recovery of pure benzoic acid.

(C) Crystallization of Acetanilide from Water.[6] Recrystallize a 5-g sample of impure acetanilide from about 100 mL of water using a 250-mL Erlenmeyer flask. Follow the general directions given in part (A). About 0.2 g of decolorizing charcoal should be sufficient. In washing the crystals, be careful to use ice-cold water and not too much of it, since acetanilide is somewhat soluble even in cold water (about 0.5 g/100 mL).

[6] Recrystallization of acetanilide is challenging because in concentrated solutions it tends to separate as an oil rather than a crystalline solid. At the same time, too much solvent must be avoided, because acetanilide is somewhat water soluble at room temperature.

(D) Crystallization from a Flammable Solvent. Weigh out 1.5 g of impure dimethyl terephthalate for recrystallization and transfer it to a 125-mL Erlenmeyer flask. Add 25 mL of 95% ethanol and warm the mixture on a steam bath until the solvent boils. Add successive small portions of ethanol (not more than 5–10 mL total) and boil gently after each addition, until the dimethyl terephthalate has dissolved; then add 1–2 mL more of the solvent. Do not attempt to dissolve admixed particles of sand, grit, and so on. Meanwhile, prepare a fluted filter (Figure 5.5) and arrange a funnel (65-mm diameter, 10–20-mm stem) and a clean dry flask to receive the filtrate.

▶ *CAUTION* Flammable solvents such as ether, alcohols, and hydrocarbon must never be heated in an open flask over a burner, or manipulated near a flame. These solvents should be heated on a water or steam bath (or an electric hot plate having no exposed hot filament). If a burner is used, the flask must be fitted with an upright reflux condenser. Take particular care to insure that no lighted burner is nearby during the filtration of the hot solution!

Remove the boiling solution from the steam bath, add gradually 0.2 g of decolorizing carbon (*caution—frothing*), and swirl the solution gently. Reheat to boiling and pour the hot solution into the fluted filter.[7] Cover the mouth of the flask containing the hot filtrate with a watch glass[8] and allow it to cool and stand undisturbed. When the product has separated completely, collect the crystals with suction on a properly fitted filter paper in a clean Büchner funnel; wash all the crystals into the funnel by rinsing the Erlenmeyer flask with part of the filtrate. Discontinue the suction and wash the crystals with two 3- to 5-mL portions of fresh ethanol (cold). Apply suction again and press the crystals firmly with a cork or flat glass stopper. Spread the crystals on a large clean paper, *allow them to dry thoroughly*, and record the weight of purified dimethyl terephthalate. Calculate the percentage recovery and determine the melting point of the purified product.

(E) Crystallization of Naphthalene from Ethanol. Recrystallize a 5-g sample of naphthalene from about 25 mL of ethanol in a 125-mL Erlenmeyer flask. Follow the general directions given in part (C). About 0.2–0.3 g of decolorizing charcoal should be sufficient. Care is required in drying the crystals, since naphthalene is volatile and will sublime slowly at room temperature.

---

[7] Occasionally a few fine particles of the decolorizing agent may pass through the pores of the filter paper. Often this may be remedied by heating the filtrate to boiling and filtering again through the same filter; or a small amount of filter aid (Filter-Cel, Super-Cel, etc.) may be added. If the filter has been damaged in folding or is defective, a fresh one must be used.

[8] If crystals have separated in the hot filtrate during the filtration, the filtrate should be heated to redissolve them. Crystals that have formed in the funnel may be recovered by placing the funnel in the mouth of the receiver and allowing condensing vapor from the hot solution to redissolve them.

(F) Sublimation. Assemble the apparatus pictured in Figure 5.7. Place 1 g of impure *p*-dichlorobenzene in the filtering flask and spread it uniformly over the bottom. Insert the test tube, attach a drying tube to the side arm of the flask, and fill the test tube with ice. Warm the flask gently in a sand bath. When no more material sublimes, remove the condenser and scrape the condensate into a tared bottle. Record the weight of purified material and determine its melting point range.

Questions

1. Define accurately the term melting point.
2. What general conclusions can be drawn from the determination of melting points of mixtures?
3. Suppose that two different organic compounds, M and N, have about the same melting point and that you are given an unknown organic compound, X, which also has the same melting point and is suspected of being identical with either M or N. Describe a procedure for identifying X and state the results you would obtain in each of the following situations.
   (a) X is identical with M.
   (b) X is identical with N.
   (c) X is not identical with either M or N.
4. (a) What physical constants other than melting point and melting points of mixtures may be used to aid in establishing the identity of an organic solid?
   (b) What constant other than boiling point may be used for an organic liquid?
5. Explain why it is essential to
   (a) pack the sample densely and tightly in the melting point tube
   (b) heat the bath slowly and steadily in the vicinity of the melting point
6. Outline the successive steps in the crystallization of an organic solid from a solvent and state the purpose of each operation.
7. What properties are necessary and desirable for a solvent to be well suited for recrystallizing a particular organic compound?
8. Not counting solubility relations and effectiveness for removal of impurities, in what respects would *n*-hexyl alcohol be a less desirable crystallization solvent than methanol or ethanol?
9. (a) Mention at least two reasons why suction filtration is preferable to ordinary gravity filtration for separating the purified crystals from the mother liquor.
   (b) Why is it desirable to release the suction before washing the crystals with small portions of the fresh solvent?
10. What means other than crystallization from a solvent may be used to purify an organic solid or to effect preliminary separation of a solid mixture?

6 Extraction with Solvents

The process of extraction with solvents is used in organic chemistry for the separation and isolation of substances from mixtures that occur in nature, for the isolation of dissolved substances from solutions, and for the removal of soluble impurities from mixtures.

6.1 Extraction of Solids

The extraction of alkaloids from leaves and bark, flavoring extracts from seeds, perfume essence from flowers, and sugar from sugar cane are examples of extractions of the first type. Solvents commonly used for this purpose are ether, methylene chloride, chloroform, acetone, various alcohols, and water. A common form of apparatus for continuous extraction of solids by means of volatile solvents is called a Soxhlet extractor. A typical laboratory setup employing this extraction device is shown, mounted and ready for use, in Figure 6.1.

Vapors from the solvent boiling in the pot rise through the vertical tube at the left into the condenser at the top. The liquid condensate drips into the filter paper thimble in the center, which contains the solid sample to be extracted. The extract seeps through the pores of the thimble and eventually fills the siphon tube at the left, where it can flow back down into the pot. If the sample being extracted is not volatile, it gradually accumulates in the pot. With the apparatus shown, the siphoning action is intermittent. No liquid will flow through the siphon until the liquid level in the thimble

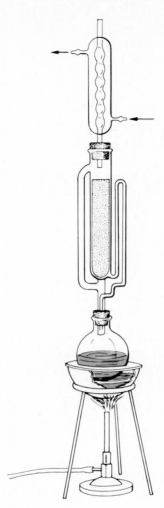

FIGURE 6.1
Soxhlet Extractor Assembly

reaches the top of the siphon tube. At that point almost all of the liquid in the siphon and the thimble drain out and the cycle of filling and draining starts again.

6.2 Extraction of Solutions

A more common application of extraction is in liquid–liquid extraction, which is used to isolate a dissolved substance by shaking a solution of it with an immiscible solvent in a separatory funnel (see Figure 6.2). In the ideal situation, the substance is extracted into the second solvent, the impurities are left behind, and after the two layers are separated, the substance is isolated by removal of the solvent.

The general principle underlying this process is known as the distribution law. In dilute solutions a substance distributes itself between two immiscible solvents so that the ratio of the concentration in one solvent to the concentration in the second solvent always remains constant (at constant temperature). This constant ratio of concentrations for the distribution of a solute between two particular solvents is called the distribution coefficient or the partition coefficient for the substance between the two solvents.

$$\text{Distribution coefficient}[1] \text{ of S between solvents A and B} = \frac{\text{concentration of S in A}}{\text{concentration of S in B}}$$

$$= K \text{ (at constant temperature)}$$

where the concentrations are expressed in grams per milliliter of the solution.[2] Actually, no two solvents are absolutely immiscible; each solvent is at least slightly soluble in the other. In practice the most common application of the distribution law is to the extraction of dissolved substances from aqueous solutions by *almost insoluble* solvents such as ether,[3] hexane, and methylene chloride.

The actual distribution coefficient is determined by bringing the solvents and solute into equilibrium at a given temperature and determining experimentally the concentration of solute per unit volume of each separate phase. A rough approximation of the distribution coefficient can be made by determining the solubility of the solute in each pure solvent independently, since the distribution coefficient is approximately the same as the ratio of the solubilities in the two solvents. The values obtained in this way are subject to several errors but are usually sufficiently good for simple laboratory calculations.

---

[1] In the case of dissolved substances that are ionized or associated in one of the solvents, the distribution law holds true for the ratio of the concentrations of the simple molecules only. To obtain an expression that holds for the total concentrations, it is necessary to introduce an expression for the ionization or association equilibrium.

[2] The values of the distribution coefficients found in the chemical literature are usually based on concentrations per volume of solution but in making rough calculations of the type illustrated in the following paragraphs, it is much simpler to use concentrations per unit volume of the solvent. For dilute solutions, the error introduced in this way is negligible. It is important to observe the manner in which the ratio is expressed; for example, at 20° the distribution coefficient of butanoic acid between ether and water is approximately 5 (conc in ether/conc in water = 5), but if the ratio is expressed as the distribution between water and ether the value is 0.2 (conc in water/conc in ether = 1/5).

[3] Ether, which is used frequently as an extraction solvent, is somewhat more soluble than other common solvents; for example, at 30° diethyl ether is soluble to the extent of 1 g in 18.8 g of water, and water is soluble to the extent of 1 g in 73 g of ether. Methylene chloride (bp 40°) has several advantages over ether and is to be preferred whenever it has satisfactory solvent properties; methylene chloride is less soluble in water (about 2 g/100 mL at 20°) and is not flammable under ordinary conditions.

In order to apply the distribution law to laboratory extractions it is convenient to use the formula shown in equation (6.1), which gives the fraction of compound S remaining in volume V_A of solvent A after extraction with a volume V_B of solvent B.

$$\text{Fraction remaining in A} = \frac{C_{\text{final}}}{C_{\text{initial}}} = \frac{1}{1 + \dfrac{V_B}{V_A K}} \tag{6.1}$$

where C_{final} is the final concentration of S in A, C_{initial} is the initial concentration of S in A, and K is distribution coefficient (concentration of S in A divided by the concentration of S in B).

As an example of the application of equation (6.1), consider the situation in which the distribution coefficient of the compound S between hexane and water (K hexane/water) is $\frac{1}{3}$ at 25°. If a hexane solution containing 8 g of S in 100 mL of hexane is extracted at 25° with 100 mL of water, the fraction of S remaining in the hexane is

$$\frac{1}{1 + \dfrac{100 \text{ mL}}{100 \text{ mL} \times \frac{1}{3}}} = 0.25$$

whence the weight of S in the hexane layer is 2.0 g and the amount in the aqueous layer is 6.0 g.

If the hexane layer in this example is separated and extracted with another fresh, 100-mL portion of water, the fraction of S remaining will be reduced by another factor of 0.25 to give only 0.5 g remaining in the hexane layer. The second aqueous extract contains 1.5 g of S, which when combined with the 6.0 g of S in the first aqueous extract gives a total of 7.5 g of S in the combined aqueous layers.

6.3 Multiple Extraction

One can ask whether, with a specified quantity of solvent, it is preferable to make a single extraction with the total quantity of solvent or to make several successive extractions (multiple extraction) with portions of the solvent. The answer, practical considerations aside, is that the amount of material extracted is greater with the second method.

The general multiple extraction formula shown in equation (6.2) gives the fraction of compound S remaining in volume V_A of solvent A after n extractions with solvent B, each of volume V_B/n.

$$\text{Fraction remaining in A} = \frac{C_\text{final}}{C_\text{initial}} = \left(\frac{1}{1 + \frac{V_B}{V_A nK}} \right)^n \tag{6.2}$$

where the terms have the same meaning as in equation (6.1).

In the limit of very large n it can be shown that equation (6.2) reduces to equation (6.3), which gives the theoretical maximum for the amount of S that can be extracted from a volume, V_A, of solvent A with a given volume, V_B, of solvent B.

$$\text{Fraction remaining in A} = e^{-V_B/V_A K} \tag{6.3}$$

for large n, where $e = 2.718$, the base for natural logarithms.

With the example used in the previous section, and assuming that both V_A and V_B equal 100 mL, the greatest possible reduction in the amount of S in the hexane is

$$e^{[-100 \text{ mL}/(100 \text{ mL} \times 1/3)]} = e^{-3} = 0.05$$

Thus, with just 100 mL of water, it is theoretically possible to extract 95% (7.6 g) of S if the 100 mL is divided into a large number of portions, which are used to consecutively extract the hexane layer.

Whether the increased efficiency of extraction is worth the trouble depends very much on the circumstances. It is worth noting that essentially the same amount of S can be extracted by successive extractions with two 100-mL portions of water. In practice, most chemists prefer to use a larger total volume of solvent and do only a few successive extractions.

Effect of Salts on Solubility. The solubility of organic substances in water is markedly affected by the presence of dissolved inorganic salts. For example, ethanol, which is perfectly miscible with pure water, is only slightly soluble in strong aqueous solutions of sodium chloride, potassium carbonate, and certain other inorganic salts. The same is true of acetone, pyridine, methanol, and many other water-soluble organic compounds. This phenomenon (*salting-out effect*) occurs commonly with salts that have ions of small radius and concentrated charge. The opposite effect of enhanced solubility in salt solutions (*salting-in*) occurs frequently when the salt has ions of large radius and diffuse charge. Benzene, for example, is about 40% more soluble in 1 M aqueous tetramethylammonium bromide than in pure water.

In the case of solutions of ionized organic substances, such as metallic salts of organic acids and salts of organic bases with mineral acids, it is possible that the *common-ion effect* may also act to decrease the solubility of the salt. Thus, sodium benzenesulfonate is quite soluble in water but is precipitated from an aqueous solution by adding sodium chloride; aniline hydrochloride is readily soluble in water but is only slightly soluble in strong hydrochloric acid solutions.

Solubility Relationships. The solubility of an organic compound in inert solvents (water, alcohols, ether, hydrocarbons, etc.) and in chemically reactive solvents is directly related to its molecular structure. Consequently, one can predict solubility relations in a qualitative way for various classes of compounds by taking into account the structure of the solute and the physical and chemical characteristics of the solvent medium. The factors to be considered are polarity, hydrogen-bonding ability, acidity, basicity, and whether it is ionic. Saturated hydrocarbons, as their other name, paraffins, suggests, do not interact significantly with species having these attributes. For this reason saturated hydrocarbons represent an extreme solvent type. By contrast, water and acids are polar, strong hydrogen-bond donors, and modest hydrogen-bond acceptors; they interact strongly with similar molecules as well as with ionic salts and basic compounds. Alcohols contain a combination of hydrocarbon and hydroxyl groups and have intermediate properties. Liquid ammonia and amines represent the analogous spectrum of behavior among basic solvents for which the hydrogen-bond accepting quality dominates. Another extreme of solvent behavior is observed with nonhydroxylic solvents such as dimethylformamide, dimethylsulfoxide, and acetonitrile for which polarity is the dominant quality.

Solubility in an inert solvent may be predicted with fair success on the basis of the following empirical solubility rules, which reflect the structural factors described previously.

1. A substance is most soluble in that solvent to which it is structurally most closely related. Thus, simple alcohols are soluble in water, esters are soluble in alcohol and ether, and so on.
2. In any homologous series, the higher members become more and more like hydrocarbons in their physical properties. In most homologous series containing only one functional group, the solubility in water falls below 4–5% when the member attains five carbon atoms.
3. A compound with branched chains is more soluble in a given solvent than the straight-chain isomer. Thus, isobutyl and *t*-butyl alcohols are much more soluble than *n*-butyl alcohol in water.

4. Liquids and low-melting solids are generally more soluble than high-melting solids in inert solvents.
5. Compounds of high molecular weight (polymers, etc.) are usually sparingly soluble in inert solvents. However, polymers often form colloidal dispersions in certain solvents.

These simple rules are a convenient guide to solubility relations, but one must expect to encounter many exceptions in actual practice.

Solubility tests are often used as a preliminary step in the identification of organic compounds. The systematic outline developed by Kamm is shown in modified form in Section 9.6. The reader is encouraged to compare the entries in this diagram with the previous discussion of extraction by chemically active solvents.

Solubility in a chemically reactive solvent depends on formation of a soluble salt and is readily predicted from the structures of the solute and the solvent. Commonly used *reaction solvents* are dilute aqueous alkalis (5% sodium hydroxide or potassium hydroxide solutions); dilute aqueous mineral acids (5% hydrochloric acid, etc.); cold concentrated sulfuric acid.

Dilute sodium hydroxide solution (also sodium carbonate or bicarbonate) can be used to extract an organic acid from its solution in an organic solvent, or to remove traces of acid that are present as an impurity in an organic preparation. The use of aqueous alkali depends upon the conversion of the free acid to the corresponding sodium or potassium salt, which is soluble in water or dilute alkali. Thus, butanoic acid may be extracted quantitatively from a benzene solution by dilute sodium hydroxide because this reagent converts the acid to sodium butanoate, which is very soluble in water or dilute alkali but insoluble in benzene. Likewise, an organic acid or mineral acid present as an impurity in a water-insoluble liquid or solid can be removed by washing with dilute alkali.

Dilute hydrochloric acid can be used in a similar way to extract basic substances from mixtures or to remove impurities. The use of dilute acids depends upon converting the base (amines, ammonia, etc.) into a water-soluble salt (amine hydrochloride, ammonium chloride, etc.). Thus in the preparation of benzanilide, unreacted aniline can be removed by washing with dilute hydrochloric acid, and the benzoic acid side product can be removed by subsequent washing with sodium carbonate solution; the anilide does not react with either the acid or the base, but the amine is converted into the water-soluble salt aniline hydrochloride, and the benzoic acid is converted into the water-soluble salt sodium benzoate.

Cold concentrated sulfuric acid can be used to remove unsaturated hydrocarbons, alcohols, ethers, esters, and so forth, present as impurities in inert compounds such as saturated hydrocarbons and alkyl halides. Alkenes and alcohols are dissolved by chemical reaction to form alkyl hydrogen sulfates; ethers, esters, and so on, form addition complexes that are soluble in concentrated sulfuric acid.

6.4 Laboratory Practice

The selection of a solvent for extraction involves considerations analogous to those for crystallization. Properties desired for a suitable solvent are (1) it should readily dissolve the substance to be extracted (favorable distribution coefficient); (2) it should be sparingly soluble in the liquid from which the solute is to be extracted; (3) it should extract little or none of the impurities or other substances present; (4) it should be capable of being easily separated from the solute after extraction (usually by distillation); and (5) it should not react chemically with the solute in an undesired way (see chemically reactive solvents in Section 6.3). Relative cost, ease of manipulation, flammability, and similar factors will be significant in choosing among various possible solvents.

Laboratory extractions are commonly carried out by shaking the solution to be extracted, together with the extraction solvent, in a glass separatory funnel. A long tapered funnel of the Squibb type, with a short stem cut off obliquely (Figure 6.2), is particularly convenient for this purpose. Cylindrical or pear-shaped funnels with a short stem may also be used. A long stem is disadvantageous, because it holds a long column of the liquid that is being drawn off and makes for difficulties in manipulation.

The separatory funnel should be of such size that it is more than three-fourths filled by the solution and solvent. The funnel is shaken to obtain good physical mixing of the insoluble liquids and is then allowed to stand undisturbed until the layers have separated completely. Vigorous shaking is desirable provided it does not produce an emulsion[4] that interfers with subsequent separation of the layers; for such systems, it is better to mix the layers by repeated gentle swirling.

[4] Emulsions of water solutions with solvents such as ether and benzene may often be broken by adding 1–2 mL of ethanol or ethyl acetate and swirling the contents of the funnel gently.

FIGURE 6.2
Extraction Funnels

Squibb   Pear   Cylindrical

During the shaking operation, it is *essential* to grasp the funnel with both hands, one hand at the top and the other on the stopcok, in such a way that the stopper and stopcock are held firmly in place. The internal pressure should be released from time to time by inverting the funnel (so that the stem points upward) and opening the stopcock momentarily until the pressure is reduced. This is particularly important when a very volatile solvent, such as ether, is used.

▶ *CAUTION*   Do not point the stem of the separatory funnel toward any one when releasing the pressure. Drops of liquid caught in the stem may be ejected forcefully.

When the liquids have separated, the stopper is loosened or removed and the lower layer is carefully drawn off into an Erlenmeyer flask. The stopcock should be manipulated with both hands to avoid loosening it, with consequent loss of material. If the liquid remaining in the funnel is corrosive or valuable, it is good practice to place a beaker or pan beneath the funnel if it is to stand for any length of time. If only one layer is to be retained, *it is a safe rule to save both layers* until you are certain which one contains the desired material.[5]

As the interface of the two liquids approaches the stopcock, the liquid should be drawn off slowly. After the separation has been made, the funnel is rotated by a twisting motion with the stopcock closed, to assist in draining droplets of the more dense liquid from the walls. This small additional quantity is drawn off into the receiving flask. The upper layer is poured through the mouth of the funnel into a clean flask; it is not allowed to flow through the stopcock, as this would lead to contamination with the liquid adhering to the stem. The organic solvent layer is usually dried by means of an appropriate solid drying agent (see Section 8.1), and the solvent is removed by distillation.

In any operation where a separatory funnel is used, the stopcock must be kept lubricated to avoid sticking or leakage during manipulations. The only exception is with Teflon or Teflon-coated stopcocks, which are self-lubricating. After it is used, the separatory funnel should be cleaned thoroughly and the stopcock freshly lubricated to inhibit "freezing" in a fixed position. If a silicone-based lubricant is used, it should be removed before cleaning with an oxidizing mixture or a siliceous film will be formed on the glass surface. The problem of "frozen" stopcocks is sufficiently severe that many workers prefer to store funnels disassembled. Stopcocks that have become "frozen" sometimes may be freed by warming the outer barrel with steam and applying gentle pressure to the stopcock plug (using a towel to protect the fingers). A vise-like stopcock plug remover is available from scientific supply firms.

[5] An aqueous layer can be distinguished from a water-insoluble layer by adding a small test portion to a few milliliters of water in a test tube; the aqueous layer will form a homogeneous solution, whereas the nonaqueous layer will form a two-layer system.

For special purposes, laboratory methods and apparatus have been developed for continuous extraction of aqueous solutions with immiscible solvents and for multiple countercurrent extraction. In some instances, an extraction solvent made up of a mixture of solvents is employed.

6.5 Representative Extractions

(A) Simple and Multiple Extraction. The purpose of this procedure is to demonstrate the effectiveness of a single extraction with a fixed volume of solvent compared with two extractions, each with half of the fixed volume. An aqueous solution of the intensely purple dye, crystal violet, will be extracted by methylene chloride (dichloromethane) in which the dye is somewhat more soluble.

1. Simple extraction. Place 30 mL of the stock aqueous solution of crystal violet (0.2 g/L) in a clean 125-mL separatory funnel and add 30 mL of methylene chloride. Stopper the separatory funnel shake it gently, and turn it upside down. While the separatory funnel is in this position, open the stopcock to release the internal pressure, close the stopcock, shake it vigorously, and again release the internal pressure.

▶ *CAUTION* Do not point the stem of the funnel at anyone when you release the pressure. Any liquid in the stem may be ejected forcefully.

Repeat this procedure four or five times, then support the separatory funnel upright in a ring and let it stand undisturbed. When the liquids have completely separated, draw off the lower methylene chloride layer into a Erlenmeyer flask (following the procedure for laboratory extractions given in Section 6.4).

▶ *CAUTION* In the manipulation of organic solvents in a separatory funnel with a glass stopcock, it is important that the stopcock be kept properly greased to avoid sticking (see Section 6.4). If this is not done, the stopcock is likely to become "frozen" in a fixed position and the separatory funnel will be rendered useless.

Transfer a portion of the remaining aqueous layer to a test tube and set it aside for later comparison.

2. Multiple extraction. Clean the separatory funnel well with water and place a second 30-mL portion of the stock solution of crystal violet in it. Extract the solution with 15 mL of methylene chloride, as described in part 1. Draw off the methylene chloride layer into the Erlenmeyer flask used in part 1, and extract the remaining aqueous layer with a second fresh 15-mL portion of methylene chloride. Draw off the methylene chloride into the Erlenmeyer flask and transfer a portion of the aqueous layer into a clean test tube of the same size used in part 1. Fill both tubes to the same height.

Compare the effectiveness of extraction by the two different procedures by noting the intensity of color remaining in the aqueous layer. The difference in color will be more noticeable if you look down the mouth of each test tube.

Pour all of the methylene chloride extracts into a bottle in the hood labeled Methylene Chloride from Extraction Experiments.

(B) Separation by Chemically Active Solvents. Obtain from your instructor 60 mL of a solution containing 5 g of benzoic acid and 5 g of acetanilide dissolved in methylene chloride. Calculate the volume of 2 N aqueous sodium hydroxide required to react completely with the benzoic acid and extract the methylene chloride solution twice with the 2 N base using the calculated volume each time (follow the procedure for laboratory extractions given in Section 6.3). Complete the extraction by washing the methylene chloride solution with 50 mL of water; combine the aqueous layers in a 250-mL beaker. Place the methylene chloride layer in a 250-mL Erlenmeyer flask, add about 1 g of anhydrous magnesium sulfate,[6] and set the flask aside to allow the residual water time to be absorbed.

Cool the combined aqueous extracts in an ice bath and add, while swirling, sufficient concentrated (12 N) hydrochloric acid to neutralize the base. Test the acidity of the solution with pH paper, and if it is not distinctly acidic add more acid dropwise until it is. Collect the precipitated benzoic acid and allow it to dry as described in Section 5.2 until the next period. Place the dried sample in a tared bottle and determine the weight of recovered benzoic acid. Determine and record the melting point of your sample.

Assemble a simple distillation apparatus (see Section 2.3) using a 250-mL round-bottom flask as the boiler. Transfer the methylene chloride solution into the boiler by passing it through a fluted filter to remove the drying agent and distill off all but 10 mL of the solvent. Pour the warm concentrated solution of acetanilide into a small beaker and rinse the flask with 5–10 mL of the recovered solvent. Evaporate the combined solutions to dryness on a steam bath in a hood. If hood space is not available, the vapors may be drawn off by means of an inverted funnel connected to a water aspirator. Transfer the solid to a tared bottle and determine the weight of recovered acetanilide. Determine and record the melting point.

[6] The anhydrous magnesium sulfate is used as a drying agent to remove droplets of water and traces of dissolved water. The amount is not critical, and only the amount needed to absorb the water should be added. Wetted drying agent is dense and clumps at the bottom of the flask, whereas unwetted excess drying agent remains fluffy and floats easily in the liquid when the flask is swirled.

If a rotary evaporator (see Section 8.6) is available, it can be used in place of the distillation assembly.

(C) Extraction of Caffeine from Tea. The stimulating effects of aqueous infusions of coffee beans, tea and mate leaves, and cola nuts are due mainly to the presence of caffeine, a nitrogen heterocycle of the molecular formula $C_8H_{10}N_4O_2$. Its structure has been established by study of its degradation products and by synthesis to be 1,3,7-trimethyl-2,6-dioxopurine. Tea leaves contain 3–5% of caffeine and a trace of theophylline, a lower homolog lacking the methyl group at position 7. These compounds are related structurally to the important purines adenine and guanine, which are present in the ribonucleic acids (RNA and DNA).

Caffeine

Guanine

In this experiment the caffeine is extracted from tea leaves by hot water, in which it is quite soluble (about 18 g/100 mL at 80°; 2.2 g/100 mL at 20°). Colored impurities such as tannic acids can be removed as calcium salts by adding calcium carbonate. From the aqueous extract the caffeine is isolated conveniently by multiple extractions with small portions of methylene chloride, in which caffeine is quite soluble. Caffeine forms a monohydrate that loses water rapidly on warming to give the anhydrous form, mp 238°.

In a 500-mL Erlenmeyer flask, place 30 g of ordinary dry tea, 300 mL of water, and 15 g of powdered calcium carbonate. After boiling the mixture gently for 20 min, occasionally swirling it, add 5 g of Celite or other filter aid, filter the hot mixture on a Büchner funnel, and press the filter cake firmly with a large cork to obtain as much of the liquid as possible.

Cool the aqueous extract to 15–20°, transfer it to a separatory funnel, and extract the caffeine with four successive 25-mL portions of methylene chloride (following the procedure for laboratory extractions given in Section 6.4).[7]

Transfer the extracts to a simple-distillation assembly (see Figure 2.10), and distill off all but 10 mL of the solvent on a steam bath. Save the recovered solvent. Pour the warm concentrated solution of caffeine into a small beaker and rinse the flask with 5–10 mL of the recovered solvent.

---

[7] Aqueous extracts of plant materials tend to form stubborn emulsions when extracted with organic solvents. An effective means for breaking difficult emulsions is to press, with the aid of a glass rod, a small wad of glass wool into the bottom of the separatory funnel and draw off the lower layer through the glass wool.

Evaporate the combined solutions to dryness on a steam bath in a hood. To purify the crude product, dissolve it in about 10 mL of hot toluene on a steam bath, add 15–20 mL of petroleum ether (bp 60–90°), and allow the product to crystallize. Collect the green-tinged crystals on a small suction filter and wash them with a little petroleum ether. The melting point of caffeine reported in the literature is 236°. This is too high to be safely determined in an apparatus using an oil bath.

The green color of the caffeine sample can be removed by sublimation.

Questions

1. What conclusions can you draw about the most efficient method of extracting acetic acid from an aqueous solution by means of an immiscible solvent?
2. What is meant by the term distribution coefficient?
3. What properties do you look for in a good solvent for extraction?
4. Explain the fact that acetic acid can be extracted quantitatively from an ether solution by dilute aqueous sodium hydroxide solution.
5. In the extraction of an organic compound from a dilute aqueous solution, will the organic solvent form the upper or lower layer when each of the following solvents is used?
    (a) chloroform
    (b) cyclohexane
    (c) *n*-heptane
    (d) methylene chloride
6. If toluene (density 0.87) is used to extract ethylene bromohydrin (density 2.41) from an aqueous solution,
    (a) Could you be certain that the organic solution would form the upper layer?
    (b) By what test could you determine which is the nonaqueous layer?
7. Name two organic compounds that cannot be extracted effectively from an aqueous solution by means of an immiscible organic solvent such as ether or cyclohexane.

7 Chromatography

The term chromatography is applied to numerous purification processes that share the principle of distributing a sample between a stationary phase and a mobile one. As with extraction, the degree of separation of a mixture is determined by differences in distribution coefficients, which are related to the same structural factors that control solubility.

7.1 Liquid–Solid Chromatography

Solid surfaces adsorb thin layers of foreign molecules as a result of forces identical in character with those operating between molecules of a liquid or a gas. Since adsorption strengths differ with the character of the solid surface, a properly chosen solid may adsorb selectively one component of a mixture. An important example of selective adsorption is the use of charcoal in crystallization to remove colored impurities. The ideal limiting law governing adsorption from a dilute solution is

$$\frac{\text{amount of solute A adsorbed per unit surface area}}{\text{concentration of solute A in solution}} = K_A$$

The factors that determine the extent of adsorption of a molecule on a solid surface are closely related to the factors that enter into solubility considerations. An additional complication in adsorption is that the solvent and the solute are competing for the active sites on the surface. For molecules containing polar functional groups the value of K_A (the adsorption coefficient) is determined principally by the relative polarities of the substance and the solvent. Highly polar solvents tend to be preferentially adsorbed, so that a low K_A results for the solute. For molecules containing hydroxyl groups, their relative abilities to form hydrogen bonds (proton bonding) to the solid or the solvent are significant.

Two solutes with different adsorption coefficients toward a certain solid can be separated by the process of liquid–solid chromatography. One of the most practical methods involves preparation of a cylindrical column of the solid (*stationary phase*) and addition of a concentrated solution (*liquid phase*) at the top of the column. As the solution penetrates the column, the solutes are adsorbed. At the moment the solution has completely penetrated the column, fresh solvent is added at the top. The solvent flows down the column and redissolves the solutes in amounts determined by the adsorption law and carries them to lower clean sections of the column, where they are readsorbed (in amounts governed by the adsorption law). As more solvent percolates through the column, the cycle of adsorption–solution continues, and the solutes gradually move down the column in concentrated bands (*development*). With solutes having different adsorption coefficients, the least tightly adsorbed material tends to move ahead more readily. If the coefficients are sufficiently different or the column is sufficiently long the faster moving component will form a separate band below the slower moving one. At the lower end the solutes are forced off the column (*elution*) and can be collected separately in successive fractions.

For satisfactory separation by liquid–solid column chromatography, it is essential to choose an appropriate combination of solid adsorbent and eluent that is compatible with the compounds to be separated. Compounds that are adsorbed very tightly require an excessive volume of eluent for development. Compounds adsorbed weakly may move too rapidly to give separation before being eluted. Table 7.1 gives some generalizations that are useful as a guide in selecting appropriate solid–solvent combinations.

A common variation of liquid–solid chromatography is the use of a thin film of solid (mixed with a binder such as plaster of paris) on a sheet of glass or plastic. The solution is added as a spot at the bottom of the plate and the plate dipped vertically into a shallow layer of solvent, which ascends by capillary action and moves the solutes with it. The particular advantage of this technique is that the solutes are exposed and can be isolated readily or treated on the plate at any moment. The method is widely used for qualitative identification of mixture components because of its exceptionally good resolution. For a fixed combination of solid, binder, and solvent, each substance will travel along the thin-layer plate a characteristic fraction of the distance traveled by the solvent. It is customary to report thin-layer chromatography data as R_f values (retention factor), defined as the distance of the spot from the starting point divided by the distance of the solvent front from the starting point. Thin-layer chromatography is restricted to small samples. A method known as "flash" chromatography, which combines the

TABLE 7.1
Adsorbants and Solvents for Liquid–Solid Chromatography

Chromatographic solids in decreasing order of adsorption strength for polar molecules	Solvents in increasing order of eluting ability[b]
Activated alumina,[a] charcoal	Saturated hydrocarbons
Activated magnesium silicate[a]	Aromatic hydrocarbons
Activated silicic acid[a]	Partially halogenated hydrocarbons
Inorganic carbonates	Ethers
Sucrose, starch	Ketones
	Alcohols
	Organic acids

[a] The adsorption strength can be diminished by adding water. Under carefully controlled conditions, this strength is reproducible.
[b] This approximate order only applies to alumina. With nonpolar solids, the order tends to be inverted.

high resolution and speed of thin-layer chromatography with the large sample capacity of regular column chromatography, is described in Section 7.5.

On a chromatographic column or a thin-layer plate, a single component usually appears as a more or less uniform spot. However, on careful analysis it will be found that the sample is distributed along the length L of the column or plate, as shown in Figure 7.1. The position of maximum concentration, when the solvent just reaches the end, occurs at length l from the starting point. By definition $R_f = l/L$; it can be demonstrated that R_f is also equal to $n/1 + K$, where K has the definition given at the beginning of this section and n is the average number of effective exchanges of a molecule of the sample on and off the surface in the time it takes the solvent front to

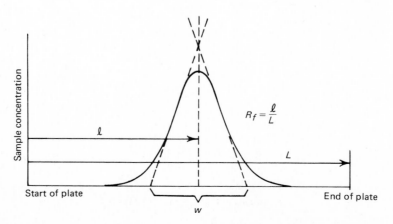

FIGURE 7.1
Chromatographic Sample Distribution

traverse the column or plate.[1] For tightly retained samples, K is very large and n is small. For the ideal case it has been shown that

$$n \approx 16\left(\frac{l}{w}\right)^2 \tag{7.1}$$

where w is the width at the base of the triangle formed by drawing straight lines through the most linear portions of the sample distribution curve, as shown in Figure 7.1.

A more recent innovation in liquid–solid chromatography is to use finely divided solids, which provide an exceptionally high surface area for a given weight of solid.[2] The effective number of exchanges in a given length of column is larger because of the small particle size, and hence for a particular value the peak width is correspondingly smaller in accordance with equation (7.1) for n. The practical drawback to packing a chromatography column with fine adsorbent is that the flow rate is then extremely slow. The problem was overcome by the development of essentially surgeless solvent pumps that allow the chromatograph to be developed under pressures as high as 1000 psi at the inlet. With such equipment remarkable analytical and preparative separations of complex mixtures have been obtained. Much less expensive low-pressure chromatographs (pressures up to about 100 psi at the inlet) show intermediate advantages and are widely used in research laboratories.

Still more useful for general laboratory use is "flash" chromatography, which employs short columns (typically 20 cm) of finely divided silica gel and pressures of only 15–30 psi at the inlet. The adjective *flash* refers to the speed with which the chromatogram is developed.

7.2 Ion-Exchange Chromatography

Ion-exchange chromatography is a special example of liquid–solid chromatography, wherein strong ionic attractions replace relatively weak polar adsorption forces.

A column of solid acidic material (such as Amberlite IR-120, a resin of polystyrene beads containing free sulfonic acid groups) can donate protons to any bases present in the surrounding liquid phase to form cations that are strongly attracted to the anions bound to the resin. The extent of proton transfer depends on the basicity of the solute and can be described

---

[1] Some workers use the reciprocal of this definition of K, in which case $R_f = nk'/1 + k'$, where $k' = 1/K$.

[2] The surface area of a given weight of spheres increases as the inverse of their radius.

by an equilibrium constant K (analogous to the previously discussed adsorption coefficient).

$$\text{Solid}^- - \text{H}^+ + \text{base} \xrightleftharpoons{K} [\text{solid}]^- \cdots [\text{H-base}]^+$$

$$K = \frac{\text{number of donated protons}}{\substack{\text{number of available} \\ \text{proton-donating sites}} \times \text{concentration of free base}}$$

Bases with larger values of K are present in lower concentration as free base and descend the column more slowly. A mixture of bases with sufficiently different constants can be separated by this method.

When all components of a mixture are held tightly, as happens frequently, it is necessary to percolate dilute acid through the column to move the components down the column (*displacement development*).

Basic columns also are available (such as Dowex 3, a resin of polystyrene beads containing free amino groups); these accept protons and can be used to separate mixtures of organic acids of different acid strengths. Special column materials that form ionic complexes with various inorganic cations or anions are useful, as are columns containing ions that form complexes with certain organic molecules.

7.3 Liquid–Liquid Chromatography

Liquid–liquid chromatography (partition chromatography) employs a liquid-moving phase and a second liquid phase held immobile by adsorption as a thin film on a solid support. Since the chemical influence of the solid support may be largely ignored, the adhering film behaves essentially like a liquid stationary phase. A substance added to such a column will be distributed (partitioned) between the two liquid phases. The distribution law is identical with that pertaining to the distribution of a solute between two immiscible solvents.

$$\text{Solid} + \text{film of solvent}_1 + \substack{\text{solution of} \\ \text{compound in solvent}_2} \xrightleftharpoons{K} \text{solid} + \substack{\text{film of solution of} \\ \text{compound in solvent}_1} + \text{solvent}_2$$

A convenient form of partition chromatography involves the use of a paper strip as the solid support. When the paper is treated with a mixture of two insoluble or slightly soluble solvents, the more polar solvent is adsorbed on the paper to form the stationary liquid phase. In one technique the sample is placed at a spot near the bottom of a dry paper strip and the strip dipped into a shallow pool of the mixed solvents. As the solvents ascend by capillary action, the more polar solvent is adsorbed, and the sample is partitioned between the moving and stationary liquid phases. Partition chromatography on paper strips is useful in separating amino acids by means of water–butanol solvent mixtures.

7.4 Gas–Liquid Chromatography

In gas–liquid chromatography (GLC or vapor-phase chromatography, VPC) the stationary phase is a film of liquid adsorbed on a solid support and the moving phase is a mixture of vaporized sample and a *carrier gas*, usually helium or nitrogen. The pertinent equilibrium is the distribution of sample between solution in the liquid film and vapor in the moving carrier gas. The rate at which the sample progresses through the column is determined principally by the rate of flow of carrier gas and the equilibrium vapor pressure of the sample in contact with the solution. A diagram of a gas chromatographic instrument is given in Figure 7.4, which shows the essential elements, consisting of the supply of carrier gas, the injection port for introducing the sample, the column for separating the sample, the detector at the end of the column, and the signal recorder attached to the detector.

$$\text{Solid} + \text{film of nonvolatile solvent} + \text{vapor of sample} + \text{carrier gas} \overset{K}{\rightleftharpoons} \text{solid} + \text{film of solution} + \text{carrier gas}$$

The fundamental data collected in gas chromatography are the periods of time (retention time) required for each of the components of the injected sample to reach the output detector. Retention time is inversely proportional to the vapor pressure of the sample in the flowing carrier gas and inversely proportional to the gas flow rate. In discussing gas-chromatographic separations, it is customary to speak of relative retention times, the ratio of the observed retention time to the retention time of some standard either already present or added to the mixture. The relative retention time is independent of the flow rate; it is determined by the equilibrium vapor pressure of the components in contact with the immobile liquid phase. From the previous discussion of fractional distillation, it can be seen that relative retention time is closely related to the relative volatility α for the sample and reference standard. The difference is that, in distillation, the liquid phase contains the same components as the vapor, whereas in gas chromatography the nonvolatile liquid phase must be considered as well. If the GLC liquid phase is similar in character to the sample components, the relative retention times will closely reflect the volatility behavior of the mixture on distillation. For example, if squalane (a fairly nonvolatile

$C_{30}H_{62}$ saturated hydrocarbon) is used as the immobile liquid phase, pentane (bp 36°), hexane (bp 68°), heptane (bp 98°), and octane (bp 125°) have relative retention times at 100° of 1.00, 2.22, 4.65, and 10.51, respectively. The retention times parallel the relative vapor pressures of the pure hydrocarbons. By contrast, if 1-propanol (bp 97°) is injected on the squalane column, the relative retention time is only 1.09; it comes off the column much faster (i.e., it is more volatile) than would be expected from its boiling point. The difference is that, with pure liquid 1-propanol, the hydroxyl group of one molecule finds itself immersed in a sea of hydroxyl groups of other propanol molecules. In the GLC column each propanol molecule is surrounded mostly by the hydrocarbon molecules of the liquid phase. In the absence of hydrogen bonding, the propanol is quite volatile. The enhanced volatility is closely related to the enhancement of volatility that occurs in steam distillation.

The situation is completely reversed if the squalane liquid phase is replaced by Carbowax, a polymer that contains ether groups (hydrogen bond acceptors) and terminal hydroxyl groups (hydrogen bond donors). The retention data, given in Table 7.2 along with the squalane retention data for comparison, show that propanol now has a very long relative retention time; the hydrocarbons show slightly enhanced volatilities (i.e., lower retention times).

Silicone oil DC-200 contains hindered silicon oxygen bonds (weak hydrogen bond acceptors) and gives retention results intermediate between the previous two liquid phases, as shown in Table 7.2. These data demonstrate that, because relative retention times can be varied by the choice of liquid phase, GLC provides a separation flexibility absent in fractional distillation.

In both GLC and distillation, separation depends on differences in vapor pressure of the components of a mixture. In fractional distillation the counterflow of rising vapor and descending liquid establishes (ideally) equilibrium between the components at each point in the column. In gas–liquid chromatography there is a unidirectional flow of carrier gas and the components move independently of each other. An expression for the number of

TABLE 7.2
Relative Retention Times for Different Liquid Phases at 100°

Compound	Bp, °C	Relative retention time[a]		
		Squalane	Silicone oil	Carbowax
Pentane	36	1.00	1.00	1.00
Hexane	68	2.22	2.04	2.00
Heptane	98	4.65	4.04	3.62
Octane	125	10.51	7.92	6.75
1-Propanol	97	1.09	1.58	40.5

[a] Relative retention time is the ratio of two times and thus is a pure number without units.

plates required to obtain a 95% pure sample with 80% recovery from a 50 : 50 mixture is

$$\text{Number of required GLC plates} = \frac{2.0}{(\log \alpha)^2} \qquad (7.2)$$

Comparison of this expression with the analogous expression for fractional distillation shows that, aside from a small difference in the constant, they differ by the exponent of the log α term. Since α, the relative volatility, is close to unity for any mixture likely to be fractionally distilled or chromatographed, the log α term is near zero. Gas–liquid chromatography, therefore requires many more theoretical plates to achieve a separation than does fractional distillation. The HETP of gas–liquid chromatography columns is usually much smaller than those of distillation columns, so that the same length of gas–liquid chromatography column contains many more theoretical plates. Moreover, since separations by gas–liquid chromatography do not depend on gravity return of a counterstream, it is possible to use long lengths of column coiled into a small volume (50-ft columns are common). Another important factor aiding separations by gas–liquid chromatography is the availability of a wide range of liquid phases, one or more of which may give a greatly enlarged value. This flexibility is inherently absent in fractional distillation. Hundreds of different liquid phases have been employed in gas chromatography, and, since these can be supported on over half a dozen different solid phases, it is apparent that the practical art of column selection is complex. Table 7.3 lists several general-purpose liquid phases with a few of their characteristics. A widely used solid support suitable for both polar and nonpolar samples is Chromosorb W, a white, flux-calcined diatomite. The liquid phase is dispersed on the surface of the support by adding the support to a solution of the liquid phase in a volatile solvent and allowing the solvent to evaporate slowly. The most serious deficiencies of gas–liquid chromatography are the related restrictions that the sample be readily vaporized and that it be small (typical sample sizes are 0.1–10 μL). Elaborate instruments are available that handle larger samples (0.1–10 mL), but they are expensive.

TABLE 7.3 Liquid Phases for Gas–Liquid Chromatography

Liquid phase	Temperature limit, °C	Application
Carbowax 20M	250	Separation of high-boiling polar compounds
Silicone oil DC-550	275	For compounds of intermediate polarity
Silicone oil DC-200	250	Separation of nonpolar compounds
Silicone gum rubber GE SE-30	375	Separation of nonpolar compounds
Apiezon (a hydrocarbon-based grease)	300	For low-boiling hydrocarbons

7.5 Laboratory Practice

Thin-Layer Chromatography (TLC). Thin-layer plates may be either prepared in the laboratory or purchased. Unless a large number of plates is to be used or some special adsorbent or binder is required, it is not much more expensive (and is considerably faster) to purchase the plates. One convenient type comes as 20×20-cm sheets consisting of a 100-μm layer of adsorbent bound to a 200-μm sheet of plastic. With reasonable care these can be cut with ordinary (but sharp) scissors into 2×10-cm strips suitable for analytical separations.

In liquid–solid chromatography, it is found that the resolution obtained depends on the ratio of solid to sample. For difficulty separated mixtures (R_f values differing by 0.1 or less), the ratio should be at least 200 : 1. With more easily separated mixtures, this ratio may be reduced proportionately. Because of the small amount of solid on a TLC plate, the sample spot should be applied with a microcapillary tube prepared by drawing out an ordinary capillary tube in a soft flame. In order to simplify the later calculation of R_f values, the plate should be marked lightly with two pencil lines 0.5 cm from each end. A microdrop of a solution of the sample in a volatile solvent is placed precisely on one of the two lines. If only one sample is being analyzed, the drop should be centered between the edges; if more are to be analyzed on the same plate, the spots should be placed symmetrically along the starting line. With 2×10-cm plates not more than three samples should be applied. When a low concentration of any component is being sought, it is necessary to superpose additional drops on the first spot until sufficient sample has accumulated. The solvent should be evaporated between additions.

A convenient developing chamber for TLC plates can be prepared from an ordinary wide-mouth screw-cap bottle. The inside of the bottle is lined with a folded circle of filter paper, which acts as a wick to transfer the developing solvent to the upper portions of the chamber. As shown in Figure 7.2, the circle of filter paper is folded to give a rectangle that is inserted in the wide-mouth bottle with the folds against the walls of the bottle. The size of filter paper should be chosen so that the folded paper rises close to the top of the bottle. It is essential that a gap in the paper be left near the top of the bottle so that the approach of the solvent front to the upper line on the plate can be seen without removing the cap. In practice, sufficient solvent is added to the bottle to saturate the liner and leave a layer 2–4 mm deep at its shallowest point. The spotted end of the plate is centered in the bottom of the chamber with the upper edge leaning against the wall; the spotted face of the plate should face the gap in the filter paper lining so that the rising spots will be visible. The bottle is capped and gently set aside until the rising solvent front has just reached the upper line. The plate is then removed and the solvent allowed to evaporate from it.

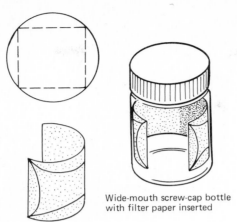

FIGURE 7.2
Developing Chamber for
Thin-Layer Chromatography

Wide-mouth screw-cap bottle with filter paper inserted

If one or more of the components to be identified is colorless, a convenient visualization technique is to place the plate in a second screw-cap bottle containing a few crystals of iodine mixed with sand. The capped bottle is held horizontally and rotated for a few seconds. Iodine vapor is absorbed on the plate wherever there is a concentration of organic material and produces a brown spot (the commercial plastic plates do not absorb a significant amount of iodine under these conditions). After the color has developed, the plate is removed and a circle penciled around each spot. On exposure to air, the brown iodine spots evaporate gradually.

Another method for visualization, which works with compounds that absorb ultraviolet light, is to use thin-layer plates that have been impregnated with a fluorescent dye. When the plate is exposed to uv light it will glow everywhere except where the organic compound has absorbed the light and quenched the fluorescence. While the plate is glowing, the dark spots should be circled carefully with a pencil so that the position of the spots can be measured and recorded after the ultraviolet light has been withdrawn.

Column Chromatography. A simple apparatus for liquid–solid column chromatography is a glass tube that has been constricted at one end (Figure 7.3). For separation of 0.1 to 0.5-g samples, a convenient tube size is 60 cm of 15-mm diameter tubing. This size will hold about 50 g of solid support and give a 100 : 1 ratio of packing to sample. Other sample sizes may be used with appropriately scaled apparatus.

An alternative apparatus widely used in research laboratories is a column assembled from a thick-walled glass tube having molded threads at the ends connected by matching threaded couplers constructed from nylon or other inert plastics. The tubes and couplers come in different lengths and diameters and allow virtually any length of column to be assembled. Special fittings are available that make it easy to attach a stopcock or a porous glass bit to the bottom of the column or to attach a solvent reservoir or a pressure inlet to the top.

A small wad of cotton or (preferably) glass wool is pushed into the tube with a wooden dowel until the wad rests on the constriction. With the tube clamped in an upright position, a 1-cm supporting layer of clean coarse sand is poured into the tube. Columns may be packed quickly by pouring in the solid support, but a packing that gives superior separations may be prepared by using a slurry of the solid in the desired eluent. The slurry is added slowly though a funnel to a column that has been stoppered temporarily at the bottom with a medicine dropper bulb. One advantage of the slurry method is that the solid settles slowly, giving a more uniform packing. Nonuniform packing usually permits channels to develop in the packing, and these seriously diminish the resolution. Another advantage of the slurry method is that any heat of adsorption of the solvent on the support will be given off before the solid is added to the tube. When solid supports containing many active adsorption sites are packed dry and then

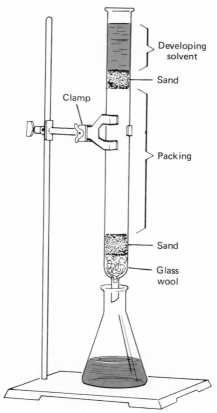

FIGURE 7.3
Apparatus for Column Chromatography

wetted with solvent, the heat evolved is frequently sufficient to expand the packed column and cause channels to develop.

Alumina frequently produces tightly packed columns that have excessively slow flow rates unless pressure is applied to force the eluent through the column. This problem can be minimized by adding about 10% Celite to the alumina to produce a coarser column. When the slurry method is applied to mixtures of adsorbents, they should be added in particularly small portions to prevent segregation of the solids as they sink through the solvent.

After the addition of the solid support has been completed, a second 1-cm layer of clean sand is added, followed by a small circle of filter paper just large enough to touch the wall of the chromatography column. The sand and filter paper prevent the upper layer of support from being disturbed during subsequent operations (this is important). Small irregularities at the top of the column produce large distortions in the shape of each band of sample by the time it reaches the bottom of the column and may cause closely spaced bands to overlap.

The sample is normally added in a solution as concentrated as possible so that narrow bands can be formed. When the solution is ready, the bulb is removed from the base of the column, and the excess solvent is allowed to drain. At the moment the solvent level reaches the top of the packing, the solution of the sample is added and allowed to penetrate into the column. After the sample has penetrated and before the column can become dry, the residue adhering to the walls is washed down with a few drops of solvent. Enough eluent is then added to fill the tube.

As the eluent flows through the column, the sample is separated gradually into bands that descend at different rates. It will be necessary from time to time to refill the tube with more eluent to keep the column from running dry. If all of the components of the sample are colored, it will be obvious when to change receivers for each component. When one or more of the components are colorless, it is necessary to collect fixed volumes of eluent in tared flasks, evaporate the solvent, and reweigh the flasks. A graph of the weights of sample eluted in successive fractions plotted against the accumulated volume of eluent reveals information about the number of components and the degree of separation. As shown in Figure 7.4, a larger number of fractions yield more detailed information about the number of components.

Flash Chromatography. Because of the superior resolution of mixtures using TLC, there have been numerous attempts to scale up the procedure to achieve separations of preparative size samples. In general such "thick-layer" methods do not separate as well, and the apparatus becomes unmanageably large for samples of more than about 0.5 g. The recently introduced technique of flash chromatography appears to combine the resolving ability

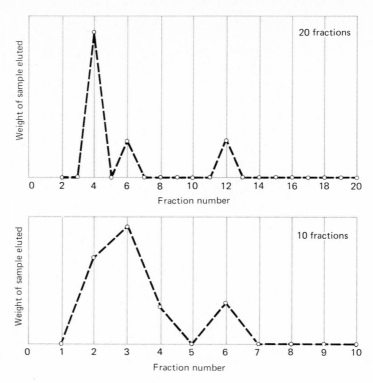

FIGURE 7.4
Effect of Number of Fractions on Chromatographic Resolution

of TLC with the large sample capacity of column chromatography and is rapidly becoming one of the most powerful separation techniques available to the organic chemist.[3]

The secret of flash chromatography is to use a very fine and uniformly sized silica gel (typically 32 to 63-μm particles), similar to that used in preparing TLC plates. The small particle size provides a large surface area and the uniformity enhances the resolution. In order to achieve the desired flow rates through a column packed with such a fine silica gel, it is necessary to apply a pressure of about 5–15 psi. For safety reasons, this, in turn, mandates heavy-walled columns. Threaded-glass columns (Ace glass) are well suited for this application and their use will be described here.

The first step is to assemble the chromatographic column using a tube of either 11 or 15 mm inside diameter to which is attached, by means of a nylon coupler, a fitting bearing a porous plastic plug and a drip tip. A stopcock is not needed. Attach the column firmly to a ring stand with clamps at both the bottom and the top of the column. Set the height to allow receiving flasks to be slid underneath the drip tip. Now add slowly, with gentle tapping, 15 g (35 mL) of 32 to 63-μm particle size silica gel. To

---

[3] Still, Kahn, and Mitra, *J. Org. Chem.*, **43**, 2923 (1978).

the top of the column, carefully add about a 1-cm layer of clean sand. Place a 100-mL Erlenmeyer flask underneath the drip tip.

Fill the column with the chosen developing solvent and, making sure that you are wearing safety goggles,[4] apply 15 lb of air pressure from a rotary vane pump (described in Section 3.2).[5] The solvent should flow down the column at a rate of 2–4 mL/min. Save the solvent for later use. When the liquid level just reaches the top of the sand, release the air pressure and transfer your sample solution to the column. In order to minimize the disturbance of the column surface, the first few milliliters of solution are best added with a long pipet that directs the stream along the walls of the column.

Adjust the vane pump to deliver 5–10 lb of pressure and force the sample solution onto the column. When the last portion has just penetrated the sand layer, interrupt the pressure and carefully rinse the walls of the column with a few milliliters of solvent. Reapply 5–10 lb of pressure until the solvent reaches the sand layer. Now fill the column with solvent and apply pressure again. Collect a series of fractions, as described in Section 7.3. Additional solvent may have to be added to the column in order to elute all of the sample components.

An extraordinary advantage of flash chromatography is that for similar adsorbents the R_f values obtained correspond closely with those obtained on a TLC plate. This correspondence permits to carry out several trial separations using TLC and then to transfer the best conditions to a flash chromatography separation of a preparative size sample.

The correspondence between the R_f values of the two techniques also can be used advantageously to calculate the volume of solvent required to elute a given component from the column. From the definition of R_f as the ratio of migration distance of the sample to the solvent front, it follows that the volume of solvent required to bring a sample to the bottom of the column is just $1/R_f$ times the volume of solvent required initially to just wet the entire column. For example, consider the case where a particular component has an R_f value of 0.2 and a volume of solvent V is required to just wet the column. After one volume of solvent has passed down the column, the component in question will be 0.2 of the way down; no solvent will yet have emerged. After a second volume V has been added, the component will have moved to a point 0.4 of the way down the column; one volume of solvent will have emerged. After five volumes of solvent have passed down the column, the component will have reached the bottom of the column; four volumes of solvent will have emerged. An additional small volume of solvent will elute the component from the column. Because of the spread in

---

[4] In the laboratory with all operations carried out under pressure or vacuum, you should wear safety goggles rather than the less protective safety glasses.

[5] The pressure can be applied conveniently by attaching a large one-holed rubber stopper to the end of the hose from the pump and holding the flat surface of the stopper against the top of the tube.

the sample along the column and the consequent lack of a unique value for R_f, one should collect the emerging solvent a little before and a little after the main peak.

Gas–Liquid Chromatography (GLC). Many different styles of GLC instruments are used in undergraduate laboratories; however, the discussion that follows is applicable to most of them with minor alterations. The major components of a gas chromatograph are shown schematically in Figure 7.5.

Recorder. The strip chart recorder makes a permanent record of the signal coming from the sample detector on a paper chart that unrolls at a constant speed while the pen slides back and forth on it in response to the applied signal. Laboratory recorders are classified by size (6-in. and 10-in. are the two most common), full-scale signal sensitivity (1–10 mV dc is used for GLC work), the minimum response time for the pen to travel the width of the chart (1 s for GLC work), and the speed at which the chart paper unrolls ($\frac{1}{2}$ in./min is convenient; other speeds are available). The electronic

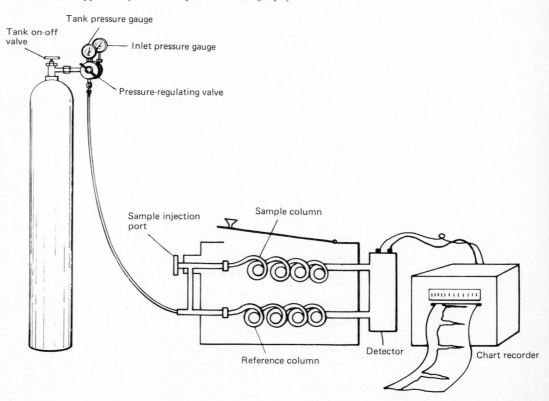

FIGURE 7.5 Apparatus for Gas–Liquid Chromatography

amplifier inside the case has adjustable gain and damping controls, but once set these controls should *not* be altered by anyone other than the service personnel. The recorder has an ON/OFF switch for power, which should be turned on at least 15 min before the recorded is used and left on during the laboratory period. There is also a chart drive switch that controls the flow of paper; you should turn on the paper drive switch shortly before a sample is injected and off when the last peak of interest has been recorded.

The signal being recorded is proportional to the amount of sample passing through the detector at any moment, so that the area under the peak on the chart paper is proportional to the total quantity of sample. Unfortunately, the proportionality constant between chart area and sample size differs slightly with the structure of the compound, so that for quantitative work standard mixtures of known composition must be analyzed first. For qualitative work the *relative* peak areas are approximately in the ratio of the weight of each component.

The areas under the recorded peaks can be determined by a number of methods; electronic integration, mechanical integration with a planimeter, or cut-and-weigh (this assumes that the density of the chart paper is uniform). In addition to these direct methods, the relative areas of the peaks can be approximated by drawing tangent lines through the sides (see Figure 7.1) and then equating the peak area to the area of the triangle thus constructed: area = $\frac{1}{2}$(base width) × (triangle height). Unfortunately, the triangle height is very sensitive to the precision with which the tangent lines are drawn and is difficult to reproduce. Another method of approximating the peak area, which avoids the problem of uncertain height, is to draw the tangent lines as before but to take the triangle height to be the height of the peak recorded on the chart paper. Because the quantities of interest are the ratios of areas, the mathematical errors introduced in this approximation tend to cancel and are more than compensated for by the increased precision of the measurement. If done carefully this method is reproducible to about 1–2% and is as precise as any other method except electronic integration, which is good to about 0.5%.

Detector. The most widely used detector for ordinary laboratory work is the thermal conductivity detector which consists of an electrically heated filament (similar to the one in a light bulb) that is immersed in the flowing gas stream. The temperature of the filament is determined by the (fixed) electrical current passing through it and the thermal losses to the cooler walls of the detector block from thermal conduction of the flowing gas. The carrier gas conducts heat at a rate characteristic of the gas, its pressure, and the rate at which it flows past the filament. If organic vapor is mixed with the carrier gas, the thermal conductivity of the mixture decreases because the main source of conductivity is from kinetic energy transfer to which the relatively massive organic molecules contribute little. The lowered thermal conductivity results in an increase in the filament temperature, which in turn leads to an increase in the electrical resistance of the filament wire. It is this increase in electrical resistance that is measured and passed on to the recorder as an electrical voltage proportional to the percentage of organic compound present in the detector gas stream.

Because of uncontrollable fluctuations in the gas flow rate and variations in the temperature of the detector walls, the background signal in the absence of organic vapor tends to fluctuate and drift. This can be controlled in practice by splitting the carried gas into two streams, one of which goes to the sample injection port and the other to a second reference filament in the detector block. If the electrical signal passed to the recorder is taken as the difference between the sample signal and the reference signal, the distortions introduced by variations in the temperature and gas flow largely cancel.

The electrical current passing through the filaments can be changed by the operator, but in practice it is best left untouched once adjusted, since each time it is changed it takes some time for the filaments to stabilize at their new equilibrium temperature. The balance control determines the electrical balance between the reference and sample filaments and is used to bring the recorder to zero when no sample is present. The "zero balance" should be adjusted before each sample is injected. The attenuator switch is a variable resistance network that determines what fraction of the detector signal is passed on to the recorder. The graduations on the dial give the factor by which the signal is reduced, thus a setting of "1" is the maximum sensitivity and "∞" gives no signal at all. The attenuator can be altered during an analysis to keep both the strong and weak signals on the recorder chart scale. Each time the attenuator is altered, some indication of the setting should be noted on the chart paper.

Carrier gas. Helium, the most commonly used carrier gas for thermal conductivity detectors, comes in large tanks pressurized to as high as 2600 psi. Because the GLC apparatus operates with a helium pressure of about 20–70 psi, a pressure-reducing and regulating value is required between the tank and the apparatus. The reducing value has two gauges: the one closest to the tank measures the tank pressure, and the other measures the pressure being delivered to the apparatus. The applied pressure, and consequently the gas flow, can be controlled by turning a bar-handle that comes out of the body of the pressure regulator. This handle should not be confused with the round ON/OFF tank valve (see Figure 7.5).

The flow rate through $\frac{1}{8}$-in. columns should be about 30 mL/min and 60 mL/min for $\frac{1}{4}$-in. columns. It can be measured by attaching the bubble flowmeter to the column exit tube on the detector and adjusted by turning the pressure-regulating valve (*not the main tank valve*) until the time measured with a stopwatch for a soap bubble to rise through a 10-mL volume of the flowmeter is close to 20 s ($\frac{1}{8}$-in. columns) or 10 s ($\frac{1}{4}$-in. columns). The flowmeter is removed after the adjustment.

Columns. Selection of a liquid phase for achieving a desired separation can be quite tedious, since there are literally dozens of combinations of liquid phases and solid supports that can be tried. The information given in Table 7.3 provides a starting point. Columns are connected to the instrument by Swagelok fittings. The diagrams provided by the manufacturer should be studied carefully before attempting to connect a column. The fittings are easily ruined if cross-threaded or overtightened; too loose a fitting will leak helium. Whenever a column is changed, it is imperative that the *detector* current *be turned off first* and then the helium flow stopped, since the filaments will overheat and possibly burn out if they are operated for any length of time without a stream of gas flowing over them.

The HETP[6] of a column depends on several factors that change with the velocity of the carrier gas through the column. If the gas flow is too fast, the moving sample does not have time to equilibrate with the stationary liquid phase and the HETP increases. If the gas flow is too slow, the sample spends too much time in the column and diffuses to give broad peaks, which from equation 7.3 corresponds to increased HETP. These two velocity-dependent contributions to HETP were combined by van Deemter with the contribution inherent in the choice of column material to give the equation for HETP as a function of gas velocity.

$$\text{van Deemter equation:} \quad \text{HETP} = A + \frac{B}{V} + CV \tag{7.3}$$

where V is the velocity of the carrier gas and A, B, and C are the parameters characteristic of the column and the factors described previously. The most important consequence of the van Deemter equation and the physical laws it represents is that there is an optimum flow rate for the lowest HETP. In critical separations, this optimum must be found; in ordinary work the standard carrier gas flow rates of 30 mL/min for $\frac{1}{8}$-in. columns and 60 mL/min for $\frac{1}{4}$-in. columns can be used.

The temperature of the oven and the column in it affects the quality of separation. Usually, the lower the temperature, the greater the difference will be in relative volatilities of the components. Lower temperatures also lead to peak broadening (increased time for diffusion), so that the increase in separation is not as marked as the difference in retention times might suggest.

---

[6] HETP = column length/number of plates. Broadened peaks correspond to a lowered number of plates.

Sample injection. The sample is injected by means of a syringe through a rubber septum into a heated injection block where it is vaporized and carried into the column by the flowing carrier gas. Typical syringes used in analytical GLC work hold up to 10 μL of sample.

The syringe is filled by pushing in the plunger to expel the air and then drawing it back slowly while the needle is immersed in the liquid to be analyzed. This operation may have to be repeated if the column of liquid drawn up is not continuous. The syringe is held vertically with the needle up, and the plunger is pressed in until the required sample volume is read on the graduated syringe barrel. The needle is wiped with a tissue to remove excess liquid, and then the plunger is pulled back slightly to draw some air into the syringe needle.

The syringe is held with both hands. One guides the needle into the septum, and the other holds the barrel and prevents the plunger from being forced out by the gas pressure inside the injection block. The needle is guided through the septum as far is it will go, the plunger is pressed in completely, and the syringe is withdrawn from the injection port. These three operations should be performed quickly but deliberately. If too much time is taken, the resulting peaks will be broaded; if the procedure is done too quickly, there is a high probability that the needle or plunger will be bent. Until experience is gained it is better to err on the slow side.

The rubber septum is self-sealing, but after a dozen or so injections it will begin to leak, particularly if a larger gauge needle is being used on the syringe. A leaky septum will result in distorted GLC traces. One test for a leaky septum is to place a few drops of water on the suspected septum; if it is leaking helium gas, bubbles will form. If the septum must be replaced, remember to turn off the detector current first so that the filaments will not burn out when the gas stream is interrupted. Also be careful in handling the injector fittings, since they are likely to be very hot.

The syringe should be cleaned after each use by drawing in and ejecting several consecutive samples of acetone and then pushing the plunger back and forth several times in the air to evaporate the acetone.

Calculation of Theoretical GLC Plates. When a sample is injected onto a GLC column, it occupies only a small portion of the column. As the vaporized sample is swept down the column by the carrier gas, it diffuses gradually so that by the time it reaches the detector, the distribution of sample resembles the peak shown in Figure 7.1. It is useful to speak of the "effective" number of times the sample is absorbed in the liquid phase and revaporized during its passage through the column. It can be shown that the total effective number of plates provided by a column is approximately

$$\text{Number of GLC plates} = 16 \left(\frac{t_{\text{ret}}}{t_{\text{width}}} \right)^2 \qquad (7.4)$$

where t_{ret} is the retention time of the peak maximum and t_{width} is the width of the peak base expressed in the same units of time. Comparison of equation (7.4) with equation (7.1) reveals the similar theoretical basis of gas–liquid and liquid–liquid chromatography.

7.6 Chromatographic Separations

The separations presented here are designed to illustrate several of the more commonly used chromatographic techniques. The materials to be separated in most of these initial experiments are highly colored because this permits the operator to see the band development and to determine immediately the consequences of careless or hasty technique.

(A) Separation of Ink Pigments by Thin-Layer Chromatography. On a 2 × 10-cm thin-layer plate[7] draw two horizontal pencil lines across the plate 7 mm from each end. On the bottom line, about 5 mm from the left-hand edge, make a single sharp dot of ink from a black Flair pen. In the center of the line make a heavy dot, and on the right-hand side make a very heavy spot. Add sufficient 95% ethanol to an 8-oz wide-mouth screw-cap bottle containing a filter paper lining (see Section 7.5) until a layer 3 mm deep is produced. Center the spotted plate in the bottle with the upper edge leaning against the side and screw the cap tightly onto the bottle. When the solvent front reaches the upper pencil line, remove the plate and allow the solvent to evaporate. Observe and record the number of colored spots.

The experiment can be repeated with other colors of Flair pens to determine if the same dyes are used that were found in the analysis of the black Flair pen ink.

(B) Separation of Plant Pigments by Thin-Layer Chromatography. In a clean porcelain mortar place 1 g of spinach, 1 g of clean sand, 5 mL of acetone, and 5 mL of petroleum ether. Grind the spinach until the green chlorophyll appears to have been extracted completely. Decant the solution into a small beaker.

Prepare a thin-layer plate as described in Part (A) and in the center of the bottom line place a microdrop (see Section 7.4) of the chlorophyll extract. Blow gently on the spot so that the solvent evaporates quickly.

---

[7] Eastman Chromagram Sheet, Type 6060 or 6061, is suitable. It can be cut conveniently with ordinary scissors. In humid climates it is desirable to activate the plates by heating them in an oven at 100° for 15–30 min.

Repeat the addition of the extract several times until a distinct green spot is visible. The additions should superpose as closely as possible.

Develop the plate with an appropriate solvent[8] as described in Part (A). It may be necessary to repeat the development to resolve the chlorophylls. Mark the spots with a pencil immediately, since the colors fade (because of air oxidation) fairly quickly.

(C) Analysis of Analgesics by Thin-Layer Chromatography. The colorless components of common analgesics (Contac, APC tablets, etc.) can be identified by comparison of their R_f values with those of the pure components. Common analgesics are usually a mixture of several drugs. Most contain aspirin and caffeine, but several contain other components as well. Because the components are colorless, ultraviolet fluorescence quenching and iodine staining must be used to make the spots visible.

Select a sample and identify it in your notebook. Pulverize your sample to a fine powder with the back of a laboratory spoon. Place a small wad of glass wool in the tip of a pipet obtained from the side shelf and transfer the powder into the pipet to form a column. With a second pipet, drain 5 mL of 95% ethanol through the column and collect the extract in a small test tube.

On a thin-layer plate that has been impregnated with fluorescent dye, draw two pencil lines 7 mm from each end. With a microdropper place a small spot of the ethanol extract in the center of the bottom line. Once the ethanol has evaporated the spot will be invisible, but it can be made visible ("visualized") by placing the plate under ultraviolet light. If the spot where the compound was applied is faint, carefully apply a second drop directly on top of the first. The spot size should be less than 5 mm in diameter.

Prepare a solution of about 0.1 g of salicylamide and 0.1 g of caffeine in 5 mL of 95% ethanol. Place spots of this solution on both sides of the central spot about halfway between the center and the edges. The spots from this mixture can be used as calibration points for the values of the sample spots.

The developing solvent is 25:1 ethyl acetate–acetic acid. Place the solvent in the developing chamber and allow the filter paper (use fresh paper) to become saturated with it. Develop the plate until the solvent has migrated to the top pencil line. Remove the plate, allow the solvent to evaporate, and then examine the plate under an ultraviolet lamp. Circle the dark spots lightly with a pencil and then compute the R_f values of the reference spots on either side of the central spots. If they are reasonably close to the values given below ($\pm 20\%$), complete the identification of the

---

[8] Suitable developing solvent is prepared by mixing 8 volumes of petroleum ether with 2 volumes of acetone. Lower percentages of acetone tend to improve the separation of chlorophylls a and b at the expense of poorer separation from the yellow xanthophylls.

unknown. If the reference values are outside this range, prepare another plate and change the proportion of acetic acid to approach the tabulated values (more acetic acid increases the values).[9]

After the analysis is complete, place the plate in an iodine chamber and compare the brown spots developed in this way with the circled spots found previously.

Salicylamide, $R_f = 0.79$

Aspirin, $R_f = 0.56$

Caffeine, $R_f = 0.31$

Phenacetin, $R_f = 0.68$

Acetaminophen (Tylenol), $R_f = 0.49$

(D) Separation of a Dye Mixture. Insert a small wad of glass wool into the constricted end of a 60-cm length of 15-mm diameter tubing and clamp the tube in an upright position (see Figure 7.3). Add a 1-cm layer of coarse sand to the tube. In a 100-mL beaker, prepare a slurry of 5 g of alumina[10] and 5 g of Celite[11] in 50 mL of warm water, and transfer the slurry in small batches to the tube (swirl between additions). The water that filters through the sand and glass wool should be collected and used to transfer any column material that remains in the beaker. After the packing has settled, add a second 1-cm layer of sand, followed by a small filter paper circle.[12]

When the last drop of water penetrates the column, add 10 drops of the dye solution[13] to the top of the column. When the dye solution has penetrated, add a few drops of water to wash down any dye adhering to the walls. After the wash water has penetrated, fill the tube with water and allow the chromatogram to develop.

---

[9] The R_f values of aspirin and acetaminophen are particularly sensitive to the activity of the plate and the proportion of acetic acid in the solvent.

[10] Merck Reagent Grade Alumina, 71707, is suitable.

[11] The 50:50 alumina–Celite mixture is much more porous than the more usual 90:10 mixture and removes the need to use pressure or vacuum.

[12] These circles can be prepared conveniently from a larger piece of filter paper using a cork borer as a cutter and a large cork as a cutting surface.

[13] A suitable dye solution can be prepared by dissolving 0.1 g each of crystal violet, auramine hydrochloride, and malachite green in 100 mL of water. Alternatively, one of the dark-colored commercial food colors may be analyzed.

(E) Flash Chromatography. Assemble a chromatography column, as described in Section 7.5, and pack it with 15 g (35 mL) of 32–63 μm silica gel topped by a 1-cm layer of sand. Prepare 5 mL of leaf extract as in Procedure B, and place it in a 10-mL graduated cylinder. Add 5 mL of water and stir the mixture with a glass rod (to extract most of the highly polar acetone in the sample). Transfer the upper dark green petroleum ether layer to the column by means of a Pasteur pipet. Force the sample onto the column with pressure, as described in Section 7.5, and develop it with 4:1 petroleum ether–acetone.[14]

Compare the pattern of bands with the spots obtained by thin-layer chromatography. Individual bands can be isolated and their purity determined by thin-layer chromatography.

(F) Gas–Liquid Chromatography. After familiarizing yourself with the GLC apparatus (compare with Figure 7.5), check to see that the carrier gas and filament current are on. With the attenuator set at "1," adjust the zero-balance control so that the recorder pen is on zero. Turn the attenuator knob to "8," and start the chart drive.

Draw up 0.5 μL of a standard hydrocarbon mixture[14] in a 1- or 10-μL syringe, and inject it into the injection port. Make a mark on the chart paper at the point of injection.[15] Calculate the relative retention times, and compare them with the data in Table 7.2. Calculate the number of theoretical plates for each of the peaks, and compare their relative areas with the known composition.

Determine the effect of altered environment on relative volatility by injecting a mixture containing three components of similar boiling points but different polarities.[16] Identify the components by their relative areas, and justify the observed relative retention times. Calculate the number of theoretical plates from each of the peaks, and compare their relative areas with the known composition.

Questions

1. Arrange the following compounds in the order of their elution from a silica gel column, with benzene as eluent.

$$CH_3(CH_2)_{10}CH_3 \quad CH_3CO_2H \quad CH_3CH_2CH_2OH \quad CH_3COCH_2CH_3$$

---

[14] A suitable mixture is 1 part by volume of *n*-pentane, 2 parts *n*-hexane, and 3 parts *n*-heptane, which on a DC-200 column have relative retention times of about 1.0, 2.0, and 4.0, respectively. Because of the broadening of the slower moving peaks, the specified relative quantities will give peaks of about the same height.

[15] A convenient technique for marking GLC charts is to momentarily hold your finger over the column exit port. *Be certain that the exit port is not too hot!*

[16] A suitable mixture is 1 part by volume of 2-propanol (bp 82°), 2 parts ethyl acetate (bp 77°), and 3 parts cyclohexane (bp 81°).

2. Suggest suitable liquid phases for separation of carboxylic acids by liquid–liquid chromatography. (*Hint:* Consider the solubility relationships discussed in Chapter 6 on extraction.)
3. Table 7.1 lists several classes of solvents arranged in order of increasing eluting ability. Give a practical example of each class.
4. Figure 7.4 illustrates the effect of the number of fractions on chromatographic resolution obtained on analysis of the same sample. Show that if the twenty-fraction chromatogram samples are combined in consecutive pairs, the ten-fraction chromatogram will be obtained.
5. Silicone oil exhibits approximately ideal behavior as a liquid phase in gas–liquid chromatography, so that relative volatility values, α, obtained from fractional distillation can be employed in estimating the number of GLC theoretical plates required.
   (a) If a mixture of two liquids requires fifty theoretical plates for adequate separation by fractional distillation, how many GLC plates would be required for separation using silicone oil as the liquid phase?
   (b) With silicone oil as the liquid phase, what would be the expected elution order of acetone, n-butyl alcohol, benzene, and pentane?

8 Accessory Laboratory Operations

This chapter describes a number of general laboratory procedures that are used throughout the experiments presented in the remainder of this manual.

8.1 Drying Agents

The removal of admixed or dissolved water from starting materials and finished preparations is an important feature of organic laboratory work. In general, water must be regarded as an objectionable impurity since it may bring about an undesired hydrolytic reaction or exert an unfavorable catalytic effect. The presence of water sometimes retards a desired reaction and may inhibit it completely, as in the formation of Grignard reagents. On the other hand, it would be superfluous to remove the last traces of water from a substance that is to be brought into contact with aqueous reagents.

An organic solid may be dried by spreading it in thin layers exposed to the air at room temperature, but this method usually allows at least a small amount of moisture to remain. More effective drying is obtained by heating the substance in thin layers in a drying oven at a temperature below the melting or decomposition point, or by placing it in a desiccator over drying agents such as anhydrous calcium chloride, solid sodium hydroxide, or phosphorus pentoxide. The use of concentrated sulfuric acid in desiccators is dangerous.

An organic liquid, or an organic solid dissolved in an organic solvent, is usually dried by placing the liquid in direct contact with a solid inorganic drying agent. After it is mixed thoroughly and the mixture is allowed to stand, the dried liquid is filtered to remove the spent drying agent and may then be distilled, and so on. *If, in any drying operation, sufficient water is present to cause the separation of an aqueous layer, the organic liquid should be separated and treated with a fesh portion of the drying agent.*

Selection of a Drying Agent. The selection of an appropriate drying agent involves consideration of the properties of the substance to be dried and the characteristics of the various drying agents. The latter must remove water efficiently and must not dissolve in the liquid or react with it in any way.

The efficiency of a drying agent is a composite of three factors: *capacity* (the amount of water that can be removed per gram of drying agent), *speed* (the rate at which the water is taken up), and *intensity* (the amount of water present even with an excess of drying agent). Ratings[1] for these three factors for the drying of diethyl ether solutions by a number of common drying agents are presented in Table 8.1. It can be seen from Table 8.1 that the three most efficient drying agents are calcium chloride, magnesium sulfate, and 4 Å molecular sieves. Their use and restrictions will now be described.

Calcium chloride is available in pellet form and is an extremely effective drying agent. Unfortunately, it has two limitations. Calcium chloride binds strongly not only to water but also to alcohols, amines, phenols, and to a lesser extent, molecules containing other polar functional groups. For this reason it is used primarily for drying hydrocarbons and halogenated hydrocarbons. Also there is the practical problem that any pellets spilled on the desk will eventually absorb enough water to form puddles of calcium chloride solution. These are both messy and corrosive; it is imperative that spilled drying agent be removed promptly.

Anhydrous magnesium sulfate is a fluffy white solid that settles slowly in an organic solution that has been swirled. Wetted drying agent, by contrast, is dense and tends to clump at the bottom of the flask. The usual practice is to add a small portion and, after a few minutes, observe its behavior when the flask is swirled. Additional portions are added until the newly added drying agent remains fluffy.

[1] The comparisons are based on a study by Pearson and Ollerenshaw, *Chemistry and Industry*, **1966**, 370.

TABLE 8.1
Common Drying Agents for Organic Compounds

Drying agent	Capacity	Speed	Intensity
Calcium chloride	High	High	Very high
Calcium sulfate	Low	High	High
Magnesium sulfate	High	Medium high	High
Molecular sieves, 4 Å	High	High	Very high
Potassium carbonate	Medium	Medium	Medium
Sodium sulfate	Very high	Low	Low

4 Å molecular sieves are available as a powder, as $\frac{1}{16}$-in. and $\frac{1}{8}$-in. pellets and as beads. The $\frac{1}{16}$-in. pellets are particularly convenient. Fresh molecular sieves are conditioned by heating at 320° for 3 hr; used sieves can be reconditioned after they are rinsed with acetone and *all* of the solvent is allowed to evaporate.

Molecular sieves possess a large number of pores; the 4 Å size allows water to penetrate but rejects almost all organic molecules, which are much larger. The most common application of molecular sieves is to dry organic solvents either by addition to the solvent or sometimes by passing the solvent down a 2-ft column packed with $\frac{1}{16}$-in. pellets. The one inconvenience of the sieves is that, unlike magnesium sulfate, they give no indication of whether their drying capacity has been saturated.

8.2 Cooling Baths

The common cooling bath is a slush of crushed ice and water. Because of the inversion of density of water at 4°, an ice bath should be stirred well if it is desired to maintain the whole bath at 0°. Temperatures below 0° can be obtained by mixing inorganic salts with ice or cold water. Table 8.2 lists the proportions of ingredients to be mixed to obtain the stated temperature. It is important in using ice–salt baths that the mixture be stirred well. Temperatures down to about −80° can be maintained by addition of pieces of dry ice to isopropanol or other low-melting heat transfer liquids.

8.3 Refluxing

In preparative organic work it is frequently necessary to maintain a reaction at an approximately constant temperature for a long period of time with a minimum of attention. The simplest procedure for reactions carried

TABLE 8.2
Cooling Baths

Ingredients	Lowest temperature obtained, °C
1 part sodium chloride 3 parts crushed ice	−20
1 part ammonium chloride 1 part sodium nitrate 1 part cold water	−20
3 parts powdered calcium chloride 2 parts crushed ice	−50

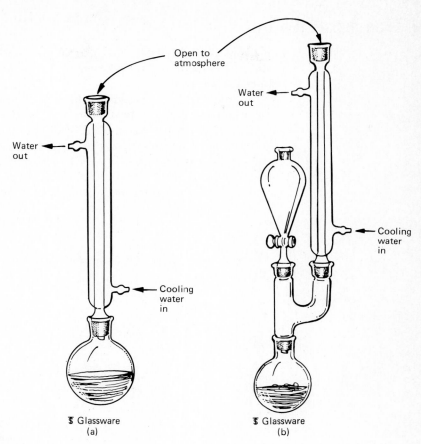

FIGURE 8.1
Reflux Assemblies

out in solution is to boil the solution and condense the vapors so that they are returned to the reaction flask (refluxing). The temperature in the flask remains nearly constant at approximately the boiling point of the solvent (the actual temperature is elevated above the solvent boiling temperature to an extent governed by the concentration of nonvolatile solutes). This operation is so common and important that the principle and technique should be understood clearly. The apparatus for a simple reflux operation is pictured at the left in Figure 8.1.

Another common requirement in preparative chemistry is to add a liquid to a reaction mixture while the mixture is refluxing. An assembly suitable for this purpose is shown at the right in Figure 8.1. In both assemblies it is essential that there be some opening to the atmosphere to prevent pressure buildup. If the reaction is sensitive to atmospheric moisture and must be protected from it, a drying tube is inserted in the top of the reflux condenser; *under no circumstances should a solid stopper be used.*

8.4 Gas Absorption Traps

In some organic preparations, noxious gases are liberated that must not be allowed to escape into the laboratory. Two common methods[2] for trapping water-soluble gases are pictured in Figure 8.2. The principal precaution to be observed in trapping gases by the inverted funnel method is to construct the apparatus so that a drop of pressure in the reaction flask will not suck water back into the reaction flask. The lower edge of the funnel should not dip into the water more than 1 or 2 mm. If the possibility of sucking water back into the reaction flask poses a hazard, a safety flask (see Figure 3.2) should be connected between the reaction vessel and the trap.

In the water-aspirator method, the flow of air sucked in by the aspirator is normally sufficient to overcome all but the most vigorous evolution of gas.

[2] Horodniak and Indicator, *J. Chem. Educ.*, **47**, 568 (1970).

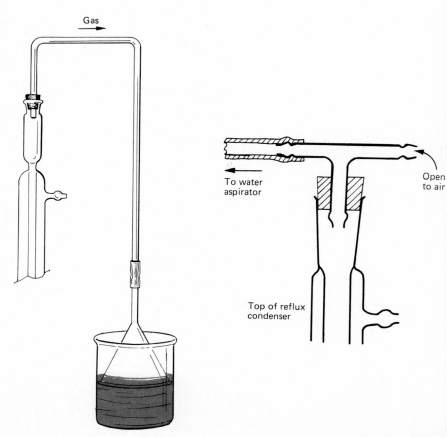

FIGURE 8.2
Gas Absorption Traps

8.5 Mechanical and Magnetic Stirring

In many organic preparations it is desirable to stir the reaction mixture. This is particularly true of heterogeneous reactions or large-scale reactions in which heat transfer from the central portions of the liquid can be slow enough to cause an excessive temperature rise. Inexpensive motors of the brush type give off sparks and are exceedingly hazardous in an organic laboratory containing vapors of volatile organic chemicals. Nonsparking motors of the induction type normally have high rotation speeds that make them unsuitable; they have low torque at speeds lower than their design speed so that attempts to slow down induction motors by adding a friction device usually are not successful. Geared induction motors are satisfactory but expensive.

An inexpensive air-driven stirrer is available. When the reaction must be stirred while it is refluxing a vapor seal, such as is shown in Figure 8.3, is necessary. In use, a lubricant (glycerin or stopcock grease) is required to prevent the rubber tubing from sticking to the stirrer shaft. One of the least

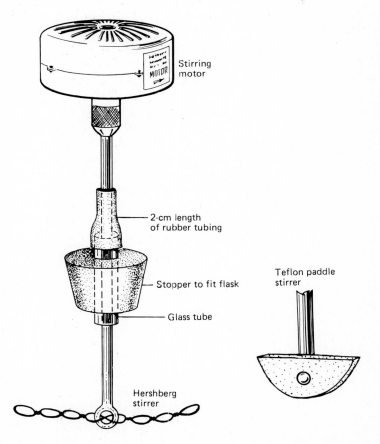

FIGURE 8.3
Stirring Assembly with Vapor Seal

expensive and most effective stirrers is of the Hershberg type, shown at the left half of Figure 8.3; a more convenient alternative is a Teflon paddle of the type shown at the right of the figure. Another particularly convenient stirring device is the magnetic stirrer, which consists of an enclosed electric motor attached to a large bar magnet. In practice, a second magnet ("spin bar") is placed in the reaction flask, which rests on the stirrer enclosure; when the motor is turned on, the enclosed magnet produces a rotating magnetic field that forces the spin bar to rotate and stir the contents. The rotating magnetic field is usually sufficiently strong to be effective even if the reaction flask is being heated by a Glass-Col heater or an oil bath. To prevent contamination of the reaction mixture, the spin bar is normally coated with Teflon or glass. Several sizes are available.

The magnetic stirrer is more convenient to use than a mechanical stirrer, but it does not have as much turning force and thus is ineffective with viscous reaction mixtures.

8.6 Rotary Evaporation

A common operation in the organic laboratory is the removal of a volatile solvent from a reaction mixture. This can be done by simple distillation, but a faster technique is to use a rotary evaporator, which consists of an electrical motor with an elongated hollow shaft. One end of the shaft is machined to accept ⚟ joints; the other end is connected through a ball joint to a vacuum pump or aspirator. In some evaporators the vacuum connection is made through a sleeve that fits around the shaft. The apparatus is assembled with the shaft at about a 45° angle to the desk top. In operation the motor turns the flask so that a thin film of the solution is continuously being exposed on the upper walls of the flask where it evaporates quickly. In practice the evaporation occurs so quickly that the flask becomes quite cold unless a warm water bath is applied. The advantages of rotary evaporation over simple distillation are the speed of the operation and its simplicity.

II IDENTIFICATION OF ORGANIC COMPOUNDS

9 Identification by Chemical Methods

9.1 Introduction

Identification of the molecular structure of organic compounds is a daily activity of practicing organic chemists. Whether the task is to identify the milligrams of toxin present in "red tide" algae or the gallons of distillate from a pilot plant catalytic hydrogenator, the challenge is to translate observations of chemical and physical behavior into a unique structure. Unlike many intellectual problems, structure elucidation usually does not yield to straightforward deductive logic but requires a more intricate interplay of facts and hypotheses. It is this complexity that makes the task always interesting and sometimes just plain fun.

There are two quite different approaches to structure determination. The traditional one, which depends on chemical reactions to identify functional groups and convert them to known derivatives, is the subject of this chapter. The modern method, involving spectrometry, is described in Chapter 10. Spectrometric methods are used extensively today because they are faster and capable of dealing with smaller amounts of compounds with more complex structures. Although the traditional methods now are seldom used alone, they are described here for a number of reasons. On occasion, parts of the traditional scheme are still quite useful. Also, the required techniques strongly reinforce fundamental chemical and physical principles and expose the beginning student to the making of chemical judgments, an essential skill for productive research. The time invested in learning to interpret chemical and physical behavior will be repaid many times over in future work.

Structure elucidation of a completely unknown substance by the traditional methods begins with the isolation of the material in a pure state followed by qualitative tests to disclose the presence of elements such as nitrogen, sulfur, or the halogens. Quantitative analysis furnishes the weight composition of the substance and permits the calculation of an empirical formula, which gives the atomic ratios of the elements present. Determination of the molecular weight permits the assignment of a definite molecular formula that expresses the actual number of atoms of each element present in the compound.

The next stage in structure elucidation is to identify the functional groups and other characteristic structural features present in the molecule. If you believe the compound has been previously prepared and characterized, convert it into several derivatives that can be compared with the properties reported in the literature. If the compound is new, the partial information obtained from the chemical tests is pieced together to give a total structure consistent with all available data. Where possible, the tentatively assigned structure is confirmed by an independent synthesis. Alternatively, the compound may be selectively degraded to simpler known substances. The structures of millions of organic compounds have been assigned by just these methods.

The full procedure outlined here is much too difficult for beginners. What will be described in this chapter is a procedure for structure determination that has been simplified by limiting the range of possible functional groups and restricting the unknowns to previously identified and well-characterized compounds. The sequence of steps to be followed is

1. Preliminary examination.
2. Purification of the unknown sample.
3. Physical constant measurements.
4. Element identification.
5. Solubility classification.
6. Functional group identification.
7. Derivative preparation.

9.2 Preliminary Examination

The first step in the identification procedure is to answer some simple but important questions about the unknown. It should be emphasized from the beginning that all observations are to be recorded immediately in your notebook. So many observations are made in the course of a structure identification that some are certain to be forgotten if recording is put off to later.

Physical Appearance. As a first step, note and record whether your sample is a liquid or a solid; the tables that will be consulted in the last stages of the analysis are subdivided into liquid and solid compounds.

Next, note the color of your unknown. Simple aliphatic and aromatic compounds with single functional groups tend to be colorless, whereas aromatic compounds with both electron donating and withdrawing groups in conjugation with each other tend to be strongly colored. Many organic compounds, particularly aromatic ones, decompose when heated and form extensively conjugated and hence colored impurities. It is particularly important to note any changes in color as the sample is purified.

Another physical property of the compound you should be aware of is its odor. Although many compounds have characteristic odors, the relationship of odor to molecular structure is not well understood, so specific structural conclusions cannot be drawn from this property. There are, however, some useful generalizations. For example, low molecular weight amines have an unmistakable (dead) fish odor, esters are fruity, and so on. There are also many surprises; benzaldehyde and hydrogen cyanide, structurally quite different species, have a similar bitter almond odor. It is a good idea to become acquainted with the odors of the common solvents. It is useful, for example, to be able to recognize quickly the presence of any residual solvent impurities in an unknown.

One should be extremely cautious in observing odors since the vapors could be obnoxious or toxic (see Section 1.1). The proper technique is to open the container well away from your nose and gently waft the vapors toward you. If the odor is not too strong, the container can be brought closer but never right up to the nose. Any odor worth noting will be detected several inches away.

Ignition Test. From the behavior of a small sample heated in a flame you can determine if a solid has an accessible melting point and whether a solid or liquid is volatile, forms volatile decomposition products, or is explosive. Combustion of the sample giving a sooty flame, indicates the presence of unsaturation, aromatic groups or long aliphatic chains. A residue indicates the presence of a metal, usually as the salt of an acid.

Ignition tests can be carried out on a laboratory spatula or on an inverted porcelain crucible cover supported by a wire triangle and ring stand over a small burner. Only a few milligrams of sample should be burned.

9.3 Purification of the Unknown Sample

Your instructor may assign a pure enough unknown that you can go directly to the next step in the identification procedure. If not, you will have to take a small detour. With a solid unknown, determine the melting-point range. If it is not sharp (less than 2°), recrystallize the solid according to the procedures described in Section 5.2. Determine the melting point again and see if it is not sharp enough to proceed. Additional recrystallizations are required until the melting point either becomes sharp or has a constant range.

An impure liquid unknown will have to be distilled and a constant-boiling fraction collected. Caution is called for here. If the ignition test indicated that the sample decomposes on heating, a vacuum distillation will be necessary. Be sure to record both the pressure and the boiling-point range of the purified sample.

An independent measure of purity for solids and liquids is by thin-layer chromatography. The developing solvent should be chosen so that the R_f of the main component is about 0.5.

9.4 Physical Constants

Melting points and boiling points are characteristic properties of pure materials. The boiling point of a liquid is related approximately to its molecular weight; the melting point of a solid is determined partly by molecular weight but more by the presence or absence of polar groups that interact strongly in the crystal lattice. With liquids the refractive index and density are characteristic properties frequently recorded in handbooks.

If you purified the sample, the melting point or boiling point is already known and you can proceed to the next section on element detection. The determination of boiling points by the method described in Section 2.4 requires that at least 5 mL of the liquid be available. Boiling points of much smaller samples, even a few drops, can be determined easily by the inverted capillary technique (Figure 9.1).

Boiling Points of Micro Samples. Place 2–5 drops of the sample in a boiler tube prepared by sealing one end of a Pasteur pipet (or a 5-cm length of 4-mm glass tubing). Seal one end of a melting-point capillary tube and break off the tube about 25 mm from the seal. Drop the capillary open end down, into the boiler tube (Figure 9.1). Attach the boiler tube to a thermometer by means of a small rubber band,[1] and support the assembly in an oil bath (the Thiele tube apparatus used for melting points is ideal) so that the sample is at least 10 mm below the bath level. The bath is heated gradually until a rapid stream of bubbles emerges from the capillary. The temperature at which rapid bubbling occurs is a few degrees above the boiling point of the sample. Immediately discontinue heating, which causes

---

[1] Tiny rubber bands can be made by cutting off 2-mm lengths of soft rubber tubing.

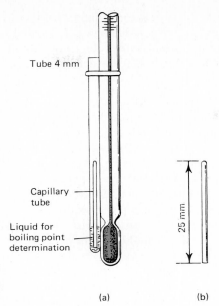

FIGURE 9.1
Apparatus for Boiling points of Micro Samples

the bubbling to cease. When the temperature reaches the boiling point of the sample, and as the temperature continues to fall, the liquid is drawn up into the capillary. The cycle of heating and cooling replaces most of the air in the capillary by the vapor of the sample. At this point, resume heating, only more cautiously, so that the temperature rises at a rate of about 2°/min until bubbles once more emerge. Remove the heat and note the exact temperature at which bubbling ceases. This is the boiling point of the liquid, since it is the temperature at which the vapor pressure inside the capillary equals the external atmospheric pressure exerted on the surface of the liquid in the boiler tube.

9.5 Element Identification

Carbon, Hydrogen, and Oxygen. Normally one assumes that a sample obtained in an organic laboratory contains carbon and hydrogen. If there is any doubt, you may detect their presence by heating a sample in a tube with dry powdered copper oxide, whereby carbon dioxide and water are formed. Carbon is detected by passing the evolved gases into an aqueous solution of calcium or barium hydroxide, in which the carbon dioxide produces a precipitate of the carbonate. Hydrogen is detected by the condensation of droplets of water in the cool upper portion of the reaction tube.

There is no satisfactory qualitative test for the presence of oxygen in organic compounds. To determine if oxygen is present, quantitative analysis is required. If the sum of the percentages of all known constituent elements does not amount to 100%, the deficit is taken as the percentage of oxygen.

Nitrogen, Halogens, Sulfur. The qualitative detection of these elements in organic compounds is more difficult than in inorganic compounds because most organic compounds are not appreciably ionized in solution. Since the tests used in qualitative analysis are based on ionic reactions, they cannot be applied directly to organic compounds. For example, sodium chloride or bromide gives an immediate precipitate of the silver halide when treated with an aqueous solution of silver nitrate; carbon tetrachloride, bromobenzene, and most organic halides do not produce a precipitate of the silver halide when treated with aqueous silver nitrate solution because they do not furnish an appreciable amount of halide ion in solution.

For qualitative detection it is necessary, therefore, first to convert elements such as nitrogen, sulfur, and halogens into ionized substances. This conversion may be accomplished by several methods, of which the most general is fusion with metallic sodium, which produces sodium cyanide, sodium halides, sodium sulfide, and so forth, as indicated in the following reaction. The resulting anions may then be identified by applying the usual inorganic tests.

$$\text{Organic compound containing C, H, O, N, S, Cl} + \text{Na} \xrightarrow{\text{high temperature}} \text{NaCN} + \text{NaCl} + \text{Na}_2\text{S} + \text{NaOH} + \text{etc.}$$

Sulfur. A fresh portion of the filtered alkaline solution is acidified with acetic acid and treated with an aqueous solution of lead acetate. If sulfide is present, a dark brown precipitate of lead sulfide results. The acetic acid neutralizes the base to prevent precipitation of lead hydroxide.

$$\text{Na}_2\text{S} + \text{Pb(OAc)}_2 \longrightarrow \underset{\text{black}}{\text{PbS}} + 2\,\text{NaOAc}$$

Nitrogen. Two tests for nitrogen are presented here. In the traditional test, a portion of the filtered alkaline solution is treated with aqueous ferrous sulfate and ferric chloride, boiled for a few moments, and acidified with hydrochloric acid. If nitrogen is present, a precipitate of prussian blue results.

$$2\,\text{NaCN} + \text{FeSO}_4 \longrightarrow \text{Fe(CN)}_2 + \text{Na}_2\text{SO}_4$$

$$4\,\text{NaCN} + \text{Fe(CN)}_2 \longrightarrow \text{Na}_4\text{Fe(CN)}_6$$

$$3\,\text{Na}_4\text{Fe(CN)}_6 + \text{FeCl}_3 \longrightarrow \underset{\text{prussian blue}}{\text{Fe}_4(\text{Fe(CN)}_6)_3} + 12\,\text{NaCl}$$

Sec. 9.5] ELEMENT IDENTIFICATION

The second test for nitrogen uses cyanide ion as a catalyst to carry out the benzoin condensation of 2-pyridinecarboxaldehyde to give a dimeric product that precipitates as a copious bright yellow flocculent solid. The new test is more sensitive and reliable than the traditional one.

2-Pyridinecarboxaldehyde → (CN⁻) → bright yellow solid

Halogens. A fresh portion of the filtered solution is acidified with nitric acid and boiled for a short while to expel any hydrogen cyanide or hydrogen sulfide that may be present. The resulting solution, containing free nitric acid, is treated with aqueous silver nitrate, and, if halides are present, a precipitate of silver halide results. The individual halides can be distinguished by oxidation to the elemental halogen and observation of the color of a methylene chloride extract.

$$X^- + Ag^+ \longrightarrow AgX$$
off-white

$$5 X^- + MnO_4^- + 8 H^+ \longrightarrow \tfrac{5}{2} X_2 + Mn^{2+}$$
purple, brown, or colorless

Sodium Fusion

The sodium fusion should be carried out in the hood. Support, a small, Pyrex test tube (about 75 × 10 mm) by inserting it through a small hole in a piece of transite board so that the tube is held by its rim, as in Figure 9.2. Drop a small piece of bright sodium metal no larger than a pea (about 3–4 mm on each edge), and heat the bottom of the tube gently with a micro burner until the sodium melts and its vapors fill the lower part of the tube. Remove the flame momentarily and then drop the sample[2] directly on the molten sodium.

▶ *CAUTION* Carry out the sodium fusion with great care; be particularly cautious in decomposing the fused mass, since the organic vapors sometimes ignite. You must wear *safety goggles* while carrying out this procedure.

---

[2] A stronger test for nitrogen can be obtained by mixing the sample with about half its volume of confectioner's sugar.

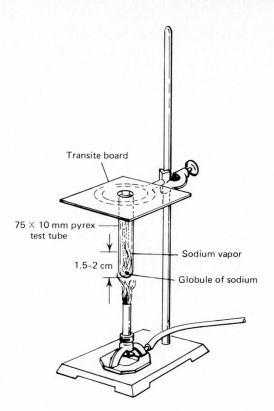

FIGURE 9.2
Apparatus for Sodium Fusion

If the sample is a solid, use only about 10 mg; if it is a liquid, use only 2 drops. A spontaneous exothermic reaction takes place, frequently accompanied by a flash of fire. Heat the tube again until the vapors start to rise and then, after removing the flame, add a second small portion of the sample. Adjust the burner to produce a hot flame and heat the tube to redness in order to complete the reactions. Extinguish the flame and allow the tube to cool to about room temperature. Slowly add about 1 mL of methanol and, after the reaction has subsided, heat the tube gently to evaporate the residual methanol. Sometimes the methanol ignites during the evaporation; the fire can be extinguished easily by placing a transite board over the top of the test tube.

Allow the tube to cool and then crush it with a clean mortar and pestle. Add about 5–6 mL of distilled water to the mortar and grind the mixture to dissolve the fusion solids. Filter the solution through a small filter paper or wad of glass wool and apply the following tests to portions of the solution.

▌▶ *CAUTION* During the crushing and grinding operation cover, the mortar and pestle with a laboratory towel.

1. Sulfur. Acidify a 1-mL portion of the fusion solution with acetic acid (litmus) and add a few drops of 5% lead acetate solution. A black precipitate indicates sulfide.

2. Nitrogen. *Traditional test:* To a 1-mL portion of the fusion filtrate add 2 drops of ferrous ammonium sulfate solution (saturated), and 2 drops of potassium fluoride solution (10%). Bring the mixture to a gentle boil and then cool it to room temperature. Acidify the mixture carefully with dilute sulfuric acid (20–30%) until the precipitate of iron hydroxide just dissolves. Avoid excess acid. A precipitate of prussian blue indicates the presence of cyanide. A faint precipitate can be detected by allowing it to stand for a short time and then filtering it through a small filter paper.

Catalytic test: To 1-mL of 1 M aqueous 2-pyridinecarboxaldehyde add a 1-mL portion of the fusion filtrate and allow the mixture to stand for 10 min. If a bright yellow precipitate forms, the test is positive for nitrogen (as cyanide ion). The test reagent does not store well at room temperature and should be prepared fresh each day, although it will keep for several days if stored in a refrigerator.

3. Halogens. Acidify a 1-mL portion of the fusion filtrate with dilute nitric acid (1 volume of concentrated acid to 1 volume of water) and if nitrogen or sulfur is present, as shown by tests 1 and 2, boil gently for 5–10 min to remove any hydrogen sulfide or hydrogen cyanide (**Caution**—*toxic vapor!*) that may have been formed. Add about 1 mL of a dilute solution of silver nitrate (5–10%) and boil gently for a few minutes. A heavy precipitate indicates the presence of halide; if there is only a faint turbidity, it is probably due to the presence of impurities in the reagents.

The halides chloride, bromide, and iodide can be distinguished by the following procedure. Acidify a 1-mL portion of the fusion solution with dilute nitric acid, as in the preceding paragraph. To this solution add 10 drops of 1% potassium permanganate and shake the test tube for about 1 min. Add just enough oxalic acid (about 20–30 mg) to discharge the color of the excess permanganate, and then add 1 mL of methylene chloride. Shake the test tube and observe the color of the methylene chloride layer. Purple indicates iodine, brown indicates bromine, and the absence of color indicates chlorine. The iodine and bromine colors may be quite faint and require close scrutiny for their detection. The presence of bromine can be confirmed by adding several drops of allyl alcohol, which will react quickly with bromine but not with iodine.

9.6 Solubility Classification

The classification of an unknown according to its solubility behavior, when combined with a knowledge of the elements present, greatly limits the number of functional groups that need be considered. These may be further differentiated either by chemical tests or, as will be discussed in Chapter 10, by spectrometry. Systematic solubility classification was introduced by Kamm,[3] and his scheme forms the basis of the flowchart presented in Figure 9.3. The classes of molecules included in this chart are restricted to the more commonly encountered functional groups.[4]

The principle behind the scheme is that although all hydrocarbons are insoluble in water, attachment of polar functional groups produces favorable interactions with the polar water molecules such that the substituted hydrocarbon becomes water soluble. The stronger the interactions, the larger the hydrocarbon chain can be. By convention, a substance is designated "soluble" in a solvent if it dissolves to the extent of 3% or more. This definition is arbitrary, but it provides a practical classification scheme.

In the abbreviated scheme shown in Figure 9.3 the primary classification solvents are water, 2.5 N NaOH (10%) and 1.5 N HCl (5%), which are used to categorize the unknown as water soluble (**S**), acidic (**A**), or basic (**B**). The acidic compounds are further divided into strong and weak acids according to their solubility in the weak base sodium bicarbonate. Those molecules not soluble in any of these aqueous solvents can be subdivided by their elemental composition and behavior toward concentrated sulfuric acid, as shown in the scheme.

Most molecules that belong to solubility class **S** have four or fewer carbon atoms combined with a polar functional group. Particularly effective water solubilizing groups such as carboxylate anions or quaternary ammonium ions confer water solubility to molecules with as many as 20–25 carbon atoms; polyfunctional molecules may have proportionately greater numbers of carbon atoms. Any functional group, except the halogens, may be present in a molecule belonging to class **S**. Compounds with five carbons have borderline solubility. Branching increases water solubility; many cyclic compounds with six carbons are soluble.

The following molecules are examples of class **S** compounds; each has a polar functional group and four or fewer carbon atoms.

$$CH_3CH_2CH_2CH_2OH \qquad CH_3\overset{\underset{\displaystyle\|}{O}}{C}CH_3 \qquad CH_3\overset{\underset{\displaystyle\|}{O}}{C}-O-CH_2CH_3$$

$$\text{\textit{n}-Butyl alcohol} \qquad\qquad \text{Acetone} \qquad\qquad \text{Ethyl acetate}$$

$$CH_3CH_2\overset{\underset{\displaystyle\|}{O}}{C}OH \qquad CH_3CHO \qquad CH_3CH_2-O-CH_2CH_3$$

$$\text{Propionic acid} \qquad \text{Acetaldehyde} \qquad \text{Diethyl ether}$$

*Water-soluble (class **S**) compounds.*

---

[3] Kamm, *Qualitative Organic Analysis* (New York: Wiley, 1922).

[4] For more exhaustive treatments of solubility classification and lists of molecular classes see Cheronis, Entrikin, and Hodnett, *Semimicro Qualitative Organic Analysis*, 3rd ed. (New York: Interscience, 1965), or Shriner, Fuson, Curtin, and Morill, *The Systematic Identification of Organic Compounds*, 6th ed. (New York: Wiley, 1980).

Sec. 9.6] SOLUBILITY CLASSIFICATION

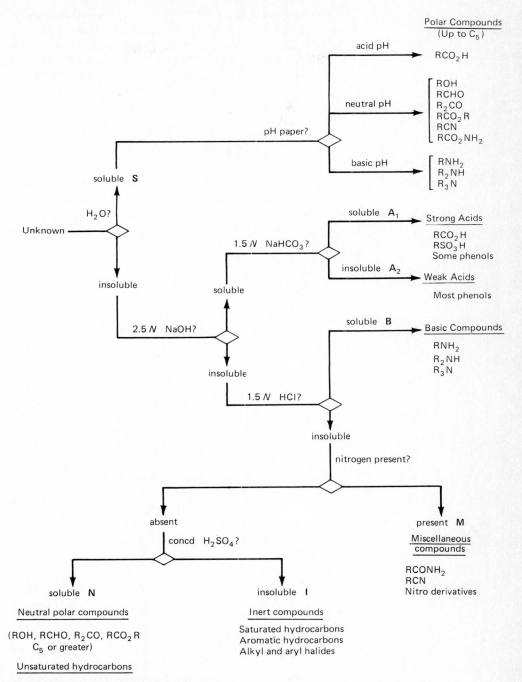

FIGURE 9.3 Solubility Classification Scheme

Note particularly the inclusion of diethyl ether, which is widely used to extract organic materials *from* water. There is no contradiction. If you shake 100 mL of diethyl ether with 100 mL of water, two layers will form. The apparent insolubility is only superficial; about 6 mL of the ether will dissolve in the water and about 1 mL of water will dissolve in the remaining 94 mL of ether. One must not confuse solubility (defined here as 3% solubility) with miscibility (complete solubility), terms that tend to be used interchangeably.

Acidic species with fewer than 20–25 carbon atoms will be soluble in 2.5 N sodium hydroxide because of the strong solubilizing effect of the negative charge on the anion. Strong acids, class A_1 (carboxylic acids and some phenols bearing electron-withdrawing groups), will also be soluble in 1.5 N aqueous sodium bicarbonate but weak acids, class A_2 (most phenols, enols, primary and secondary nitro compounds, and primary sulfonamides), will not dissolve in this weak base. Examples of A_1 and A_2 compounds are shown below.

$CH_3(CH_2)_{14}CO_2H$ $(CH_3)_3C$—⟨ ⟩—OH

Hexadecanoic acid (A_1)   *p-tert*-Butylphenol (A_2)

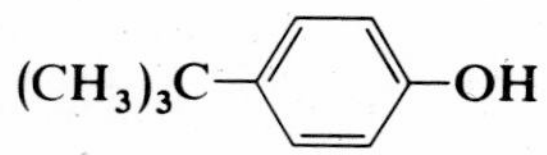

Base-soluble (class *A*) molecules.

Molecules soluble in 1.5 N hydrochloric acid, class **B**, will be primary, secondary, or tertiary amines.

$CH_3(CH_2)_8NH_2$ $(CH_3CH_2CH_2CH_2)_2NH$ $(CH_3CH_2CH_2)_3N$

Decamine   Di-*n*-butylamine   Tri-*n*-propylamine

Acid-soluble (class *B*) molecules.

In classifying the remaining molecules, which are insoluble in all of the aqueous test solvents, it is useful to distinguish between those that contain nitrogen or sulfur and those that do not. Molecules containing these elements are assigned to class **M**, which includes nitro compounds, amides, mercaptans, sulfides, and sulfonyl derivatives.

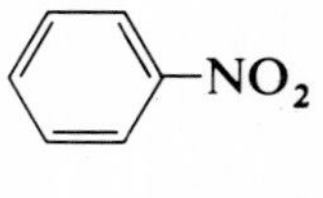

⟨ ⟩—NO_2 $CH_3\overset{O}{\overset{\|}{C}}-N\underset{CH_2CH_3}{\overset{CH_2CH_3}{\diagup\diagdown}}$ CH_3—⟨ ⟩—$\overset{O}{\underset{O}{\overset{\|}{\underset{\|}{S}}}}-NH_2$

Nitrobenzene 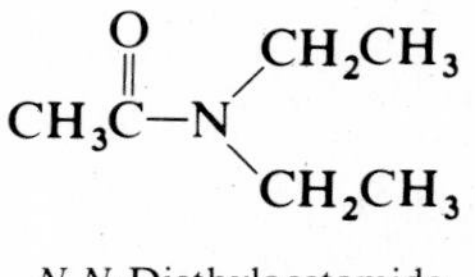   *N,N*-Diethylacetamide 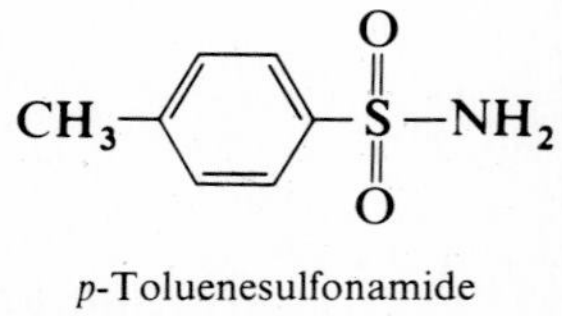   *p*-Toluenesulfonamide

Nitrogen- and sulfur-containing (class *M*) molecules.

When nitrogen and sulfur are absent, the remaining molecules are classified according to their solubility in cold, concentrated sulfuric acid. Soluble molecules, class **N**, may contain any of the functional groups found in class **S** since each of them is capable of being protonated to produce an ionic compound that is soluble in the acid. Another group of molecules falling into class **N** are the alkenes and alkynes, which react with concentrated sulfuric acid to produce polar species that are soluble in that solvent. Compounds that react with sulfuric acid, as evidenced by darkening or heat evolution, should be assigned to class **N** even if an insoluble precipitate is formed. Class **N** molecules will have more than four carbons and nitrogen and sulfur will be absent.

$$CH_3CH_2CH_2\overset{\overset{O}{\|}}{C}CH_2CH_3 \qquad C_6H_5\overset{\overset{O}{\|}}{C}-OCH_3 \qquad CH_3C{\equiv}CCH_3$$

3-Hexanone   Methyl benzoate   2-Butyne
*Class **N** molecules.*

The inert molecules, class **I**, do not dissolve in any of the classification solvents and include the saturated hydrocarbons, aromatic hydrocarbons (weakly reactive toward sulfuric acid because of their delocalization energy), and alkyl and aryl halides.

$$CH_3(CH_2)_4CH_3$$

n-Hexane   Adamantane
*Class **I** molecules.*

Solubility Classification Tests

In a small test tube place 0.1 mL of a liquid or 0.1 g of a solid compound[5] and add, in small portions, a total of 3 mL of water. Between each addition stir the sample vigorously with a rounded stirring rod; with solids it is desirable to crush the crystals to increase their surface area. If the sample has dissolved, test the aqueous solution with a wide-range pH paper to determine if the solution is acidic, neutral, or basic. Record your observations directly in your notebook.

Follow the solubility scheme illustrated in Figure 9.3 using fresh samples and 3-mL portions of the appropriate solvents. Keep a careful record of each test as you make it. Note that in following the scheme, not all the test solvents need be tried; use only those that are logically required

---

[5] It is desirable to use precise amounts of sample. A convenient procedure for solids is to weigh out 0.4 g on a weighing paper and divide it visually into four equal portions. With liquids you can use a dropper that delivers a known number of drops per milliliter.

to classify the sample. Also note that the strongly basic and acidic solutions are quite corrosive and should be cleaned up immediately if spilled.

The most serious difficulty encountered in these tests is with compounds having borderline solubility. You should observe the sample carefully and note whether any of it appears to dissolve. If partial solubility is observed, give the sample a longer time to dissolve because the rate of solution of borderline compounds can be quite slow.

9.7 Functional Group Identification

Any molecule other than a saturated hydrocarbon has at least one functional group, which can be identified by carrying out a series of "classification tests" that serve to narrow the range of possibilities until only one remains. When the functional group is identified, an appropriate table of characterized compounds containing this group is consulted, and those compounds having chemical and physical properties consistent with the sample are selected. In favorable cases only a few molecules will be found; rarely will there be more than 10.

The lists of molecules containing each functional group give not only the physical properties of the molecules but also the properties of solid substances (derivatives) that can be prepared from them by test procedures. Since the melting points of these derivatives are usually distinctive, the combination of properties of the original substance and of its derivatives is sufficient to identify it.

Table 9.1 lists the different functional groups that will be considered in this text with the classification tests and derivatives appropriate for each. The list of functional groups is restricted but does include the most commonly encountered types. If your instructor wishes to broaden the range, advanced texts devoted to organic qualitative analysis will have to be consulted.[4] Discussions of the chemistry of each test and its structural significance, as well as the experimental details, are given for each functional group in this section. Once the functional group has been identified, the final identification can be made by the preparation of derivatives as described in Section 9.8.

In the laboratory it is important to perform the classification tests in a sequence consistent with the accumulated evidence, never at random. A good guide is the solubility classification scheme (Figure 9.3), which lists the possible functional groups for each solubility class. For example, if the elemental analysis reveals nitrogen and the compound falls in solubility class **B**, the amine tests should be performed directly.

As a second example, if a neutral compound falls in class **S** or **N** and does not contain nitrogen, sulfur, or halogens, the functional group must be an alcohol, aldehyde, ketone, or ester. In this case, the recommended next

Sec. 9.7] FUNCTIONAL GROUP IDENTIFICATION

TABLE 9.1 *Classification Tests and Derivatives of Functional Groups*

Functional group	Classification tests	Derivatives
Alcohol	Oxidation test Lucas test Iodoform test	*p*-Nitrobenzoate 3,5-Dinitrobenzoate Phenylcarbamate
Aldehyde and ketone	Tollens' test Fuchsin aldehyde test 2,4-Dinitrophenylhydrazine test Iodoform test	Methone derivatives 2,4-Dinitrophenylhydrazone Semicarbazone Oxime
Ester	Hydroxamate test	Saponification equivalent Hydrolysis of ester to acid and alcohol, which are suitable derivatives if solid
Carboxylic acid	Solubility classification (pH of solution if **S**)	Acid amides Neutralization equivalent
Sulfonic acid	Elemental analysis Solubility classification (pH of solution if **S**)	*S*-Benzylthiouronium salts Sulfonamides
Acid amide and nitrile	Elemental analysis Hydroxamate test	Hydrolysis to carboxylic acid and amine or ammonia, which may be identified separately
Amine	Elemental analysis Solubility classification (pH of solution if **S**) Hinsberg test	Acetamide Benzamide Benzenesulfonamide *p*-Toluensulfonamide Quarternary ammonium salt
Phenol	Solubility classification Ferric chloride test	3,5-Dinitrobenzoate Aryloxyacetic acid Bromination product
Unsaturated hydrocarbons	Solubility classification Bromine test Permanganate test	No generally applicable derivatives Bromine titration
Aromatic and saturated hydrocarbons and halide	Solubility classification Elemental analysis Alcoholic silver nitrate test	No generally applicable derivatives

step is to test with 2,4-dinitrophenylhydrazine for an aldehyde or ketone. If the test result is positive, further structural distinctions can be made with the tests described in the procedures for aldehyde and ketone tests. A negative 2,4-dinitrophenylhydrazone test should be followed by the hydroxamate test for esters. If that test is negative, only the alcohol class remains, and this can be confirmed by the classification tests for alcohols. Functional groups of compounds that fall into other solubility classes can be identified by analogous strategies.

To insure satisfactory results for the tests, it is recommended that the specified quantities of liquid reagents be measured in a graduated cylinder or a calibrated dropper. If a test is being done for the first time, it is important to practice on materials of known structure.

It is in the area of identifying functional groups that infrared (IR) analysis is so powerful, since a single IR spectrum reveals much about the nature of all of the functional groups present. However, the IR spectrum usually does not provide a total answer, and one must resort either to other instrumental techniques or to the chemical methods described here.

Alcohols.[6] The chromic acid test is a general test for alcohols or other readily oxidizable functional groups such as aldehydes. The *Lucas test* and the *iodoform test* provide further structural information about the alcohol.

(A) Chromic Acid Oxidation Test

This test, based on the ability of primary and secondary alcohols to be oxidized by chromic acid, distinguishes these alcohols from tertiary alcohols. For low-molecular-weight alcohols, aqueous chromic acid can be used.

To 5 mL of a 1% solution of sodium dichromate, add 1 drop of concentrated sulfuric acid and mix thoroughly by shaking. (Use a cork, not your thumb, to stopper the tube!) Add 2 drops of the liquid (or about 40 mg of a solid) to be tested, and warm gently. Observe any change in the color of the solution.

$$H_2Cr_2O_7 + R-CH_2OH \text{ or } R_2CHOH \xrightarrow{H_2SO_4} Cr_2(SO_4)_3 + R-CO_2H \text{ or } R_2C=O$$

orange — blue-green

For higher molecular weight alcohols that are insoluble in the aqueous reagent, the oxidation test can be carried out by using acetone and chromic acid test reagent. Under these conditions, the blue-green chromic reduction product precipitates and is quite visible in the orange solution.

In 2 mL of acetone in a small test tube dissolve 2 drops of the liquid (or about 20 mg of a solid), and to the solution add 2 drops of *chromic acid test reagent*.[7] Observe any change that occurs within 5 sec; ignore any change

---

[6] A suitable set of alcohols with which to practice the alcohol tests is allyl alcohol, *sec*-butanol, *t*-butanol, and *n*-propanol.

[7] Chromic acid test reagent is prepared by dissolving 10 g of chromic anhydride (CrO_3) in 40 mL of 25% sulfuric acid (10 mL of sulfuric acid in 30 mL of water). (**Caution:** CrO_3 is *corrosive* and *poisonous*. In contact with organic materials it may cause inflammation!)

that occurs later. It is advisable to run a control test on a sample of the acetone.

(B) Lucas Test

The reagent used is concentrated hydrochloric acid containing 1 mole of fused zinc chloride to 1 mole of the acid.[8] The Lucas test distinguishes among primary, secondary, and tertiary alcohols and is based on the rate of formation of the insoluble alkyl chloride. For the test to be reliable, the alcohol should be soluble in water (class **S**).

The ease of conversion of alcohol to chloride follows the stability of the corresponding carbocation, modified by the solubility of the alcohol in the test reagent. Allyl alcohol, $CH_2=CH-CH_2OH$, which yields a stabilized charge-delocalized cation, acts like a tertiary alcohol. Isopropyl alcohol sometimes fails to give a positive test because the chloride product is volatile (bp 36°) and may escape from the solution.

$$ROH + H^+ \xrightarrow{ZnCl_2} R^+ + H_2O \xrightarrow{Cl^-} RCl$$

insoluble in aqueous reagents

To 0.5 mL of the alcohol add quickly 3 mL of the hydrochloric acid–zinc chloride reagent at 25–27°. Close the tube with a cork and shake it; then allow the mixture to stand. Tertiary alcohols give an immediate separation (emulsion) of the chloride, and secondary alcohols require about 5 min, but most primary alcohols do not react significantly in less than 1 hr. If the result is positive, carry out a second test using concentrated hydrochloric acid alone, instead of the test reagent. This less reactive reagent will give chloride emulsions within 5 min only with tertiary alcohols.

(C) Iodoform Test

This is a test for the specific structural feature $R-CHOH-CH_3$ (R may also be H). The test depends on initial oxidation of the alcohol to $R-CO-CH_3$, which is iodinated and then cleaved to give a bright yellow precipitate of iodoform.

---

[8] Lucas, reagent is prepared by dissolving 34 g of anhydrous (fused) zinc chloride in 27 g of concentrated hydrochloric acid, with stirring and external cooling to avoid loss of hydrogen chloride. The resulting solution has a volume of about 35 mL and is sufficient for about 10 tests. To obtain reliable results, the reagent should be reasonably fresh.

$$I_2 + NaOH \rightleftharpoons NaOI + NaI + H_2O$$

$$\underset{H}{\overset{OH}{R-\underset{|}{\overset{|}{C}}-CH_3}} \xrightarrow{NaOI} R-\overset{O}{\underset{\|}{C}}-CH_3 \xrightarrow[base]{I_2} R-\overset{O}{\underset{\|}{C}}-CH_2I \xrightarrow[base]{I_2} R-\overset{O}{\underset{\|}{C}}-CHI_2 \xrightarrow[base]{I_2}$$

$$R-\overset{O}{\underset{\|}{C}}-CI_3 \xrightarrow{OH^-} \left[R-\underset{OH}{\overset{O^-}{\underset{|}{\overset{\|}{C}}}}-CI_3 \right] \longrightarrow R-\overset{O}{\underset{\|}{C}}-OH + {}^-CI_3 \rightleftharpoons R-\overset{O}{\underset{\|}{C}}-O^- + HCI_3$$
yellow

In a clean (acetone-free) 150-mm test tube, mix 3 drops of the liquid (or about 50 mg of solid) with 2 mL of water and 2 mL of 10% aqueous sodium hydroxide solution.[9] Add dropwise with shaking a 10% solution of iodine in potassium iodide,[10] until a definite brown color persists (indicating an excess of iodine).

With some compounds a precipitate of iodoform appears almost immediately in the cold. If it does not appear within 5 min, warm the solution to 60° in a beaker of water. If the brown color is discharged, add more of the iodine solution until the iodine color persists for 2 min. Add a few drops of sodium hydroxide solution to remove excess iodine, dilute the mixture with 5 mL of water, and allow it to stand for 5 min at room temperature.

For compounds that are not appreciably soluble in water, the sample may be dissolved in *pure* methanol instead of water. Before the test is started, the solvent should be tested to see if iodoform-producing impurities are present.

Iodoform crystallizes as lemon yellow hexagons that have a characteristic odor. Their identity can be confirmed by collecting the solid with suction and taking the melting point (119°).

Aldehydes and Ketones.[11] The 2,4-dinitrophenylhydrazone test is positive for both aldehydes and ketones. These may be distinguished by either the *silver mirror test*, which depends on the ease of oxidation of aldehydes, or *Schiff's fuchsin test*, which depends on the ease of formation of SO_2 adducts

---

[9] This test is suitable for alcohols with significant water solubility. For less soluble compounds add 1 mL of pure methanol (free of ethanol and acetone).

[10] Iodine–potassium iodide solution is prepared by dissolving 10 g of iodine crystals in a solution of 20 g of potassium iodide in 80 mL of water and stirring until the iodine has dissolved.

[11] A suitable set of aldehydes and ketones with which to practice is acetone, benzaldehyde, and cyclohexanone.

of aldehydes but not ketones. Another test that will distinguish aldehydes from ketones is the *chromic acid test*, described earlier under alcohols.

The *iodoform test*, also described earlier under alcohols, is specific for molecules containing a methyl group adjacent to a carbonyl group or to any other structure that can form such a methyl carbonyl combination. The only aldehyde that gives a positive iodoform test is acetaldehyde.

(D) 2,4-Dinitrophenylhydrazone Test

Most aldehydes and ketones react with 2,4-dinitrophenylhydrazine reagent to give precipitates of the 2,4-dinitrophenylhydrazones. Esters and amides generally do not respond and can be eliminated on the basis of this test.

The color of the precipitate depends on the degree of conjugation in the aldehyde or ketone. Unconjugated aliphatic carbonyl groups, such as butyraldehyde or cyclohexanone, give yellow precipitates;

conjugated carbonyls, such as benzaldehyde or methyl vinyl ketone, give red precipitates.

Unfortunately, the reagent is orange-red; one should therefore establish that a reddish precipitate is not just the starting reagent that has been made insoluble by the addition of the unknown.

In a clean small test tube, place 1 mL of 2,4-dinitrophenylhydrazine reagent[12] and add a few drops of the unknown liquid (or about 50 mg of solid dissolved in the minimum amount of 95% ethanol). A positive test is the formation of a yellow to red precipitate. Most aldehydes and ketones will give a precipitate immediately, although some that are sterically hindered may take longer. If no precipitate appears within 15 min, heat the solution gently for 5 min; examine the test tube after it has cooled to room temperature.

---

[12] The reagent is a 4% solution of 2,4-dinitrophenylhydrazine in acidified ethanol–water.

(E) Tollen's Reagent (Silver Mirror Test)

This test involves reduction of an alkaline solution of silver ammonium hydroxide to metallic silver and oxidation of the aldehyde, but *not* a ketone, to the carboxylic acid.

$$R-\underset{\underset{H}{\|}}{\overset{\overset{O}{\|}}{C}}-H \xrightarrow[H_2O]{Ag(NH_3)_2} R-\underset{}{\overset{\overset{O}{\|}}{C}}-O^- + Ag$$
silver mirror

This is an extremely mild oxidation, and alcohols do not respond. Fehling's or Benedict's solution (alkaline cupric tartrate or citrate) also may be used to test for aldehydes, but the Tollens' test is more sensitive.

In a thoroughly clean 75-mm test tube, place 1 mL of a 5% solution of silver nitrate and add a drop of 10% aqueous sodium hydroxide. Add a very dilute solution of ammonia (about 2%), drop by drop, with constant shaking until the precipitate of silver oxide just dissolves. To obtain a sensitive reagent, it is necessary to avoid a large excess of ammonia.

▶ *CAUTION* The silver ammonium hydroxide reagent should be freshly prepared just before use and *should not be stored*. On standing, the solution may decompose and deposit an explosive precipitate of silver nitride, Ag_3N.[13]

Add 2 drops of the unknown to be tested, shake the tube, and allow it to stand for 10 min. If no reaction has occurred in this time, place the tube in a beaker of water that has been heated to about 40° and allow it to stand for 5 min. A positive result is the formation of a silver mirror (if the tube is clean) or a black precipitate of finely divided silver.

Water-insoluble compounds give weak or negative tests. With such unknowns it is helpful to dissolve them in 0.5 mL of pure acetone.

(F) Schiff's Fuchsin Test

The intensely colored triphenylmethane dye fuchsin reacts with bisulfite to produce the colorless "leuco" form of the dye. Aldehydes, but not ketones, react with this "leuco" dye to produce a new triphenylmethane dye that possesses a similar fuchsia color.

---

[13] As soon as the test has been completed, pour the contents into the "silver waste" container and wash the tube with water. A freshly formed silver mirror can be removed with soap and a test tube brush; residual silver stains can be removed with dilute nitric acid.

fuchsia   colorless   fuchsia

To a few drops of the unknown to be tested, in 4–5 mL of water, add about 1 mL of the fuchsin test reagent[14] and observe any development of purple color.

Ketones do not respond to this test *when perfectly pure*, but the color reaction is very sensitive and responds to mere traces of an aldehyde.

Esters. An extremely sensitive test for esters is the formation of the intensely colored ferric hydroxamate derivative. The first step of the test is the base-catalyzed conversion of the ester to a hydroxamic acid. In the next step the hydroxamic acid is treated with ferric chloride, which produces the red-violet octahedral ferric hydroxamate.

$$3\ R-\overset{O}{\underset{\|}{C}}-OR' + H_2NOH \longrightarrow R-\overset{O}{\underset{\|}{C}}-NHOH + R'OH$$
Hydroxamic acid

$$3\ R-\overset{O}{\underset{\|}{C}}-NHOH + FeCl_3 \longrightarrow \left[R-C\underset{\underset{H}{N-O}}{\overset{O:}{\diagdown}} \right]_3 Fe + 3\ HCl$$
red-violet

All carboxylic acid esters produce the ferric hydroxamate. Acid chlorides, anhydrides, and imides react with hydroxylamine to give hydroxamic

---

[14] The fuchsin test solution is prepared by dissolving 100 mg of pure fuchsin (*p*-rosaniline hydrochloride) in 100 mL of distilled water and adding 4 mL of saturated sodium bisulfite solution. After about an hour, 5 mL of water to which 2 mL of concentrated hydrochloric acid has been added is added slowly. This produces a practically colorless, sensitive reagent. If the solution is not colorless, shake it with a little decolorizing carbon and filter it.

acids, and thus they also give positive tests. Amides and nitriles generally do not react sufficiently under the specified reaction conditions to give more than pale colorations. Free carboxylic acids (except for formic acid) given negative tests.

Some phenols react with ferric chloride to give a somewhat similar color. For this reason, if a positive test is obtained, it is necessary to run a "blank" test in which all of the steps are carried out except the addition of the hydroxylamine.

(G) Hydroxamate Test[15]

In a small test tube, place a few drops of a liquid (or about 50 mg of a solid) unknown and 1 mL of 7% methanolic hydroxylamine hydrochloride ($NH_2OH \cdot HCl$) containing a pH indicator.[16] Add 10% methanolic potassium hydroxide until the mixture just turns blue and then 0.5 mL more. Heat the solution to boiling and, after allowing it to cool slightly, acidify it with 7% methanolic hydrochloric acid[17] until the solution turns rose color. Then, add 2 drops of 3% ferric chloride solution. If the color is weak, add a few more drops of ferric chloride solution. A positive test is the development of an intense red-violet color.

Carboxylic Acids. The presence of a carboxylic acid functional group is revealed by the solubility behavior of the molecule. If the solubility class is **S**, the aqueous solution will be acidic. The only possible confusion here would be with a low-molecular-weight sulfonic acid, which requires a positive elemental analysis for sulfur. If the solubility class is **A**, and sulfur is absent, the molecule is either a carboxylic acid or one of the few phenols that are substituted with strongly electron-withdrawing groups, so that they move from their normal classification as weak acids (A_2). Tests for phenols are described later in this section.

Sulfonic Acids. There is no specific test for detecting a sulfonic acid functional group. Its presence is indicated by a positive elemental analysis for sulfur, combined with solubility evidence: either an acidic solution (if the molecule is water soluble, class **S**) or solubility in sodium bicarbonate (class A_1). The only confusion that might arise would be the rare circumstance

---

[15] A suitable set of compounds on which to practice the ester test is ethyl acetate, benzoic acid, and *p*-cresol.

[16] The indicating reagent is prepared by dissolving 70 g of hydroxylamine hydrochloride, 100 mg of thymolphthalein, and 15 mg of methyl yellow in 1 L of methanol. The solution is neutralized by careful addition of 10% methanolic potassium hydroxide until the rose color just turns orange.

[17] This is approximately 2 *M* acid prepared by dissolving 17 mL of concentrated hydrochloric acid in 83 mL of methanol.

that the molecule was a carboxylic acid substituted by some neutral, sulfur-containing substituent. Fortunately, there are good derivatives that can be made from sulfonic acids, and therefore its suspected presence can be confirmed.

Carboxamides and Nitriles. These two functional groups can be detected by their reaction with hydroxylamine and ferric chloride to form a red-violet complex. The general structure of the complex from a carboxamide or nitrile is the same as that described under the ester test reactions; it differs only by substitution of an imino group for the ester carbonyl group.

$$R-CN + H_2NOH \longrightarrow R-\underset{\underset{NHOH}{\|}}{\overset{NH}{C}}-NHOH \xrightarrow{FeCl_3} \left[R-C \underset{\underset{H}{\diagdown}}{\overset{\diagup N \diagdown}{\diagdown N-O \diagup}} \right]_3 Fe + 3\ HCl$$

red-violet

Amides and nitriles are less reactive than esters, therefore more vigorous conditions are required for the reaction with hydroxylamine. This is achieved by substituting propylene glycol (bp 187°) for methanol (bp 65°) as a solvent.

(H) Amide and Nitrile Test[18]

Add 30–50 mg of the unknown to 2 mL of 1 N propylene glycol solution of hydroxylamine hydrochloride. Add 1 mL of 1 N potassium hydroxide in propylene glycol, and boil the mixture gently for 2 min. Cool the solution to room temperature and add 0.5–1 mL of 5% aqueous ferric chloride. A red-violet color is a positive test result.

Amines. All classes of alkyl and aryl amines—primary, secondary, and tertiary (1°, 2°, 3°)—have an unshared electron pair on nitrogen and are thus basic and nucleophilic. The unshared electron pair is responsible for the ability of these amines to form salts with acids to form coordination complexes with metal cations, and to undergo alkylation with alkyl halides (nucleophilic displacements).

The availability of the unshared pair for combination with an electrophile (measured by base strength or nucleophilicity) is strongly influenced by the nature of the groups attached to the nitrogen atom. In general, alkyl

---

[18] A suitable set of compounds with which to practice this test is benzamide and benzonitrile.

groups, through electron release, enhance the base strength, whereas electron-withdrawing groups such as aryl and acyl diminish it. Base strength and nucleophilicity are also influenced by the steric bulk of the substituents, both being diminished by large groups.

All of the amines considered here will fall into either the **S** or **B** solubility class and thus are easily recognized by their elemental composition and basicity. Primary and secondary amines differ from tertiary amines in their behavior toward acyl and sulfonyl halides. A particularly useful reagent is benzenesulfonyl chloride and excess base (Hinsberg test).[19] All three classes of amines react with benzenesulfonyl chloride, but they differ in how the intermediate products respond to base. With primary amines the sulfonamide that is formed is acidic and dissolves in the excess base used to yield a solution of the corresponding anion. Addition of excess hydrochloric acid converts the anion into the water-insoluble sulfonamide.

$$RNH_2 + C_6H_5-SO_2Cl \xrightarrow{KOH} C_6H_5-SO_2NHR + KCl + H_2O \underset{\text{excess acid}}{\overset{\text{excess base}}{\rightleftarrows}}$$

1° amine, water insoluble

$$C_6H_5-SO_2NR^- K^+ + H_2O$$

water soluble

Secondary amines react with benzenesulfonyl chloride, but the sulfonamide lacks an amide hydrogen and is insoluble in the basic reagent.

$$R_2NH + C_6H_5-SO_2Cl \xrightarrow{KOH} C_6H_5-SO_2NR_2 + KCl + H_2O \xrightarrow{\text{excess base}} \text{no reaction}$$

2° amine, water insoluble

Tertiary amines behave differently with benzenesulfonyl chloride; the intermediate ammonium ion does not have a proton to lose and instead reacts rapidly with hydroxide ion to displace the benzenesulfonate anion and regenerate the tertiary amine.

$$R_3N + C_6H_5-SO_2Cl \longrightarrow C_6H_5-SO_2\overset{+}{N}R_3\ Cl^- \xrightarrow[H_2O]{OH^-} C_6H_5-SO_3^- + NR_3 + Cl^-$$

3° amine, water soluble, water soluble

---

[19] For a detailed discussion of the Hinsberg test, see Gambill, Roberts, and Shechter, *J. Chem. Educ.* 49: 287 (1972).

The overall reaction amounts to an amine-catalyzed hydrolysis of the benzenesulfonyl chloride. With tertiary amines there can be a side reaction between the amine and the intermediate ammonium ion to produce a complex mixture of water-insoluble products,[19] which could lead to confusion with the results for a secondary amine. This complication is minimized by keeping the concentration of the amine low (as specified in the test procedure).

(I) Hinsberg Test

To 8–10 drops of the amine in a large test tube, add 10 mL of 10% aqueous potassium hydroxide and 10 drops of benzenesulfonyl chloride. Shake the tube thoroughly and note any reaction. Warm the mixture very gently with shaking (*do not boil*) until the *disagreeable odor* of benzenesulfonyl chloride can no longer be detected. The reaction mixture should still be strongly alkaline at this point. Cool the tube to room temperature, shake well, and note whether any new solid or liquid separates. Do not confuse it with unreacted benzenesulfonyl chloride.

If the mixture has formed two liquid layers (not counting any unreacted benzenesulfonyl chloride layer), separate them and determine if the upper, organic layer is soluble in 5% hydrochloric acid. If the organic material is soluble, it indicates a tertiary amine (acid soluble, unreactive toward benzenesulfonyl chloride). However, if the organic material does not dissolve in the hydrochloric acid, it indicates a secondary amine (acid- and base-insoluble secondary sulfonamide). Add hydrochloric acid to the aqueous phase until the pH is 4 or less. The formation of a precipitate indicates that the unknown was a primary amine.

Failure of the original basic mixture to separate indicates the presence of a primary sulfonamide. This can be confirmed by adjusting the pH to 4 and noting the formation of a precipitate.

Phenols.[20] Many phenols and related compounds form colored coordination complexes with ferric ion, in which six molecules of a monohydric phenol are combined with one atom of iron to form a complex anion. Most phenols produce red, blue, purple, or green colors. Sterically hindered phenols give negative tests. Aliphatic enols (ethyl acetoacetate, acetylacetone) give a positive test.

---

[20] A suitable set of phenols on which to practice this test is 2-naphthol, *p*-nitrophenol, and salicylic acid.

$$\text{C}_6\text{H}_5\text{OH} + \text{Fe}^{3+} \rightleftharpoons [(\text{C}_6\text{H}_5\text{O})_6\text{Fe}]^{3-} + 6\,\text{H}^+$$

$$\underset{\text{Acetylacetone}}{\text{CH}_3\text{-CO-CH}_2\text{-CO-CH}_3} \qquad \underset{\text{Enol form}}{\text{CH}_3\text{-CO-CH=C(OH)-CH}_3}$$

(J) Ferric Complex

To 2 mL of ethanol in a test tube, add 2 drops of a liquid (or 20 mg of a solid) unknown and a few drops of a 3% aqueous solution of ferric chloride. Shake well and observe the color.

▶ *CAUTION*   Phenol, the cresols, and other phenolic compounds in the pure state or in concentrated solution are toxic and cause painful burns. If any of these come in contact with the skin, wash the area quickly and thoroughly with soap and water.

Hydrocarbons.[21] There are four classes of hydrocarbons: (1) the saturated hydrocarbons, (2) the alkenes (olefins), (3) the alkynes (acetylenes), and (4) the aromatic hydrocarbons. Of these only the alkenes and alkynes will be soluble in cold sulfuric acid (class **N**); the saturated hydrocarbons will fall into class **I**. A test for an aromatic hydrocarbon was described earlier in this section. There are no tests for saturated hydrocarbons; these substances must be detected by their failure to give positive tests for either an aromatic ring or unsaturation. Saturated hydrocarbons are best detected by nuclear magnetic resonance, as described in Section 10.3.

The suspected presence of unsaturation can be confirmed by *cis* hydroxylation with aqueous permanganate (Baeyer test) and by *trans* addition of bromine in carbon tetrachloride. Almost all alkenes and alkynes react with these reagents.

$$3\,\text{C=C} + 2\,\text{MnO}_4^- + 4\,\text{H}_2\text{O} \longrightarrow 3\,\text{C(OH)-C(OH)} + 2\,\text{MnO}_2 + 2\,\text{OH}^-$$

$$\text{C=C} + \text{Br}_2 \xrightarrow{\text{slow}} \left[\text{C}\overset{+}{\underset{\text{Br}}{-}}\text{C}\right] \xrightarrow{\text{fast}} \text{BrC-CBr}$$

---

[21] A suitable set of hydrocarbons on which to practice these tests is cyclohexane, cyclohexene, and toluene.

The only exceptions are molecules that have a strongly electron-withdrawing group on the multiple bond; they fail to react with bromine because the intermediate bromonium ion is formed too slowly.

Another complication of the bromine test is the tendency for C—H bonds adjacent to a double bond (allylic C—H) to discharge the bromine color by a free-radical substitution reaction accompanied by the evolution of hydrogen bromide.

$$C=C-C{\overset{H}{\diagdown}} + Br_2 \longrightarrow C=C-C{\overset{Br}{\diagdown}} + HBr$$

and some ketones also substitute bromine and evolve hydrogen bromide, but by an ionic mechanism. These substitution reactions can be detected by the evolved hydrogen bromide vapor, which is not soluble in the carbon tetrachloride solvent and tends to form an "acid fog" when one blows across the top of the reaction vessel.

All aliphatic amines and some pyridine derivatives discharge the bromine color by reversible formation of colorless bromine adducts.

The Baeyer permanganate test is superior to the bromine test, but it also has complications. All easily oxidized molecules, such as aldehydes and phenols, give positive Baeyer tests. Fortunately, the two tests are largely complementary. It is recommended that the permanganate test be tried first; then, if it is positive, the bromine test should be tried.

(K) Permanganate Test (Baeyer Test)

In a small test tube, dissolve 3 drops of the liquid (or 30 mg of a solid) unknown in 1 mL of pure *alcohol-free* acetone. The solvent must be tested beforehand for purity. Add dropwise, with vigorous shaking, a 1% aqueous solution of potassium permanganate. A positive test is the loss, within 1 min, of the purple permanganate ion color and formation of the insoluble brown hydrated oxides of manganese. Record the number of drops necessary to develop a persistent purple color; do not be deceived by a slight reaction caused by impurities in the unknown.

(L) Bromine Test

This test should be carried out in the hood. In a small test tube dissolve 3 drops of the liquid (or 30 mg of a solid) unknown in 1 mL of carbon tetrachloride and add dropwise, with shaking, a 2% solution of bromine in carbon tetrachloride. Record the number of drops necessary to develop a persistent (for 1 min) bromine color.

> **CAUTION** Bromine can cause painful burns. If any of the solution is spilled on the skin, wash the area quickly and thoroughly with water and then apply a dressing soaked in 10% sodium thiosulfate solution; see a physician.

Prolonged exposure to carbon tetrachloride vapor should be avoided because of its toxicity.

Aromatic Hydrocarbons. Molecules falling into solubility class **I** include saturated hydrocarbons, aromatic hydrocarbons, and their derivatives. The flame test carried out in the preliminary examination may have suggested the presence of an aromatic ring by the appearance of a yellow, sooty flame. Confirmation can be obtained by the Friedel–Crafts alkylation test described here.

Aromatic hydrocarbons (and many of their derivatives) react serially with chloroform in the presence of anhydrous aluminum chloride to produce triarylmethanes.

$$ArH + CHCl_3 \xrightarrow{AlCl_3} ArCHCl_2 \xrightarrow[ArH]{AlCl_3} Ar_2CHCl \xrightarrow[ArH]{AlCl_3} Ar_3CH$$

The intermediate chlorohydrocarbons react with aluminum chloride to produce carbocations that abstract a hydride ion from the triarylmethane to yield highly colored triarylmethyl cations. For example,

$$Ar_2CHCl + AlCl_3 \longrightarrow Ar_2CH^+ + AlCl_4^-$$
$$Ar_2CH^+ + Ar_3CH \longrightarrow Ar_2CH_2 + Ar_3C^+$$
$$\text{highly colored}$$

The color depends on the number of rings in the hydrocarbon. Benzene and its derivatives give an orange-red color; naphthalene and phenanthrene and their derivatives give blue-purple colors; an anthracene ring produces a green color. In general, the observed color depends on the nature of the substituents, but in the classification scheme described here the substituents will be either alkyl groups or halogens, which do not change the colors significantly.

In carrying out the test, it is essential that the aluminum chloride be anhydrous. This is accomplished in the test procedure by freshly subliming a sample of aluminum chloride, which drives off any water that may be present.

(M) Friedel–Crafts Test[22]

Place about 100 mg of anhydrous aluminum chloride in a small dry Pyrex test tube and heat it strongly with the tube held almost horizontally in order to sublime the chloride onto the cooler wall of the tube. While the tube is cooling, prepare (in the hood) in a separate small test tube a solution of about 20 mg of unknown in 10 drops of chloroform. (**Caution**—Chloroform is toxic; see Section 1.1). Add this solution to the test tube containing the freshly sublimed aluminum chloride by dropping it directly onto the salt, and note the color, if any, where they meet.

Alkyl and Aryl Halides.[23] Alkyl halides can be distinguished from aryl halides by a combination of two tests. The first is with alcoholic silver nitrate, which forms a precipitate of silver halide with alkyl halides that undergo S_N1 reactions. The order of reactivity for R groups is allyl and benzyl > tertiary > secondary ≫ primary. The order for the halide leaving group is I > Br > Cl. Secondary and primary halides give no reaction within 5 min; secondary halides react only when the solution is boiled. Primary, aromatic, and vinyl halides usually do not react even after 5 min of heating under reflux.

Primary chlorides and bromides can be distinguished from the aromatic and vinyl halides by the reaction with sodium iodide in acetone. Primary bromides undergo S_N2 displacement reactions within 5 min at room temperature to produce sodium bromide, which is insoluble in acetone. The same reaction occurs with primary chlorides at 50° to produce sodium chloride, which also precipitates.

$$R-X \ (X = Cl, Br) + KI \longrightarrow R-I + \underset{\text{white precipitate}}{KX}$$

Secondary and tertiary bromides and some secondary chlorides also react at 50°.

(N) Alcoholic Silver Nitrate

In a small test tube place 2 mL of a 2% solution of silver nitrate in ethanol and add 1 drop of a liquid (or 10 mg of a solid) unknown. A positive test is the formation of a whitish precipitate of silver halide within 5 min. If no reaction occurs in that time, boil the solution gently for 5 min more.

---

[22] A suitable set of hydrocarbons on which to practice this test is toluene, naphthalene, and cyclohexane.

[23] A suitable set of halides on which to practice these tests is chlorobenzene, *n*-butyl bromide, *t*-butyl chloride and *sec*-butyl chloride.

If a precipitate forms, either at room temperature or on heating, it is advisable to verify that it is not the silver salt of an organic acid by adding 2 drops of dilute nitric acid (20 : 1 water : acid). The acid salts will dissolve; the halides will not.

(O) Sodium Iodide in Acetone

In a small test tube dissolve 2 drops of a liquid (or 20 mg of a solid) unknown in the minimum volume of acetone and add 1 mL of the sodium iodide solution (15 g of sodium iodide in 100 mL of pure acetone). A positive test is the formation of a white precipitate within 5 min at room temperature. If no reaction occurs, place the test tube in a beaker of water at 50°. After 5 min cool the test tube to room temperature and note if a precipitate has formed.

9.8 Derivatization of Functional Groups

At this stage in the analysis, the functional group has been identified. It is time to consult a table of characterized compounds containing the functional group and to select those having consistent physical properties. Extensive tables of organic molecules organized by functional group are available;[4, 24] abbreviated tables containing many of the more common compounds are included in this section. Your instructor will tell you which set of tables to consult in identifying your unknown.

For most functional groups there is at least one derivative that can be prepared, and when information on the melting points of derivatives is available it is included in the table. For those functional groups without generally applicable derivatives, one must rely on the physical properties of the unknown to make an identification. It is for such compounds that the spectrometric tools discussed in the next chapter are of particular value.

This section is organized by procedures for derivative preparation listed according to functional group. Where more than one derivative is given for a functional group, only those derivatives required to distinguish among the possible compounds need be prepared.

Most of the procedures are described for sample sizes of 0.25 g. Beginners may wish to scale up the reactions by a factor of 4 (remember, this increases the quantities of all reagents but not the reaction time). Students with more laboratory experience may wish to reduce the scale by a factor of

---

[24] Rappoport, Ed., *Handbook of Tables for Organic Compound Identification* (Boca Raton, FL: CRC Press, Inc.).

4. The smaller the scale, the greater the attention that must be paid to careful technique; separations are a particular source of trouble on a small scale. If the scale is small enough, it may be better to substitute a large corked test tube for a separatory funnel and use a pipet to draw off the upper layer.

Alcohols. Three alcohol derivatives are described here: the esters of *p*-nitrobenzoic acid, of 3,5-dinitrobenzoic acid, and of phenylcarbamic acid (also known as phenylurethans). The first two are formed from the carboxylic acid chloride in the presence of 4-methylpyridine as both a catalyst and a base to neutralize the acid produced. The mechanism, in skeletal form, is believed to be

This reaction works well with primary and secondary alcohols, but with tertiary alcohols the product tends to undergo E2 elimination to give the carboxylic acid and an alkene. This competing reaction is minimized by use of milder conditions (for longer times) and a more powerful catalyst, 4-dimethylaminopyridine. In most published procedures the parent compound, pyridine, is used as the base and catalyst. However, one study of the rates of reaction using different bases gave the following order.

Pyridine	4-Methylpyridine (Picoline)	4-Dimethylaminopyridine	Triethylamine
1	5×	5000×	0.04×

It is interesting to note that triethylamine, although it is the strongest base of the compounds shown, is the poorest catalyst, at least in part because of the greater steric bulk of the three ethyl groups. This is consistent with the proposed mechanism.

The third derivative, the phenylcarbamate, is prepared from phenylisocyanate. The mechanism for its formation is

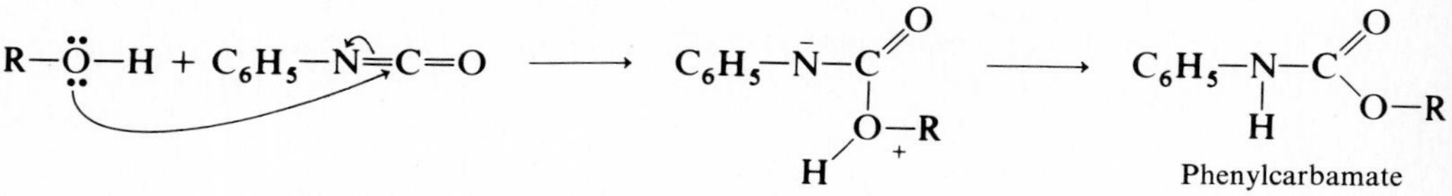

$$R-\ddot{O}-H + C_6H_5-N{=}C{=}O \longrightarrow C_6H_5-\bar{N}-C(=O)(O-R)(\overset{+}{H}) \longrightarrow C_6H_5-N(H)-C(=O)(O-R)$$

Phenylcarbamate

This reaction is base catalyzed, a reflection of the greater nucleophilicity of the alcoholate anion.

$$\underset{\text{Alcohol}}{R-O-H} + \text{base} \rightleftharpoons \underset{\text{Alcoholate anion}}{R-O^-} + H-\text{base}^+$$

It has been suggested that 4-dimethylaminopyridine is a superior base for the isocyanate reaction, but the evidence is less well documented than for the acid chloride reaction.

Here, also, tertiary alcohols tend to give an elimination reaction. The by-product, water, hydrolyzes an equivalent of phenylisocyanate to produce phenylcarbamic acid, which is extremely unstable and decomposes with loss of CO_2 to produce aniline, $C_6H_5-NH_2$. The aniline reacts rapidly with a second equivalent of phenylisocyanate to produce the crystalline, but very insoluble, diphenylurea.

$$H_2O + C_6H_5-N{=}C{=}O \longrightarrow C_6H_5-N(H)-C(=O)(O-H) \xrightarrow{\text{fast}}$$

$$C_6H_5-NH_2 + CO_2 \xrightarrow{C_6H_5-N=C=O} C_6H_5-N(H)-\underset{\parallel\,O}{C}-N(H)-C_6H_5$$

Diphenylurea
(mp 238°)

(A) 3,5-Dinitrobenzoates and p-Nitrobenzoates

1. Primary alcohols. In a large 150-mm test tube containing a boiling chip place 1 mL of cyclohexane, 0.5 mL of 4-methylpyridine (**Caution**—*lachrymator!*) and 0.25 g of the alcohol. Cautiously add 0.5 g of fresh

3,5-dinitrobenzoyl chloride[25] (or *p*-nitrobenzoyl chloride) and, after the initial reaction has subsided, heat the mixture to a gentle reflux for about 10 min. Allow the mixture to cool for a few minutes and then, while stirring thoroughly, *slowly* pour it into 10 mL of ice cold 20% hydrochloric acid. If the product does not solidify immediately, cool the mixture in an ice bath for several minutes.

Remove the cyclohexane layer with a Pasteur pipet and place it in a small beaker. If any solid has separated, remove it by filtration and add it to the cyclohexane layer. Evaporate the cyclohexane crush the solid, and stir it well with 5 mL of 10% sodium carbonate solution. Collect the product on a Hirsch filter and recrystallise it from ethanol–water. It may be necessary to repeat the crystallization to achieve a sharp, precise melting point (Table 9.2).

2. Tertiary alcohols. Follow the same procedure described for primary and secondary alcohols except substitue 0.5 g of 4-dimethylaminopyridine for the 4-methylpyridine and, instead of heating the solution, allow it to stand, well stoppered, for at least 24 hr at room temperature.

(B) Phenylcarbamates (Urethans)

In a large test tube mix 0.25 g of the alcohol with 0.3 mL of phenyl isocyanate (**Caution**—*lachrymator!*) and add a few crystals of 4-dimethylaminopyridine as catalyst. With primary and secondary alcohols warm the test tube in a beaker of boiling water for 5 to 10 min; with tertiary alcohols stopper the tube and set it aside at room temperature for at least 24 hr. When the reaction is complete, cool the tube in an ice bath and scratch the inside wall with a glass rod to induce crystallization. Dissolve the product in 2 mL of hot ligroin, bp 100–120° (**Caution**—*flammable solvent!*), which leaves any diarylurea as undissolved residue. Filter the hot solution, allow the filtrate to cool, and collect the crystals. Recrystallize the product from hot ligroin and take the melting point (see Table 9.2).

Aldehydes and Ketones. Four carbonyl derivatives are described here. The first, the methone derivative, is suitable only for aldehydes; the others are suitable for both aldehydes and ketones.

Aldehydes undergo a base catalyzed condensation with the reactive methylene group of 5,5-dimethyl-1,3-cyclohexanedione (also called methone

---

[25] Carboxylic acid chlorides are rapidly hydrolyzed by moisture. If a bottle of the chloride has been left unsealed for some time, the acid chloride should be recrystallized from cyclohexane and the melting point checked (3,5-dinitrobenzoyl chloride, 70°; *p*-nitrobenzoyl chloride, 75°). The carboxylic acids are essentially insoluble in cyclohexane.

TABLE 9.2 *Derivatives of Alcohols*

Alcohol	Bp, °C	Melting point of derivative, °C[a]		
		Phenyl-carbamate	3,5-Dinitro-benzoate	p-Nitro-benzoate
Methyl	65	47	108	96
Ethyl	78	52	93	57
Isopropyl	82	88	122–123	110
t-Butyl	83	136	142	116
Allyl	97	70	48–49	(28)
n-Propyl	97	51	74	35
s-Butyl	99	65	76	(26)
t-Pentyl	102	42	116	85
Isobutyl	108	86	87	69
3-Pentanol	116	48	101	(17)
n-Butyl	118	63	64	36
2,3-Dimethyl-2-butanol	118		111	82
2-Pentanol	119		61	
2-Methyl-2-pentanol	121	239	72	70
3-Methyl-3-pentanol	123	50	62; 97	
2-Methoxyethanol	125			51
2-Methyl-1-butanol	129		70	
2-Chloroethanol	131	51	95	
4-Methyl-2-pentanol	132	143	65	(26)
3-Methyl-1-butanol	132	55	61	(21)
2-Ethoxyethanol	135		75	
3-Hexanol	136		77	
2,2-Dimethyl-1-butanol	137		51	
1-Pentanol	138	46	46	(11)
2-Hexanol	139		39	
2,4-Dimethyl-3-pentanol	140			155
Cyclopentanol	141	132	115	62
2-Ethyl-1-butanol	148		52	
2-Methyl-1-pentanol	148		51	
4-Heptanol	156		64	35
1-Hexanol	158	42	58	(5)
2-Heptanol	159		49	
Cyclohexanol	161	82	113	50
2-Furfuryl	172	45	81	76
1-Heptanol	177	68	47	(10)
Tetrahydrofurfuryl	178	61	83–84	46–48
2-Octanol	179	114	32	(28)
1-Octanol	195	74	61	(12)
Benzyl	205	78	113	85
2-Phenylethanol	221	79	108	62
Benzohydrol (mp 66–67°)	297	140	141	132

[a] Two values are given for certain derivatives that may be encountered in polymorphic forms. Blanks mean data are not available. Melting points too low to be useful are enclosed in parentheses.

Sec. 9.8] DERIVATIZATION OF FUNCTIONAL GROUPS

or dimedone) to give an intermediate product that reacts with a second equivalent of the ketone to furnish the crystalline methone derivative.

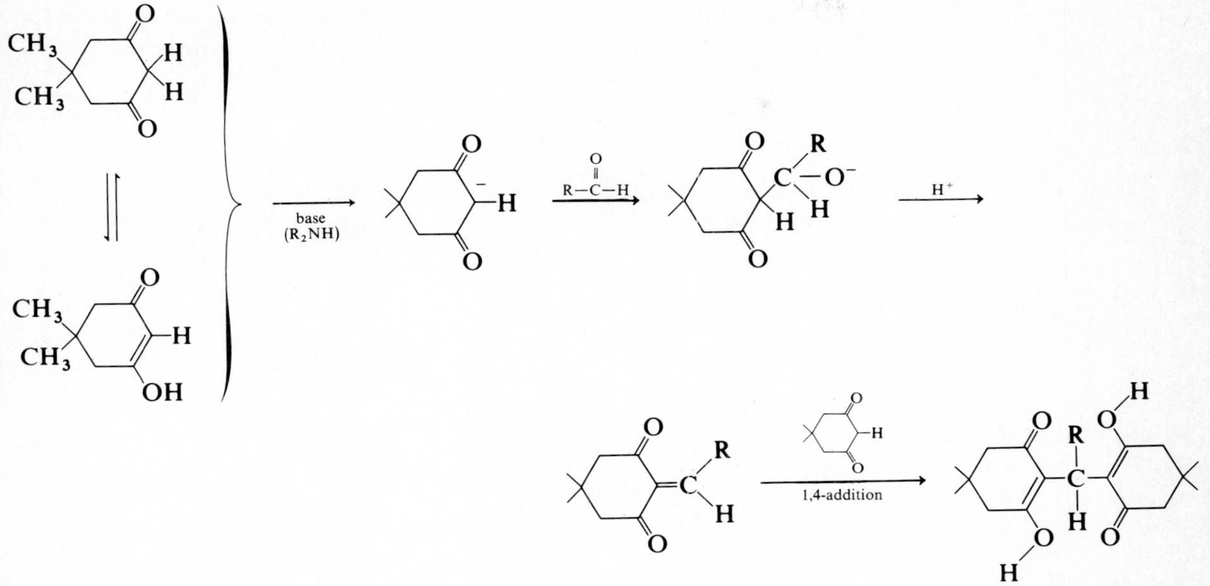

Methone derivative I

The methone derivatives (but not those of formaldehyde and *o*-hydroxybenzaldehydes) are cyclized readily to xanthene derivatives (II) by heating with dilute acid in ethanol.

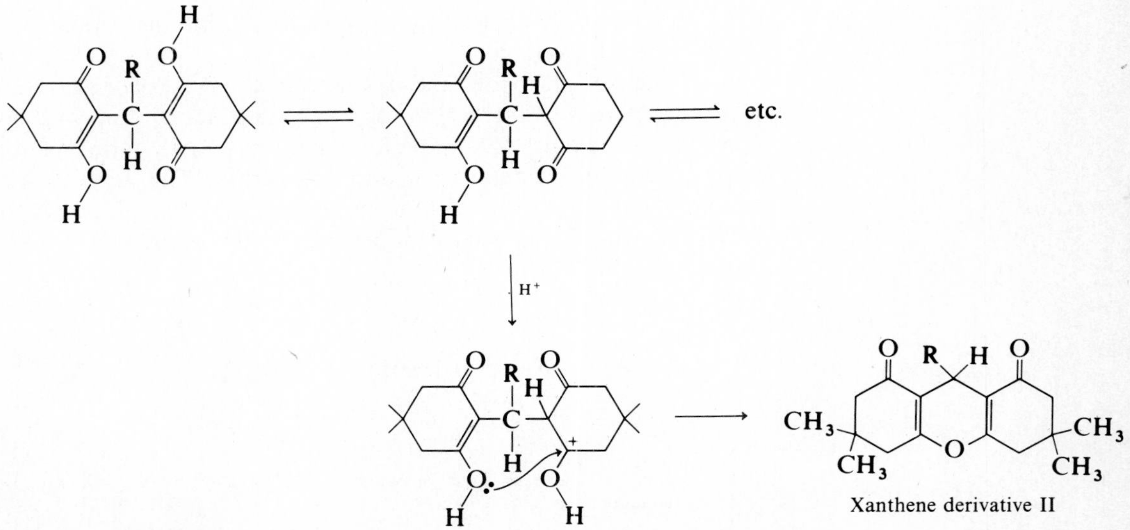

Xanthene derivative II

The methone derivatives are particularly good derivatives for small amounts of aldehyde because of the large increase in molecular weight.

Perhaps the single best derivative of both aldehydes and ketones are the 2,4-dinitrophenylhydrazones. They offer a large increase in weight, form quickly, and crystallize easily. The chemistry of their formation was discussed in Section 9.7 under aldehyde tests.

The semicarbazone and oxime derivatives are chemically analogous to the 2,4-dinitrophenylhydrazones.

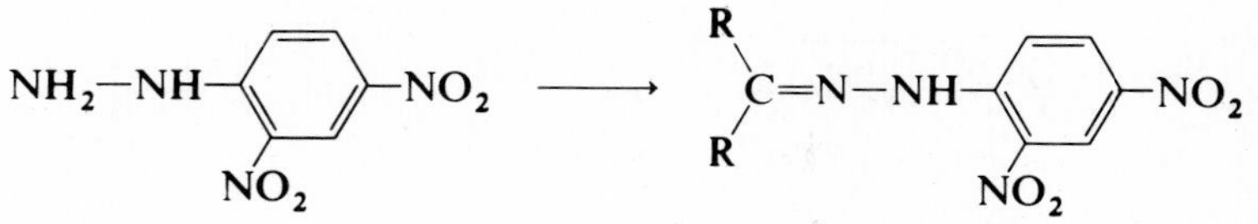

$$NH_2-NH-C_6H_3(NO_2)_2 \longrightarrow R_2C=N-NH-C_6H_3(NO_2)_2$$

2,4-Dinitrophenylhydrazone

$$R_2C=O + NH_2-NH-\underset{\underset{O}{\|}}{C}-NH_2 \longrightarrow R_2C=N-NH-\underset{\underset{O}{\|}}{C}-NH_2$$

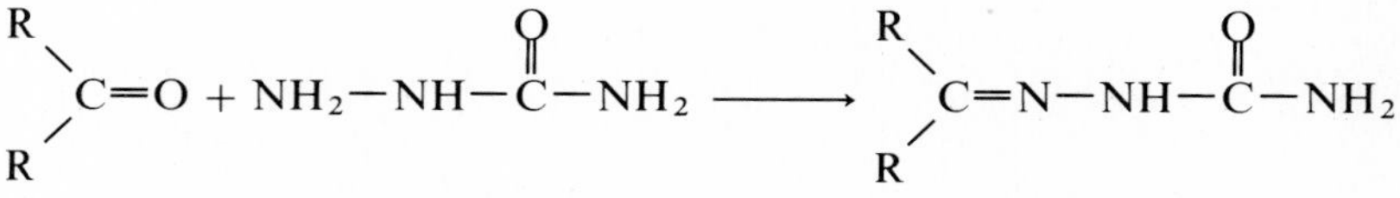

Semicarbazide   Semicarbazone

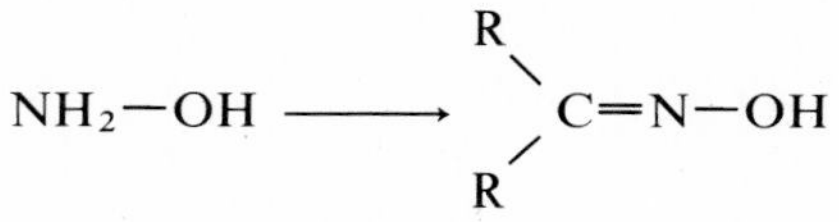

$$NH_2-OH \longrightarrow R_2C=N-OH$$

Hydroxylamine   Oxime

Many aldehydes and ketones give oximes that are liquids at ordinary temperatures (e.g., *n*-butyraldehyde, methyl ethyl ketone) or are very difficult to obtain in crystalline form. For such compounds the semicarbazones or 2,4-dinitrophenylhydrazones are likely to be suitable crystalline derivatives for characterization.

The oximes of aldehydes and of unsymmetrical ketones are capable of exhibiting geometrical isomerism, and may exist in *syn* and *anti* configurations. Benzaldoxime, for example is known in two forms of different melting points and different chemical properties. The acetate of the β-oxime, on warming with sodium carbonate solution, gives benzonitrile (C_6H_5—CN), but the acetate of the α-form merely regenerates the oxime by this treatment.[26]

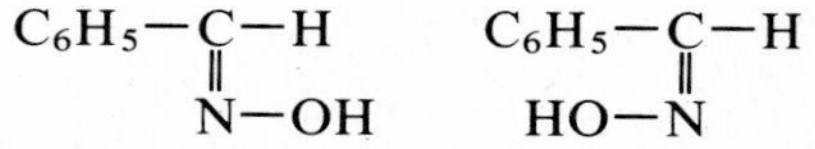

$$\underset{\underset{N-OH}{\|}}{C_6H_5-C-H} \qquad \underset{\underset{HO-N}{\|}}{C_6H_5-C-H}$$

α-Benzaldoxime   β-Benzaldoxime
(*syn*—mp 35°)   (*anti*—mp 130°)

[26] Stereoisomeric ketoximes, when subjected to the Beckmann rearrangement, give different carboxylic amides; see Blatt, *Chem. Rev.*, **12**, 215 (1933); Popp and McEwen, *ibid.*, **58**, 370 (1958).

TABLE 9.3 *Methone (I) and Xanthenedione (II) Derivatives of Aldehydes*

Aldehyde	Melting point, °C		Aldehyde	Melting point, °C	
	I	II		I	II
Formaldehyde	190–191	(171)[27]	Isovaleraldehyde	154–155	170–172
Acetaldehyde	141–142	176–177	n-Hexaldehyde	107–108	
Propionaldehyde	157–158	141–143	Benzaldehyde	194–195	204–205
n-Butyraldehyde	134–135	135–136	Anisaldehyde	142–143	241–243
Isobutyraldehyde	153–154	154–155	Piperonal	177–178	218–220
n-Valeraldehyde	107–109	112–113	2-Furaldehyde	159–160	

(C) Methone Derivative

In a small flask, place 0.50 g (3.6 mmole) of methone, 5 mL of 50% aqueous ethanol, and *not more than* 0.15 g of the aldehyde ($\sim$ 1.5 mmole). Add *one drop* of a secondary amine as catalyst (diethylamine or piperidine), and introduce two small boiling chips. Attach a small reflux condenser, and heat the flask in a beaker about one-fourth filled with water. Apply heat gently until the reaction mixture reaches its boiling point. After the mixture has boiled gently for 5–10 min, remove the flask and add water dropwise until turbidity develops. Allow the solution to cool, occasionally shaking it. Methone derivatives often crystallize slowly; if necessary, stopper the flask and allow it to stand overnight or longer. Collect the crystals with suction and wash them with cold 50% aqueous ethanol. The yield is about 0.30 g. After drying the crystals, take the melting point (see Table 9.3).

The methone derivative (except that of formaldehyde)[27] undergoes cyclization very readily. For this purpose dissolve about 0.2 g in 6–10 mL of 80% ethanol, add 1 drop of concentrated hydrochloric acid, and boil the solution gently for 5 min. Add water dropwise to the hot solution until a faint turbidity develops and allow the liquid to cool. Collect the crystals of the cyclized product (II) and take its melting point (see Table 9.3).

(D) 2,4-Dinitrophenylhydrazones

To prepare the reagent, place 0.3 g of moist 2,4-dinitrophenylhydrazine[28] in a 50-mL Erlenmeyer flask and add 1 mL of water; then add 1 mL of concentrated sulfuric acid dropwise while swirling the solution. Allow the solution to cool and then add 15 mL of 95% ethanol.

---

[27] For cyclization the formaldehyde derivative requires 6–8 hr heating with 10 times its weight of concentrated sulfuric acid. After it is poured into water and neutralized with sodium carbonate, the product is collected with suction and recrystallized from ethanol.

[28] Transportation regulations require that 2,4-dinitrophenylhydrazine must be shipped in a moist condition, containing 20% water. A procedure for the preparation of this compound from 2,4-dinitrochlorobenzene is given by Fieser and Williamson, *Organic Experiments*, 4th ed. (Lexington, MA: Heath, 1979).

Prepare a solution of the carbonyl compound in a 50-mL flask by adding 0.25 g of the compound to 10 mL of 95% alcohol. After the carbonyl compound has dissolved completely, slowly add the 2,4-dinitrophenylhydrazine test reagent and allow the combined solutions to stand at room temperature. Crystallization of the hydrazone usually occurs within 5–10 min. If the product does not separate in this time, attach a reflux condenser and heat the flask on a steam bath for 15 min. If the hydrazone seems to be too soluble in ethanol, add water dropwise at the boiling point of the solution until the product just begins to separate. Allow the solution to cool slowly, since rapid chilling may cause the hydrazone to separate as an oil.

Collect the crystals by suction filtration and wash them with a little cold ethanol. It is usually not necessary to recrystallize the product. However, if the melting range is broad, try recrystallization from ethanol. If the hydrazone does not dissolve completely in ethanol add ethyl acetate dropwise to the hot mixture until solution occurs. The melting points of the 2,4-dinitrophenylhydrazones of many aldehydes and ketones are listed in Table 9.4.

(E) Semicarbazones

In a large test tube prepare a solution of 0.25 g of semicarbazide hydrochloride ($NH_2-CO-NHNH_2 \cdot HCl$) and 0.4 g of sodium acetate in 3 mL of water. Add 0.25 g of a carbonyl compound, close the tube with a cork, and shake it *vigorously*. Allow the reaction mixture to stand (shake it occasionally) until the product has crystallized completely. If necessary, cool the tube in an ice bath. Collect the crystals with suction and wash them with a little cold water. After they have dried in the air, take their melting point.

Reactive carbonyl compounds (butyraldehyde, cyclohexanone, etc.) may be converted to semicarbazones by the same method. With less reactive compounds it is advantageous to heat the tube in a beaker of boiling water for a few minutes and allow it to cool slowly. Dissolve a water-insoluble compound in 3 mL of ethanol; add water dropwise until the solution becomes turbid, and then add 0.25 g of semicarbazide. The melting points of the semicarbazones of many aldehydes and ketones are listed in Table 9.4.

(F) Oximes

In 3 mL of water dissolve 0.5 g of hydroxylamine hydrochloride, $NH_2OH \cdot HCl$, add 2 mL of 10% aqueous sodium hydroxide, and introduce 0.25 g of the aldehyde or ketone. If the compound does not dissolve completely, add *just enough* ethanol, drop by drop while shaking the solution, to obtain a clear or only faintly turbid solution. Heat the solution in a boiling water bath for 10 to 20 min, cool it in an ice bath, and induce crystallization by

TABLE 9.4 Derivatives of Aldehydes and Ketones

	Bp (mp), °C	Oxime[b]	Semi-carbazone	2,4-Dinitrophenyl-hydrazone[c]
Aldehydes[d]				
Acetaldehyde	21	47	162	168; 157
Propionaldehyde	50	40	89; 154	154
Isobutryaldehyde	64	oil	125	187
n-Butyraldehyde	74	oil	104	123
Isovaleraldehyde	92	48	107	123
n-Valeraldehyde	103	52	108	98; 107
n-Hexaldehyde	131	51	106	104; 107
n-Heptaldehyde	153	57	109	108
2-Furaldehyde	161	89; 74	202	230; 212
Benzaldehyde	179	35; 130	222	237
Salicylaldehyde	197	57	231	252 dec
p-Tolualdehyde	204	79; 110	221	239
Citral	228	oil	164	116
Chloral hydrate	(53)	56	90 dec	131
Ketones				
Acetone	56	59	187	126
2-Butanone	80	oil	146	117
3-Methyl-2-butanone	94	oil	113	120
2-Pentanone	102		106; 112	143
3-Pentanone	102	69	139	156
Methyl t-butyl	106	75; 79	157	125
4-Methyl-2-pentanone	119	58	134	95
2,4-Dimethyl-3-pentanone	124	34	160	88; 94
2-Hexanone	129		122	106
Cyclopentanone	131	56	205	142
4-Heptanone	145	oil	133	75
2-Heptanone	151	oil	127	89
Cyclohexanone	155	90	166	162
Acetophenone	200	59	198	240
Benzalacetone	(41)	115	187	223
Benzophenone	(48)	141	164	239
Benzalacetophenone	(58)	116; 75	168; 180	245
Benzil (mono)	(95)	137; 108	175; 182	189
Benzil (di)		237	244	
Benzoin	(133)	151; 99	206 dec	245
dl-Camphor	(176)	118	235	

[a] Two values are given for certain derivatives that may be encountered in polymorphic forms or as *syn* and *anti* geometrical isomers.

[b] Oximes of many aliphatic carbonyl compounds separate as oily liquids that resist efforts to induce crystallization. The simpler oximes are appreciably soluble in water and in organic solvents.

[c] A few dinitrophenylhydrazones exist in red and yellow forms which have different melting points. Mixtures of the two have lower or intermediate melting points.

[d] Methone derivatives are useful for some aliphatic aldehydes.

scratching the walls of the container with a glass rod. If necessary, allow the solution to stand overnight or longer. Collect the crystals on a small suction filter, and wash them with 1 mL of ice cold water. Recrystallize the oxime from a little water or aqueous ethanol, reserving a few tiny crystals for seeding. The melting points of the oximes of many aldehydes and ketones are listed in Table 9.4.

TABLE 9.5
Boiling Points of Liquid Esters and Melting Points of Solid Esters

Liquid ester	Bp, °C	Liquid ester	Bp, °C
Methyl formate	32	Pentyl formate	132
Ethyl formate	54	Ethyl 3-methylbutanoate	135
Methyl acetate	57	Isobutyl propanoate	137
Isopropyl formate	68; 71	Isopentyl acetate	142
Ethyl acetate	77	Propyl butanoate	143
Methyl propanonate	80	Ethyl pentanoate	146
Methyl propenoate	80	Butyl propanoate	147
Propyl formate	81	Pentyl acetate	149
Isopropyl acetate	91	Isobutyl 2-methylpropanoate	149
Methyl 2-methylpropanoate	93	Methyl hexanoate	151
sec-Butyl formate	97	Isopentyl propanoate	160
t-Butyl acetate	98	Butyl butanoate	165
Ethyl propanoate	99	Propyl pentanoate	167
Propyl acetate	101	Ethyl hexanoate	168
Methyl butanoate	102	Cyclohexyl acetate	175
Allyl acetate	104	Isopentyl butanoate	178
Ethyl 2-methylpropanoate	110	Pentyl butanoate	185
sec-Butyl acetate	112	Propyl hexanoate	186
Methyl 3-methylbutanoate	117	Butyl pentanoate	186
Isobutyl acetate	117	Ethyl heptanoate	189
Ethyl butanoate	122	Isopentyl 3-methylbutanoate	190
Propyl propanoate	122	Ethylene glycol diacetate	190
Butyl acetate	126	Tetrahydrofurfuryl acetate	194
Diethyl carbonate	127	Methyl octanoate	195
Methyl pentanoate	128	Methyl benzoate	200
Isopropyl butanoate	128	Ethyl benzoate	213

Solid ester	Mp, °C	Solid ester	Mp, °C
d-Bornyl acetate (bp 221)	29	Ethyl 3,5-dinitrobenzoate	93
Ethyl 2-nitrobenzoate	30	Methyl 4-nitrobenzoate	96
Ethyl octadecanoate	33	2-Naphthyl benzoate	107
Methyl cinnamate (bp 261)	36	Isopropyl 4-nitrobenzoate	111
Methyl 4-chlorobenzoate	44	Cyclohexyl 3,5-dinitrobenzoate	112
1-Naphthyl acetate	49	Cholesteryl acetate	114
Ethyl 4-nitrobenzoate	56	Ethyl 4-nitrobenzoate	116
2-Naphthyl acetate	71	t-Butyl 4-nitrobenzoate	116
Ethylene glycol dibenzoate	73	Hydroquinone diacetate	124
Propyl 3,5-dinitrobenzoate	74	t-Butyl 3,5-dinitrobenzoate	142
Methyl 4-bromobenzoate	81	Hydroquinone dibenzoate	199; 204

Sec. 9.8] DERIVATIZATION OF FUNCTIONAL GROUPS

Esters. There are no universal ester derivatives. One identifying characteristic that will serve instead is the saponification equivalent, which is the weight of the ester (in grams) that reacts with 1 mole of alkali. For mono esters the saponification equivalent is the molecular weight of the ester. This is determined by heating a weighed sample of ester with an excess of standardized potassium hydroxide solution (usually in aqueous methanol or ethanol) and titrating the excess alkali with standardized hydrochloric acid, using phenolphthalein as indicator.

$$\underset{\substack{\| \\ O}}{R-C-OR} + {}^-OH \longrightarrow \underset{\substack{\| \\ O}}{R-C-O^-} + HOR$$

Since the saponification equivalent[29] expresses the molecular weight of the ester divided by the number of ester groups in the molecule, for esters of dibasic acids or of bifunctional alcohols, it is one-half the molecular weight; for trifunctional esters, one-third the molecular weight; and so on.

If either the acid or alcohol fragment of the ester is a solid, the ester can be hydrolyzed (saponified) with base and the solids isolated. These, after purification, are suitable derivatives without further transformation. The boiling points and melting points of many common esters are listed in Table 9.5.

(G) Saponification Equivalent of an Ester

Prepare an ethanolic solution of potassium hydroxide by dissolving 1 g of potassium hydroxide pellets in 5 mL of water and adding 20 mL of ethanol; if sediment is present, allow the solution to stand until it has settled. Withdraw *carefully*, by means of a pipet, two 10-mL portions of the solution and place them in separate 125-mL Erlenmeyer flasks. Add about a 0.250-g sample of the ester, weighed accurately to ± 0.001 g, to one of the flasks. Attach a reflux condenser,[30] add a boiling chip, and boil the solution gently

---

[29] The saponification equivalent of an ester is different from the saponification number, commonly used in industry for fats and fatty oils, which is defined as the number of milligrams of potassium hydroxide required to saponify *one gram* of the fat or fatty oil (mixtures of glyceryl esters of higher aliphatic acids).

$$\text{Saponification number} = \frac{\text{mL of } 1 \, N \text{ alkali} \times 56.1}{\text{grams of ester}}$$

The factor 56.1 is the molecular weight of potassium hydroxide.

[30] The condenser is best attached by means of a cleanly drilled cork. A flask with a ground glass joint may be substituted, but the joint must be greased carefully or it might freeze from contact with traces of base.

for 30 min. Meanwhile, titrate the 10-mL portion of potassium hydroxide solution in the other flask, using standardized (about 0.2 N) hydrochloric acid, with phenolphthalein as an indicator.

When the saponification has been completed, cool the solution and rinse the condenser tube with 5–10 mL of water (collect the rinsing water directly in the flask). Titrate the alkali remaining in the solution against standardized hydrochloric acid, as in the previous titration. The difference in the volumes of acid required in the two titrations represents the amount of alkali consumed in the saponification. Calculate the saponification equivalent by the following formula. The result should correspond to the molecular weight (within about 5%) if the sample of ester was fairly pure.

$$\text{Saponification equivalent} = \frac{\text{weight of ester (in grams)}}{\text{mL of KOH consumed} \times N \text{ of the KOH}} \times 1000$$

(H) Hydrolysis of an Ester

In a 100-mL round-bottom flask provided with a reflux condenser (greased joints), place 1.0 g of an ester. To this add 5 mL of 10% aqueous sodium hydroxide and about 10 mL of water. (If the ester boils above 150°, substitute 10 mL of ethylene glycol for the water.) Add two small boiling chips and boil for about 2 hr. Cool the solution in the flask and carefully transfer it to a small separatory funnel. Extract the strongly basic solution twice with 10-mL portions of methylene chloride and combine the extracts (containing the alcohol) in a clean flask. If the alcohol is a solid it can be isolated by concentration of the methylene chloride solution by distillation followed by evaporation of any residual solvent. Recrystallization of the alcohol may be desirable. Isolation of a liquid alcohol in the quantity specified here requires special techniques.

The acid can be isolated from the residual basic solution by acidification with dilute sulfuric acid (about 10%). If the acid is a solid, it can be isolated as described previously for the alcohol fragment of the ester.

Carboxylic Acids. Amides and N-substituted amides of carboxylic acid make excellent derivatives because of their high melting points and easy preparation and purification. The preparation of any of these derivatives starts with conversion of the carboxylic acid to the acid chloride using thionyl chloride catalyzed by dimethylformamide (DMF).

$$\text{R}-\overset{\text{O}}{\underset{\|}{\text{C}}}-\text{OH} + \text{Cl}-\text{S}-\text{Cl} \xrightarrow{\text{DMF}} \text{R}-\overset{\text{O}}{\underset{\|}{\text{C}}}-\text{Cl} + \text{SO}_2 + \text{HCl}$$

Sec. 9.8] DERIVATIZATION OF FUNCTIONAL GROUPS

The DMF catalyst is particularly effective. The proposed mechanism involves formation of a chloroimmonium ion.

[Reaction scheme showing $SOCl_2$ + DMF ⇌ intermediates ⇌ chloroimmonium ion + SO_2 + Cl^-]

It is this chloroimmonium ion that is the active chlorine transfer agent.

[Reaction scheme showing $R-C(=O)-OH$ + chloroimmonium ion → intermediates → DMF + $R-C(=O)-Cl$ + H^+]

Through all of these transformations the dimethylamino group serves as a reversible electron source, releasing electrons to stabilize the adjacent cationic center and taking them back again when that center encounters better electron sources (e.g., nucleophiles and the C=O group). In the overall sequence the DMF is regenerated and thus is a true catalyst.

Acid chlorides react rapidly with ammonia, amines, alcohols, and *water*. In the procedure described here the reaction is moderated by adding the inert solvent methylene chloride, chosen for its easy removal. One equivalent of HCl is liberated, which produces copious white fumes on contact with ammonia and water. The reaction should be carried out in a hood.

$$R-\overset{O}{\overset{\|}{C}}-Cl + \ddot{N}H_2R \longrightarrow R-\overset{O^-}{\underset{Cl}{\overset{|}{C}}}-\overset{+}{N}H_2R \longrightarrow R-\overset{O}{\overset{\|}{C}}-\overset{+}{N}H_2R \rightleftharpoons R-\overset{O}{\overset{\|}{C}}-NHR$$
Amide

A useful alternative to a solid derivative is a determination of the neutralization equivalent (equivalent weight) of the acid, which is defined as the weight of acid required to neutralize 1 mole of base. For mono acids the neutralization equivalent is the molecular weight of the acid; for di or tri acids it is one-half or one-third of the molecular weight.

(I) Acid Amides, Anilides, and p-Toluidides

1. Conversion of acid to acid chloride. The preparation of these derivatives can release hazardous fumes and should be carried out **in a hood**! In a small dry flask place 0.250 g of the acid, 1 mL of thionyl chloride, and 1 drop of pyridine or dimethylformamide. Attach a reflux condenser, add two small boiling chips, and boil the mixture gently in a water bath for 30 min. During this time the acid should react and dissolve completely. Small amounts of sulfur dioxide and hydrogen chloride are evolved during the heating. The excess thionyl chloride could be removed by warming the mixture under reduced pressure, but that is not necessary.

▶ *CAUTION* Handle thionyl chloride carefully. The liquid burns the skin, and the vapor is an irritant and harmful to breathe.

2. Conversion to acid amide. Cool the acid chloride mixture and add 5–10 mL of methylene chloride. Pour the solution of the acid chloride (and excess thionyl chloride) *very cautiously* and *in small portions* into 5 mL of cold, concentrated aqueous ammonia contained in a small Erlenmeyer flask. The reaction mixture should be mixed thoroughly by swirling it as the solution is added. The acid amide may separate as a crystalline precipitate during the reaction.

After allowing the mixture to stand for 20 min, occasionally shaking it, separate the methylene chloride layer and remove the solvent from it by distillation. Wash the residue with water, collect the crystals by suction filtration, and press them as dry as possible on the filter. Recrystallize the amide from ethanol–water and determine its melting point.

3. Conversion to anilide. Prepare the acid chloride as described earlier and add 5–10 mL of methylene chloride to the cooled mixture. Dissolve 0.5 g of aniline in 10 mL of methylene chloride in a small Erlenmeyer flask and add the acid chloride solution to it *cautiously* and *in small portions*. The reaction mixture should be mixed thoroughly by swirling it between additions.

After allowing the mixture to stand for 20 min, occasionally shaking it, transfer the methylene chloride solution to a separatory funnel and wash it in sequence with 5 mL of water, 5 mL of 5% hydrochloric acid, 5 mL of 5% sodium hydroxide, and finally 5 mL of water. Remove the solvent by distillation, recrystallize the residue from ethanol–water, and determine its melting point.

TABLE 9.6 Derivatives of Carboxylic Acids

		Melting point of derivative, °C		
	Bp, °C	Amide	Anilide	*p*-Toluidide
Liquid carboxylic acids				
Methanoic (formic)	101		50	53
Ethanoic (acetic)	118	82	114	153
Propenoic (acrylic)	141	84	104	141
Propanoic (propanoic)	141	81	106	126
2-Methylpropanoic (isobutyric)	155	128	105	109
Butanoic (butyric)	163	115	96	75
3-Methylbutanoic	177	135	110	106
Pentanoic (valeric)	186	106	63	74
2,2-Dichloroethanoic	194	98	118	153
Hexanoic (caproic)	205	100	94	74
Heptanoic (enanthic)	223	96	65; 70	81
Octanoic (caprylic)	239	106; 110	57	70
Nonanoic (pelargonic)	254	99	57	84
	Mp, °C			
Solid carboxylic acids				
Octadecanoic (stearic)	70	109		96
Phenylethanoic	77	156	65	117
2-Benzoylbenzoic	90; 128	165	195	
Pentandioic (glutaric)	97	175	223	218
Ethanedioic (oxalic)	101	219	148	169
2-Methylbenzoic	105	143	125	144
3-Methylbenzoic	112	94	126	118
Benzoic	122.4	130	160	158
trans-Cinnamic	133	147	109; 153	168
Propanedioic (malonic)	135	106	132	156
2-Acetoxybenzoic (aspirin)	135	138	136	
cis-Butenedioic (maleic)	137	172	187; 198	142
2-Chlorobenzoic	140	142; 202	114; 118	131
3-Nitrobenzoic	140	143	154	162
2-Nitrobenzoic	146	176	155	
Diphenylacetic	148	168	180	172
2-Bromobenzoic	150	155	141	
Benzilic	150	153	175	190
Hexanedioic (adipic)	153	125; 230	151; 249	241
2-Hydroxybenzoic (salicylic)	158	142	136	156
2-Iodobenzoic	162	110	141	
4-Methylbenzoic (*p*-toluic)	179	160	144	160; 165
4-Methoxybenzoic (*p*-anisic)	185	162; 167	170	186
2-Naphthoic	186	192	171	192
Phthalic (mono)[a]	200; 230	149	170	150
Phthalic (di)[a]		220	254	201
3,5-Dinitrobenzoic	205	183	234	
4-Nitrobenzoic	241	198	204; 211	192; 204

[a] Phthalic acid is a dicarboxylic acid that forms both mono- and dicarboxylic acid derivatives.

The melting points of the common carboxylic acid derivatives are listed in Table 9.6.

4. Conversion to *p*-toluidide. Use the same procedure described for anilides, except substitute *p*-toluidine for aniline.

(J) Neutralization Equivalent of an Acid

Place about 0.250 g of the acid, weighed accurately to ± 0.001 g, in a 125-mL Erlenmeyer flask, and add 5 mL of water and 20 mL of ethanol. Swirl the flask to dissolve the acid. Titrate the solution using standardized aqueous sodium hydroxide (about 0.2 N) with phenolphthalein as an indicator. The neutralization equivalent is calculated as

$$\text{Neutralization equivalent} = \frac{\text{weight of acid (g)}}{\text{mL of NaOH consumed} \times N \text{ of NaOH}} \times 1000$$

Sulfonic Acids. Alkylsulfonic and arylsulfonic acids and their metallic salts can be characterized by conversion to salts of organic bases, such as the S-benzylthiouronium salts, which are crystalline and have suitable melting points.

$$\left[C_6H_5CH_2-S-C{\overset{\displaystyle NH_2}{\underset{\displaystyle NH_2}{\big<}}} \right]^+ \, ^-[O_3S-C_6H_4CH_3]$$

TABLE 9.7 *Derivatives of Sulfonic Acids*

Sulfonic acid	Mp, °C	Melting point of derivative, °C	
		S-Benzylthiouronium salt	Sulfonamide
Methane-	20		90
3-Nitrobenzene-	48	146	167
Hexadecane-1-	54		97
2-Methylbenzene-	57	170	156
2,4-Dimethylbenzene-	62	146	139
3,4-Dimethylbenzene-	64	208	144
Benzene-	66	148	153
2-Carboxybenzene-	68	206	194
Naphthalene-1-	90	137	150
Naphthalene-2-	91	190	217
4-Chlorobenzene-	93	175	144
4-Methylbenzene-	104	181	105; 139
Naphthalene-1,6-di-	125	81	297

Another, more complicated method involves reaction of the sulfonic acid with phosphorus pentachloride to form the arylsulfonyl chloride, which is treated with ammonia (or an amine) to furnish a crystalline arylsulfonamide. These reactions are analogous to those discussed for formation of amide derivatives of carboxylic acids (see page 168).

$$R-SO_3H + PCl_5 \longrightarrow RSO_2Cl + POCl_3 + HCl$$
$$RSO_2Cl + NH_3 \longrightarrow RSO_2NH_2 + HCl$$
$$\text{Sulfonamide}$$

(K) *S*-Benzylthiouronium Salts

In a small test tube, dissolve 0.25 g of the sodium or potassium salt of the sulfonic acid in a minimum amount of water (warm if necessary). If only the sulfonic acid is available, dissolve it in 0.5 mL of 10% sodium hydroxide, add 1 mL (or more) of water and a drop of phenolphthalein solution, and neutralize the excess base by adding hydrochloric acid drop by drop.

In a separate test tube prepare a concentrated solution of 0.25 g of *S*-benzylisothiuronium chloride[31] in water. Mix the solutions together, shake well, and cool the mixture in an ice–water bath. If crystals do not form within a few minutes, scratch the inside of the tube with a glass rod to promote crystallization. Collect the product on a small suction filter, wash sparingly with cold water, and recrystallize it from 50% aqueous ethanol (reserving a seed crystal for inoculation).

(L) Sulfonamides

This reaction can release hazardous vapors and should be carried out in a hood. In a small round-bottom flask place 0.25 g of the sulfonic acid (or its sodium or potassium salt) and about 1.0 g of phosphorus pentachloride. (**Caution**—*Corrosive material*, do not inhale vapors!) Attach a reflux condenser and heat the mixture in a sand bath at 150° for 45 min. Cool the mixture to room temperature, add 5–10 mL of methylene chloride, and break up any solid masses present. Filter the solution through a dry filter

---

[31] *S*-Benzylthiuronium chloride is a relatively expensive reagent. Suitable material may be readily prepared from thiourea and benzyl chloride. In a flask fitted with a reflux condenser, a mixture of 25 g (16 mL) of benzyl chloride (**Caution**—*lachrymator!*), 15 g of thiourea, and 40 mL of ethanol (or methanol) is warmed on a steam bath. Soon a vigorous exothermic reaction occurs and the thiourea dissolves completely. The pale yellow solution is refluxed for 30 min, transferred while hot to a beaker, and cooled in an ice–water bath. The mass of white crystals is collected on a suction filter, washed with several 15-mL portions of cold ethyl acetate or ethanol, and pressed well. The dried product is stored in an amber bottle. The yield is 30–35 g of the salt. The crude product is satisfactory for the preparation of derivatives. It may be purified by recrystallization from ethanol or from 15–20% aqueous hydrochloric acid. The compound is dimorphic: stable form, mp 174–176°; metastable form, mp 142–145° (with reversion to the stable form).

paper into a small Erlenmeyer flask containing 5 mL of cold, concentrated aqueous ammonia. Swirl the flask during the addition. Occasionally, the sulfonamide will separate as a crystalline precipitate, but usually it will not.

After allowing the mixture to stand for 20 min, occasionally shaking it, separate the methylene chloride layer and remove the solvent by distillation. Recrystallize the sulfonamide from ethanol or ethanol–water.

Nitriles and Amides. There are no good derivatives of nitriles and amides; the best strategy is to hydrolyze them to the corresponding carboxylic acid and ammonia (or amine). The hydrolysis can be achieved with either acid or base, but only the base procedure will be described here.

$$R-C\equiv N \xrightarrow[H_2O]{NaOH} R-\overset{O}{\underset{\|}{C}}-NH_2 \longrightarrow R-\overset{O}{\underset{\|}{C}}-O^- + NH_3$$

$$R-\overset{O}{\underset{\|}{C}}-NR_2 \xrightarrow[H_2O]{NaOH} R-\overset{O}{\underset{\|}{C}}-O^- + HNR_2$$

For purely physical reasons one must use a different procedure for substituted (on nitrogen) and unsubstituted amides. With nitriles and

TABLE 9.8 Boiling Points and Melting Points of Nitriles and Amides

Nitrile	Bp, °C	Mp, °C	Amide[b]	Bp, °C	Mp, °C
Acrylonitrile[a]	77		N,N-Dimethylformamide	153	
Acetonitrile	81		N,N-Diethylformamide	176	
Propionitrile	97	−93	N-Methylformamide	185	
Isobutyronitrile	108		N-Formylpiperidine	222	
n-Butyronitrile	117		N,N-Dimethylbenzamide		41
Benzonitrile	191	−13	N-Benzoylpiperidine		48
2-Methylbenzonitrile	205	13	N-Propylacetanilide		50
2-Cyanopyridine	212	26	N-Benzylacetamide		54
Phenylacetonitrile	234		N-Ethylacetanilide		54
3-Cyanopyridine	240	52	N,N-Diphenylformamide		73
1,4-Dicyanobutane	295	2	N-Methyl-4-acetotoluidide		83
2-Chlorobenzonitrile		41	N,N-Diphenylacetamide		101
Diphenylacetonitrile		75	N-Methylacetanilide		102
4-Cyanopyridine		80	N-Ethyl-4-nitroacetanilide		118
			N-Phenylsuccinimide		156
			N-Phenylphthalimide		205

[a] Cancer suspect agent.
[b] Table 9.6 lists boiling points and melting points of many unsubstituted amides, anilides, and 2-methylanilides. Table 9.9 lists these data for a number of N-substituted acetamides, benzamides, benzenesulfonamides, and 4-methylbenzenesulfonamides.

unsubstituted amides, only ammonia is produced, and this need not be characterized. With substituted amides for which the amine fragment is *volatile*, a trap containing acid must be attached to the apparatus in order to collect the amine for identification. With substituted amides for which the amine fragment is *nonvolatile*, no trap is required. The amine can be recovered from the basic hydrolysis mixture by extraction; acidification of the mixture then yields the acid. The boiling points and melting points of many nitriles and amides are listed in Table 9.8.

(M) Base Hydrolysis of Nitriles and Unsubstituted Amides

In a small flask place 2 g of potassium hydroxide, 5 mL of ethylene glycol, and 0.5 g of the nitrile or unsubstituted amide. Add two small boiling chips, attach a reflux condenser, and heat the mixture under reflux for 1 hr. Cool the mixture to room temperature, dilute with 5 mL of water, and extract twice with 5 mL of methylene chloride to remove any unreacted nitrile or amide. The methylene chloride layer should be set aside. Acidify the aqueous solution with 6 N (50%) hydrochloric acid until the pH is about 1 to 2, and extract again with methylene chloride. This extract should be evaporated or distilled to recover the acid.

(N) Base Hydrolysis of Substituted Amides

1. Volatile amines. The procedure is the same as that described for unsubstituted amides, except that a glass tube, leading to an Erlenmeyer flask containing 5 mL of 6 N hydrochloric acid, should be attached to the top of the reflux condenser to trap the volatile amines. At the end of the reaction, the acid should be neutralized with 10% sodium hydroxide solution and extracted several times with 5-mL portions of methylene chloride. The amine can be recovered from the methylene chloride by distillation.

2. Nonvolatile amines. The procedure is identical to that described for unsubstituted amides. The nonvolatile amine will be in the first methylene chloride extract (of the basic solution), and the acid will be in the second methylene chloride extract (of the acidic solution).

Amines. The most effective derivatives for primary and secondary amines are amides. Four are described here. The chemistry of their formation is essentially the same as that described in connection with the tests for amines. They have in common an acyl or sulfonyl group that is activated by using a chloride or anhydride leaving group in place of the hydroxyl group of the acid. When a chloride is used, some base must be present to neutral-

TABLE 9.9 Derivatives of Primary and Secondary Amines

Amine	Bp, °C	Melting point of amide, °C			
		Acetyl	Benzoyl	Benzene-sulfonyl	p-Toluene-sulfonyl
Methylamine	−6	oil	80	30	75
Ethylamine	17	oil	71	58	63
Isopropylamine	33		100		26
n-Propylamine	49	47	84	36	52
t-Butylamine	45	98	134		
sec-Butylamine	63		76	70	55
Isobutylamine	69	107	57	53	78
n-Butylamine	77	oil	42		44
Ethylenediamine	116	172 (di)	244 (di)	168 (di)	160 (di)
Cyclohexylamine	134	104	149	89	
Benzylamine	185	60	105	88	116
Aniline	185	114	163	112	103
2-Methylaniline	199	112	144	124	108
4-Methylaniline (mp 45°)	200	148	158	120	117
3-Methylaniline	203	66	125	95	114
2-Chloroaniline	208	87	99	129	105; 193
2-Ethylaniline (mp 47°)	210	111	147		
2,4-Dimethylaniline	212	130	192	129	
2,5-Dimethylaniline	215	139	140	139	119
2,6-Dimethylaniline	216		168		
2,4-Dimethylaniline	217		192	130	
N-Ethyl-3-methylaniline	221		72		
2-Methoxyaniline	225	85	60	89	127
4-Chloroaniline (mp 70°)	232	172; 179	192	122	95; 119
4-Methoxyaniline (mp 57°)	243	130	157	95	114
2-Ethoxyaniline	229	79	104	102	164
4-Ethoxyaniline	254	135	173	143	107
3-Nitroaniline (mp 114°)		152	155	136	
4-Nitroaniline (mp 147°)			199	139	
Dimethylamine	7	oil	41	47	79
Diethylamine	55	oil	42	42	60
Diisopropylamine	84			94	
Piperidine	105	oil	48	93	96
Di-n-propylamine	109			51	
Morpholine	130	oil	75	118	147
N-Methylaniline	192	102	63	79	94
N-Ethylaniline	205	54	60	oil	87
N-Methyl-2-methylaniline	208	83	53	64	60

TABLE 9.10
Derivatives of Tertiary Amines

Tertiary amine	Mp, °C	Melting point of derivative, °C	
		Methiodide	Ethiodide
Triethylamine	89	280	
Pyridine	116	117	90
2-Methylpyridine (α-picoline)	129	230	123
2,6-Dimethylpyridine (2,6-lutidine)	142	233	
3-Methylpyridine (β-picoline)	143	92	
4-Methylpyridine (γ-picoline)	143	167	
Tripropylamine	157	207	238 (dec)
Dimethylaniline	193	220 (dec)	136
Tributylamine	216	180; 186	
Diethylaniline	216	102	
Quinoline	237	133	

ize the hydrogen chloride produced, which would otherwise react with amine to produce an unreactive hydrochloride salt.

$$Ar-\overset{O}{\underset{\|}{C}}-X + NHR_2 \longrightarrow Ar-\overset{O}{\underset{\|}{C}}-NR_2 + HX$$

$$ArSO_2X + NHR_2 \longrightarrow Ar-SO_2NR_2 + HX$$

The melting points of these derivatives are listed in Table 9.9.

The arylsulfonyl derivatives of aliphatic and aromatic amines are usually the better crystalline derivatives for identification purposes. If the benzenesulfonamide is a liquid, the *p*-toluenesulfonamide may serve. For the nitroanilines, halogenated anilines, and similar weak bases, which frequently give poor results in an aqueous medium, the use of pyridine in an anhydrous system is advantageous.

With tertiary amines, amide derivatives are impossible and one must turn to formation of a quarternary ammonium salt using either methyl or ethyl iodide. The iodide salts are nicely crystalline but tend to decompose near their melting points or on prolonged exposure to light. Many rapidly absorb moisture from the air. Melting points of the salts of a few tertiary amines are listed in Table 9.10.

(O) Acetylation with Acetic Anhydride in Water (Lumière–Barbier Method)

In a large test tube dissolve 0.25 g of the 1° or 2° amine in 5 mL of 5% hydrochloric acid. Prepare a solution of 0.5 g of sodium acetate crystals (trihydrate) in 5 mL of water and set this aside.

Warm the solution of the amine hydrochloride to 50°, add 0.3 mL of acetic anhydride, and swirl the mixture to dissolve the anhydride. *At once* add the sodium acetate solution and mix the reactants thoroughly. After a few minutes, cool the reaction mixture and stir it vigorously to induce crystallization. Collect the crystals by suction filtration.

(P) Acylation with Acyl Halides and Aqueous Sodium Hydroxide (Schotten–Baumann Method)

To 10 mL of 5% aqueous sodium hydroxide in a large test tube, add 0.25 g of the amine and 0.5 g (0.6 mL of a liquid) of the acyl or sulfonyl chloride (**Caution**—*irritating vapor*). Stopper the tube firmly and shake it vigorously; release any internal pressure by cautiously removing the stopper. Continue to shake for about 10 min. Collect the precipitate by suction filtration and wash it with water and then with a little dilute hydrochloric acid. Recrystallize the derivative from methanol or aqueous ethanol; a few crystals should be reserved for seeding.

(Q) Acylation in Pyridine

In a small round-bottom flask, place 0.25 g of the amine, 0.5 g (0.6 mL of a liquid) of the acyl or sulfonyl chloride (**Caution**—*irritating vapor*), and 2 mL of pyridine (**Caution**—*disagreeable odor*). Reflux the mixture gently for 30 min and then pour it cautiously into 10–15 mL of cold water. Stir the product until it crystallizes, collect it on a suction filter, and recrystallize it from aqueous ethanol.

(R) Quaternary Ammonium Salts

In a small flask place 0.3 mL (0.25 g) of the 3° amine and add, in the hood, 0.5 mL of methyl iodide.

▶ *CAUTION* Methyl iodide is a cancer suspect agent and toxic. Attach a reflux condenser and heat the mixture gently for 30 min on a steam bath. Collect the crystals of the methiodide with suction and wash them with a little methylene chloride. Without delay, place the product in a stoppered vial. Methanol or ethanol can be used for recrystallization.

The same procedure can be used with ethyl iodide to prepare the ethiodide salt.

Phenols. Phenols possess two functional groups joined together: a benzene ring and a hydroxyl group. The chemistry of phenols is largely the chemistry of these two groups. However, because of the interaction of the π-electrons, their properties are somewhat modified. The proton of the hydroxyl group is more acidic than that of an aliphatic alcohol; the benzene ring is more susceptible to attack by electrophilic reagents.

Sec. 9.8] DERIVATIZATION OF FUNCTIONAL GROUPS

Of the three derivatives described here, the first depends on the alcohol-like properties of a phenol and the chemistry is much the same as that described in the alcohol derivative section.

The aryloxyacetic acid derivative formation depends on the enhanced acidity of the phenol proton, which easily yields the highly nucleophilic phenolate anion.

$$S\text{-}C_6H_4\text{-}OH + OH^- \rightleftharpoons S\text{-}C_6H_4\text{-}O^- + H_2O$$

Phenolate anion

This strong nucleophile rapidly displaces the Cl of chloroacetate anion in an S_N2 reaction to yield the aryloxyacetate anion, which, upon acidification, gives the aryloxyacetic acid.

$$S\text{-}C_6H_4\text{-}O:^- \overset{Cl}{\underset{}{\frown}} CH_2\text{-}CO_2^- \longrightarrow S\text{-}C_6H_4\text{-}O\text{-}CH_2\text{-}CO_2^- \overset{H^+}{\rightleftharpoons} S\text{-}C_6H_4\text{-}O\text{-}CH_2\text{-}CO_2H$$

The bromination of phenol is very fast in any available *ortho* or *para* position. Phenol itself gives a tribromide; Phenols with one or more of these positions occupied will give correspondingly less bromine uptake. The mechanism involves electrophilic aromatic substitution with a bromonium ion or its equivalent (such as Br_3^+).

$$H\text{-}C_6H_4\text{-}OH + Br^+ \longrightarrow \left[\begin{array}{c}H\\Br\end{array}C_6H_4^+\text{-}OH\right] \longrightarrow Br\text{-}C_6H_4\text{-}OH \xrightarrow{\text{repeat twice}} Br\text{-}C_6H_2(Br)_2\text{-}OH$$

The physical properties and derivatives of many phenols are listed in Table 9.11.

(S) 3,5-Dinitrobenzoates

Follow the procedure given for the preparation of 3,5-dinitrobenzoates of alcohols.

(T) Aryloxyacetic Acids

In a large test tube dissolve 0.25 g of the unknown phenol in 5 mL of 10% aqueous sodium hydroxide and add 0.25 g of chloroacetic acid with vigorous shaking. If any of the sodium salt of the phenol separates, add another few milliliters of water to dissolve it. Heat the solution in a gently boiling water bath for 1 hr and, after cooling, add 1 mL of water. Acidify the

TABLE 9.11 Derivatives of Phenols

Phenol	Mp, °C	Bp, °C	Melting point of derivative, °C		
			3,5-Dinitrobenzoate	Aryloxyacetic acid	Bromo derivative
5-Isopropyl-2-methyl- (carvacrol)	1	238	77; 83	151	46
2-Chloro-	7	175	143	145	48 (mono)
					76 (di)
3-Methyl- (*m*-cresol)	10	203	165	103	84 (tri)
2,4-Dimethyl-	26	212	164	141	
2-Methoxy- (guaiacol)	28	205	141	116	116 (tri)
2-Methyl-	32	192	138	152	56 (di)
4-Methyl-	35	203	188	135	49 (di)
					108 (tetra)
2-Nitro-	45	216	155	158	95 (tri)
2,6-Dimethyl-	48	212	158	139	79
2-Isopropyl-5-methyl- (thymol)	50	232	103	149	55
4-Methoxy-	57	243		110	
3,4-Dimethyl-	63	225	181	162	171 (tri)
3,5-Dimethyl-	64	219	195	81; 111	166 (tri)
4-Bromo-	66	222	191	157	95 (tri)
2,5-Dimethyl-	73	212	137	118	178 (tri)
1-Naphthol	95	278	217	193	105 (di)
3-Nitro-	96		159	156	91 (di)
4-*t*-Butyl-	98	237	156	86	50 (mono)
					67 (di)
1,2-Dihydroxy- (catechol)	105	246	152 (di)	137	192 (tetra)
1,3-Dihydroxy- (resorcinol)	110	281	201 (di)	175; 195	112 (tri)
4-Nitro-	112		186	187	142 (di)
2-Naphthol	123	285	210	154	84
1,4-Dihydroxy- (hydroquinone)	171	286	317 (di)	250	186 (di)

solution to pH 4 (universal pH paper) and extract it with two 10-mL portions of diethyl ether. Wash the combined ether layers with cold water. To isolate the aryloxyacetic acid, shake the ether solution with 5 mL of 5% sodium carbonate solution. Draw off the aqueous layer and add it to a flask containing 2 mL of concentrated hydrochloric acid diluted with 10 mL of water. Filter the product with suction and recrystallize it from water or 95% ethanol.

(U) Bromo Derivatives

In a large test tube dissolve 0.25 g of the phenol in 3 mL of methanol and add 3 mL of water. Add dropwise, while swirling the solution, aqueous brominating solution[32] until a yellow color persists even after thorough

[32] Aqueous brominating solution is prepared by adding 2 mL (6 g) of bromine (**Caution—severe harard!** See Section 1.1) to a solution of 8 g of potassium bromide in 50 m of water.

mixing. Add 15 mL of water to precipitate the product; collect it on a small suction filter and wash it thoroughly with about 10 mL of 5% aqueous sodium bisulfite solution. Finally, wash the bromophenol with about 10 mL of water and press it as dry as possible on the filter.

Dissolve the crude product in about 25 mL of hot ethanol, by warming on a steam bath, and filter the hot solution through a fluted filter into a clean 125-mL Erlenmeyer flask. To the hot filtrate add about 50 mL of water in small portions until the bromophenol begins to separate. Mix the solution thoroughly, heat to redissolve the solid, and set the solution aside to cool undisturbed. After the product has crystallized, collect it by suction filtration, wash it with a few milliliters of cold 50% aqueous ethanol, and spread it on a clean filter paper to dry.

Hydrocarbons and Chlorocarbons. There are no universal derivatives for hydrocarbons; the recommended procedure is usually to functionalize the compound in some way and then identify the product. For example, alkenes

TABLE 9.12
Boiling Points of Saturated Hydrocarbons

	Bp, °C		Bp, °C
Pentane	36	2,2,4-Trimethylpentane	99
Cyclopentane	49	trans-1,4-Dimethylcyclohexane	119
2,2-Dimethylbutane	50	Octane	126
2,3-Dimethylbutane	58	Nonane	151
2-Methylpentane	60	Decane	174
3-Methylpentane	63	Eicosane (mp 37°)	343
Hexane	69	Norbornane (mp 87°, subl)	—
Cyclohexane	81	Adamantane (mp 268, sealed)	—
Heptane	98		

TABLE 9.13
Boiling Points of Unsaturated Hydrocarbons

	Bp, °C		Bp, °C
1-Pentene	30	3-Hexene	82
2-Methyl-1,3-butadiene (isoprene)	34	Cyclohexene	84
		2-Hexyne	84
trans-2-Pentene	36	1-Heptene	94
cis-2-Pentene	37	1-Heptyne	100
2-Methyl-2-butene	39	2,4,4-Trimethyl-1-pentene	102
Cyclopentadiene	41	2,4,4-Trimethyl-2-pentene	104
1,3-Pentadiene (piperylene)	41	1-Octene	123
		Cyclooctene	146
3,3-Dimethyl-1-butene	41	1,5-Cyclooctadiene	150
1-Hexene	63	d,l-α-Pinene	155
cis-3-Hexene	66	(−)-β-Pinene	167
trans-3-Hexene	67	Limonene	176
1-Hexyne	71	1-Decene	181
1,3-Cyclohexadiene	80		

TABLE 9.14 Boiling and Melting Points of Aromatic Hydrocarbons

Liquids	Bp, °C	Solids	Mp, °C
Benzene	80	Diphenylmethane	25
Toluene	111	Bibenzyl	53
Ethylbenzene	136	Biphenyl	69
p-Xylene	138	Naphthalene	80
m-Xylene	139	Triphenylmethane	92
o-Xylene	144	Acenaphthene	96
Isopropylbenzene	152	Phenanthrene	96
n-Propylbenzene	159	Fluorene	116
1,3,5-Trimethylbenzene	165	Pyrene	148
t-Butylbenzene	169	1,1'-Binaphthyl	160
Isobutylbenzene	173	Hexamethylbenzene	165
sec-Butylbenzene	173	Anthracene	216
Indene	182		
n-Butylbenzene	183		
Tetrahydronaphthalene	206		

can be acetylated under Friedel–Crafts conditions and the resulting ketone identified in the usual manner; aromatic hydrocarbons can be nitrated and the nitro product reduced to an amine, which can be derivatized. The situation is no better with chlorocarbons, which usually must be converted by a Grignard reaction to a carboxylic acid. This approach is lengthy, it frequently gives mixtures, and the yields can be low. With the development

TABLE 9.15 Boiling Points of Halogenated Hydrocarbons

	Bp, °C		Bp, °C
Chlorides		**Bromides**	
n-Propyl	47	Ethyl	38
t-Butyl	51	Isopropyl	60
sec-Butyl	68	Propyl	71
Isobutyl	69	t-Butyl	72
n-Butyl	78	Isobutyl	91
Neopentyl	85	sec-Butyl	91
t-Pentyl	86	Butyl	101
Cyclohexyl	143	t-Pentyl	108
Chlorobenzene	132	Neopentyl	109
2-Chlorotoluene	159	Bromobenzene	156
1,4-Dichlorobenzene (mp 53°)	173	1-Bromoheptane	174; 180
Hexachloroethane (mp 187° subl)	185	1,4-Dibromobenzene (mp 89°)	219
2,4-Dichlorotoluene	200	**Iodides**	
1-Chloronaphthalene	259	Methyl	43
Triphenylmethyl (mp 113°)		Ethyl	72
		Isopropyl	90
		Propyl	102

of NMR spectroscopy, it has become easy to identify ordinary hydrocarbons and chlorocarbons directly. For that reason, the cumbersome chemical procedures will not be described. If a hydrocarbon unknown is encountered, it should be identified by the spectroscopic methods described in Chapter 10. The boiling points and melting points of representative hydrocarbons and chlorocarbons are listed in Tables 9.12 through 9.15.

Questions

1. Predict the solubility class of each compound.
   - (a) aniline
   - (b) benzamide
   - (c) butylamide
   - (d) citric acid
   - (e) cyclohexanone
   - (f) diethyl ether
   - (g) methylene chloride
   - (h) p-nitrobenzoic acid
   - (i) n-pentane
   - (j) 2-propanol

2. Using the solubility classification solvents, indicate how you would separate each binary mixture and recover the components.
   - (a) benzoic acid and benzamide
   - (b) n-butyl bromide and di-n-butyl ether
   - (c) nitrobenzene and aniline
   - (d) 1-pentene and 2-pentanone

3. A white solid, mp 110–112°, was insoluble in water but soluble in both aqueous sodium hydroxide and sodium bicarbonate. Elemental analysis was negative for N, S, and halogens. What is the probable structure of the solid?

4. A colorless liquid, bp 177–179°, fell into solubility class N. Treatment with phenylisocyanate gave a solid derivative, mp 112–114°. What is the probable structure of the liquid?

5. What test could be used to distinguish between each pair of compounds?
   - (a) 1-butanol and 2-butanol
   - (b) 2-methyl-2-propanol and 1-butanol
   - (c) butanal and 2-butanone
   - (d) acetic acid and acetamide
   - (e) triethylamine and dibutylamine
   - (f) cyclohexene and cyclohexane

6. What derivative would you prepare to distinguish between each pair of compounds?
   - (a) isobutyl alcohol and 3-pentanol
   - (b) 2-pentanone and 3-pentanone
   - (c) propanedioic acid and 2-acetoxybenzoic acid
   - (d) benzenesulfonic acid and 2-carboxybenzenesulfonic acid
   - (e) 4-methylaniline and 3-methylaniline

10 Identification of Structure by Spectrometric Methods

The classical methods for organic structure determination involve a series of test reactions and chemical transformations. The newer spectrometric methods probe the sample with electron beams (mass spectrometry), with electromagnetic radiation (infrared, ultraviolet, and X-ray spectroscopy), or with a combination of electromagnetic radiation and a strong magnetic field (nuclear magnetic resonance).[1] The classical methods are destructive; that is, the sample is gone when the reaction is complete. Mass spectrometry is also destructive, but on such a small scale it is usually ignored. The other spectrometric methods are essentially nondestructive. Historically, it was this feature as much as anything that led organic chemists to welcome spectrometry when the instruments first became available commercially. It soon became apparent, however, that spectrometry offered much more; with this technique the environment of the atoms, the bonds, and the combinations of bonds could be probed with an ever-increasing variety and specificity denied to chemical reactions.

Spectrometry has entered a new era with the exploitation of microelectronic devices and computers. Complex sequences and combinations of nondestructive electromagnetic probes can be applied automatically with the massive outpouring of data reduced and analyzed by a computer.

Elaborate computer programs that compare an observed spectrum with a large file of spectra are available. In a matter of minutes a file of 20,000–100,000 spectra is searched and any match reported; even when no match-

---

[1] This list does not include the less common techniques such as photoelectron spectroscopy, electron spin resonance, Fourier transform ion cyclotron resonance, and electron diffraction.

ing spectrum is found, the program reports the structures of the compounds with the nearest fitting spectra—information that frequently is sufficient to identify the unknown.

The discussion of spectrometry presented here is much less ambitious. Only the fundamentals of the four most widely used methods will be considered. Even though abbreviated, the material presented here is suitable for the many routine applications of spectrometry in the organic laboratory.

10.1 Mass Spectrometry

The mass spectrometer bombards a stream of vaporized sample with a high-energy electron beam, which, on collision with a molecule of sample, ejects an electron from it to produce a vibrationally excited radical-cation species.

$$\text{Molecule} \xrightarrow{\text{electron beam}} \underset{\text{radical-cation}}{[\text{molecule}]^{\ddagger}} + \text{ejected electron}$$

$$\swarrow \downarrow \searrow$$

$$\text{neutral and ionized fragments}$$

The collisions are so energetic (typically 1400 kcal/mole) that the radical-cations formed behave as though they have been heated to thousands of degrees and begin to decompose almost immediately to form a variety of neutral and ionized fragments. The collection of ionized fragments along with the remaining parent ions are drawn out of the ionizing chamber by a set of charged electrodes and toward an ion-collecting target that generates a signal in proportion to the number of ions reaching it.

A simplified diagram of a mass spectrometer is shown in Figure 10.1. The ionization chamber is at the left of the diagram; the target (detector) is at the bottom. Along the way the rushing flow of ions passes through a strong magnetic field that bends the ion beam by an amount that depends on both the mass of the ion and the strength of the magnetic field. For a particular magnetic field strength, only those ions possessing a characteristic mass-to-charge ratio (conventionally abbreviated m/e) will reach the narrow opening of the target. As the magnetic field is altered, species with different m/e will be focused on the target. Once the calibration between m/e and magnetic field strength has been determined, it is possible to record a mass spectrum that shows the relative abundance of each ionic species produced in the fragmentation process. A representative mass spectrum, that for ethyl acetate, is shown in Figure 10.2.

Mass spectra are both fingerprints of compounds and, if interpreted in terms of the stability and ease of forming of different ions, sets of clues to the

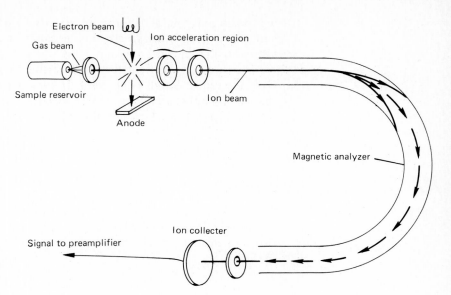

FIGURE 10.1
Diagram of Mass Spectrometer

structures. One of the most valuable pieces of information provided by mass spectrometry is the molecular weight of the parent ion, which will normally be observed as one of the group of peaks at the highest m/e, 88 for ethyl acetate (Figure 10.2). Not visible in Figure 10.2 are the low-intensity peaks at m/e 89 and 90, which arise from trace amounts of ^{13}C in the original sample. There is also a weak peak at m/e 87, which arises from loss of a hydrogen atom from the parent ion.

From close examination of the peaks around the parent peak, it is possible to obtain some idea of the elemental composition. Ordinary organic chemicals are not isotopically pure because of the 1.08% natural abundance of ^{13}C. For the saturated hydrocarbon $C_{10}H_{22}$, this means that there is a $10 \times 1.08 = 11\%$ probability that a particular molecule will

FIGURE 10.2
Mass Spectrum of Ethyl Acetate

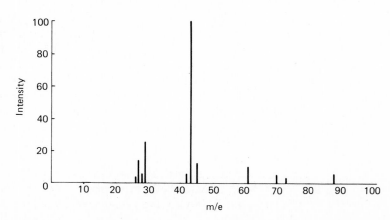

contain one ^{13}C. The mass spectrum of the C_{10} hydrocarbon will show parent peaks at m/e 142 and 143 in the ratio of 100:11. The appearance of isotopic peaks is quite helpful in identifying the presence of chlorine, since the ^{35}Cl and ^{37}Cl isotopes occur naturally in an easily recognized ratio of 3:1. The appearance of two peaks in this ratio, separated by two mass units, is strong evidence for one chloride atom in the molecule. If two chlorine atoms were present, three peaks would occur in the ratio of 9:6:1 (Why?) with each peak separated from the next by two mass units. Naturally occurring bromine contains ^{79}Br and ^{81}Br in the ratio of 1:1; fluorine and iodine are monoisotopic.

With special mass spectrometers of extremely high mass resolution, it is possible to make use of the nonintegral values of atomic weights of even the pure isotopes (^{12}C = 12.00000 basis, ^{1}H = 1.007825, ^{14}N = 14.00307, ^{16}O = 15.99491, etc.) to determine empirical formulas directly. For example, ethyl acetate with an empirical formula $C_4H_8O_2$ has a molecular weight of 88.0524. Ethyl n-propyl ether, $C_5H_{12}O$, has a molecular weight of 88.0888. These two compounds, although they have the same low-resolution molecular weight of 88, could easily be distinguished by a high-resolution mass spectrometer, which will give molecular weights to ± 0.002 mass unit.

Interpretation of the clues provided by the fragment ions requires considerable practice and study but can provide complete structures, particularly if auxiliary data from other spectroscopic sources or chemical studies are available. The details of the process will not be described here; the interested reader is referred to the excellent book by McLafferty.[2] The mass spectrum of ethyl acetate reproduced in Figure 10.2 gives an idea of the structural information available by this technique. The largest peak in the spectrum (*base peak*) at m/e 43 is the fragment ion $[CH_3CO]^+$ that arises from the parent ion (m/e 88) by loss of a neutral $CH_3CH_2O^\bullet$ radical. The peak at m/e 45 arises from the same bond cleavage, but this time with loss

---

[2] McLafferty, *Interpretation of Mass Spectra*, 3rd ed. (New York: Wiley, 1981).

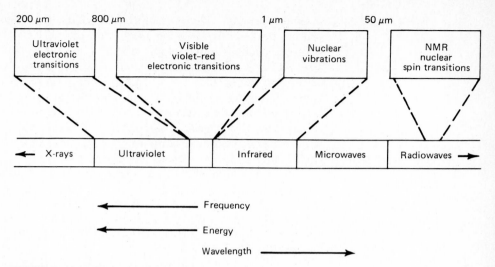

FIGURE 10.3 *Electromagnetic Spectrum*

of a neutral $CH_3CO^\bullet$ radical. The peak at m/e 29 is from the $[CH_3CH_2]^+$ ion. The other significant peaks arise from more complex rearrangements and eliminations. The relative abundance of the peaks reflects the relative stabilities of the cations and radicals formed.

A particularly powerful organic analytical technique is the computer-controlled gas chromatograph–mass spectrometer combination with which microgram or even nanogram quantities of complex mixtures can be analyzed as rapidly as the peaks come off the chromatograph. Such instruments form the backbone of environmental pollution analysis.

10.2 Infrared Spectroscopy

The electromagnetic spectrum runs continuously from X-rays, through light, to radiowaves. Electromagnetic radiation consists of bundles of energy (photons) surrounded by oscillating electric and magnetic fields. At the X-ray end of the spectrum, the frequency and energy are high and the wavelength is short; at the radiowave end of the spectrum, the converse is true. Molecules respond in various ways to electromagnetic radiation, depending on the frequency. Figure 10.3 displays the electromagnetic spectrum and several of the regions of particular interest to organic chemists.

A useful mechanical model for discussing the infrared (IR) region of absorption (wavelengths of about 1–50 μm) is to treat the atoms as masses connected by springs (bonds). A property of mechanical systems is their characteristic vibrational frequencies (resonances) that absorb energy from applied oscillatory forces. Small masses or stiff springs give rise to high vibrational frequencies; as a consequence of the small weights of atoms and the tightness of bonds, molecular vibrations occur in the infrared region (6×10^{12} to 3×10^{14} Hz). The probability of absorption depends on the change of dipole moments during the vibration; a large change in dipole moment accompanies intense absorption.

There are two basic types, or *modes*, of motion of nuclei relative to each other: stretching and bending.

$$\underset{\text{stretching}}{C-X} \quad \underset{\text{bending}}{\overset{C}{\underset{X \quad Y}{\diagup \diagdown}}}$$

These basic modes combine to form more complex vibrational modes. For example, a methylene group has six modes, each with a characteristic frequency, as shown in Figure 10.4.

Even for simple molecules the number of vibrational frequencies is large (for nonlinear molecules, $3N - 6$ vibrations, where N is the number of atoms). Not all of these will appear in the spectrum because many are highly symmetrical vibrations that produce small changes in dipole moment. A further complication is that overtones (analogous to those of a vibrating violin string) and combination bands occur with moderate intensity. Fortunately, certain functional groups and structural units have characteristic absorption frequencies that change little from molecule to molecule. For many structural units, the frequency shifts that do occur can be related to variations in the neighboring structure. Table 10.1 lists a few

FIGURE 10.4
Vibrational Modes of the Methylene Group

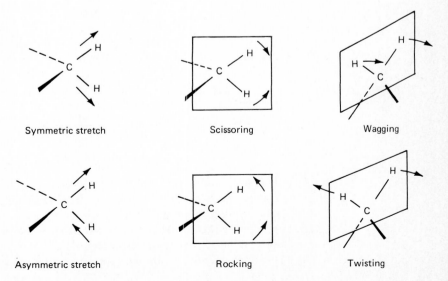

TABLE 10.1
Infrared Fundamental Absorption Frequencies

	Frequency, cm^{-1}	Wavelength	Intensity[a]
C=O stretching vibrations			
Aldehydes	1740–1720	5.75–5.81	s
Ketones			
Saturated	1725–1705	5.80–5.87	s
Aromatic	1700–1680	5.88–5.95	s
Carboxylic acids	1725–1680	5.80–5.95	s
Esters	1750–1720	5.71–5.81	s
Amides	1700–1630	5.81–6.13	s
C—H, N—H, and O—H stretching vibrations			
C—H			
Alkanes	2960–2850	3.38–3.51	m–s
Alkenes	3095–3010	3.23–3.32	m
Alkynes	near 3300	near 3.03	s
Aromatic	near 3030	near 3.30	v
Aldehyde	2960–2820 and	3.45–3.55 and	w
	2775–2700	3.60–3.70	w
N—H			
Amine	3500–3300	2.86–3.03	m
Amide	3500–3140	2.86–3.18	m
O—H			
Alcohols and phenols			
Not hydrogen bonded	3650–3590	2.74–2.79	v, sharp
Hydrogen bonded	3550–3200	2.82–3.13	v, broad
Carboxylic acids	2700–2500	3.70–4.00	w
Triple-bond stretching vibrations			
Nitriles	2260–2215	4.42–4.51	m
Alkynes	2260–2100	4.42–4.76	v
C=C stretching vibrations			
Alkenes	1680–1620	5.95–6.17	v
Conjugated diene	near 1650 and	near 6.06 and	w
	near 1600	near 6.25	w
Aromatic	near 1600 and	near 6.25	v
	near 1500	near 6.67	v

[a] w = weak absorption; m = medium absorption; s = strong absorption; v = variable intensity absorption.

structural units and their characteristic absorption ranges in the infrared region. The frequencies listed arise from stretching modes. The stronger the bond joining the atoms, the higher the frequency. The apparent exceptions are the various X—H bonds, which are normal single bonds but absorb at high frequencies because of the lightness of the hydrogen atom (the analogy here is to the vocal cords of a soprano and a basso).

Further information on characteristic frequencies can be found in specialized texts.

Frequencies in the range of 1250–600 cm^{-1}, the so-called fingerprint region, contain most of the bending vibrations. The vibrational motions occurring here can be quite complex mixtures of the basic modes and the student is cautioned not to attempt to interpret every band or wiggle found in this region.

Assignment of Principal Functional Groups. This section presents a scheme for the systematic identification of the principal functional groups. A flowchart of the scheme is given in Figure 10.5; the circled numbers correspond to the numbered steps that follow. Further details for selected functional groups along with representative spectra are presented in the next section.

The assignment scheme starts with the carbonyl group since, when present, this group in aldehydes, ketones, acids, esters, acid chlorides, and amides is usually responsible for the strongest band in the spectrum. The next most characteristic absorptions are those for O—H groups and N—H bonds. Triple bonds, though less common, also have characteristic frequencies. The scheme is not infallible, but it is much superior to random guessing; although the conclusions of the scheme are stated firmly, it should be understood that confirmation is always desirable.

1. Is there strong absorption in the region of 1820–1650 cm^{-1}? If not, proceed to **7**. If there is strong absorption (see Figures 10.10 and 10.11 for examples), the molecule contains a carbonyl group, which can be further specified by looking for other characteristic absorptions starting with **2**. The strong carbonyl band arises from the large change in dipole moment as the carbon–oxygen bond stretches and contracts.

2. Is there a pair of strong bands at 1850–1800 and 1790–1740 cm^{-1}? If not, proceed to **3**. If a pair is present, the molecule contains an anhydride functional group.

3. Is there broad absorption in the region of 3300–2500 cm^{-1} (Figure 10.6 shows an example)? If not, proceed to **4**. If this is present, the molecule is a carboxylic acid. The absorption arises from an O—H stretching vibration and is broad because the different molecules of a carboxylic acid occur with varied degrees of hydrogen bonding, each with its characteristic absorption frequency.

4. Is there a medium-intensity absorption near 3400 cm^{-1} (an example appears in Figure 10.7)? If not, proceed to **5**. If the 3400 cm^{-1} band is present, the molecule is a primary or secondary amide. Primary amides show two bands ($\sim$ 3500 and $\sim$ 3400 cm^{-1}), whereas secondary amides show only one band near 3400 cm^{-1}. These absorptions are associated with N—H stretching vibrations. Tertiary amides do not absorb in the region.

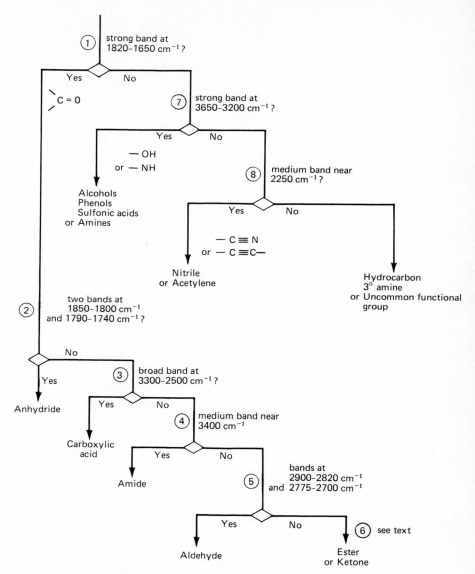

FIGURE 10.5 *Infrared Assignment Scheme for the Principal Functional Groups*

5. Is there a pair of weak bands in the 2900–2820 and 2775–2700 cm^{-1} regions? If not, proceed to **6**. If such a pair of bands is present, the molecule is an aldehyde. The bands have been associated with the mixing of the C—H and C—O vibrations. The identification of aldehydes and ketones is discussed more completely in the following text.

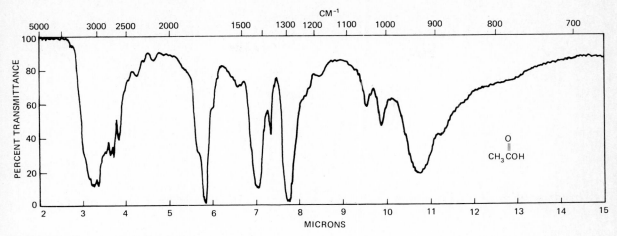

FIGURE 10.6 *Infrared Spectrum of Glacial Acetic Acid*

6. For the restricted classes of functional groups being considered here only two possibilities remain: esters and ketones.

    Saturated esters absorb in the range of 1750–1735 cm^{-1} (Figure 10.8), which increases to 1820–1760 cm^{-1} with strained cyclic esters. Aryl and α,β-unsaturated esters absorb at 1730–1717 cm^{-1}.

    Saturated ketones absorb in the range of 1725–1705 cm^{-1}. Small-ring cyclic ketones and α,β-unsaturated ketones absorb at 1685–1665 cm^{-1}.

7. If a carbonyl group is absent, look for absorption in the range of 3650–3200 cm^{-1}. If none is present, go to **8**. Alcohols and phenols absorb strongly in this range; primary and secondary amines show medium-

FIGURE 10.7 *Infrared Spectrum of Acetamide (Melt)*

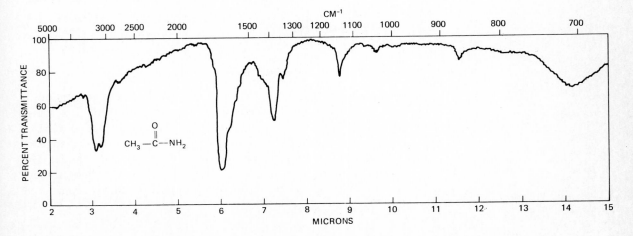

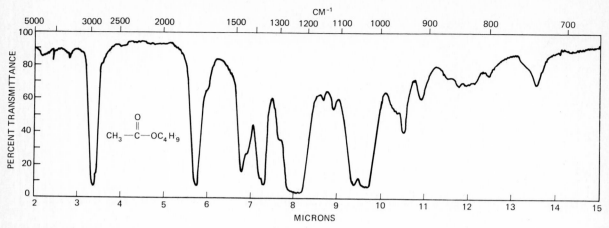

FIGURE 10.8 *Infrared Spectrum of Neat* n-*Butyl Acetate*

strength absorptions. The amines can sometimes by distinguished by the presence of weak absorptions near 1600 cm^{-1} arising from N—H bending vibrations. Hydroxyl groups show O—H bending vibrations in the fingerprint region. The spectroscopic identification of alcohols and amines is discussed more fully in the text that follows.

8. If O—H and N—H absorptions are absent, look near 2250 cm^{-1}. If no absorption is present there, go to **9**. Nitriles show a medium-intensity sharp absorption near 2200 cm^{-1} characteristic of the C≡N stretch. Alkynes absorb near 2150 cm^{-1}, but the band is usually weak.

9. If none of the above infrared bands is present, the molecule either is a hydrocarbon or contains one of the less common functional groups. The spectroscopic identification of hydrocarbons is discussed in this text, as are sulfonic acids and their derivatives. If another uncommon functional group is suspected, a specialized text on infrared spectroscopic identification should be consulted.

Spectroscopic Details for Selected Functional Groups

Alcohols. The infrared spectra of alcohols typically show absorption in the range 3200 to 3600 cm^{-1} from the O—H stretching of the polymeric species formed by intermolecular hydrogen bonding. Under conditions of high dilution, a weaker band at higher energy (3600–3650 cm^{-1}) appears for the free hydroxyl. Strong absorptions near 1050–1150 cm^{-1} and 1250–1410 cm^{-1} are characteristic of C—O stretching vibrations. These features can be seen in the IR spectrum of neat (pure liquid) *n*-butanol reproduced in Figure 10.9.

Sec. 10.2] INFRARED SPECTROSCOPY

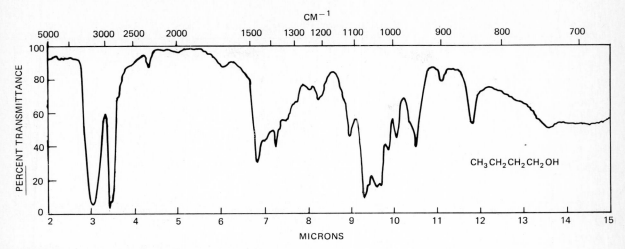

FIGURE 10.9 Infrared Spectrum of Neat n-Butanol

Aldehydes and ketones. The carbonyl group is one of the most easily identified peaks in an infrared spectrum. Saturated acyclic ketones absorb strongly in the range of 1705–1725 cm^{-1} (see the spectrum of cyclohexanone, Figure 10.10). If α,β-unsaturation is present, the peak shifts to 1680–1685 cm^{-1}. Cyclic ketones with five-membered or smaller rings are shifted to higher fields: five-membered, 1740–1750 cm^{-1}; four-membered, 1775 cm^{-1}.

Aldehydes absorb in the range of 1720–1740 cm^{-1}. As with ketones,

FIGURE 10.10 Infrared Spectrum of Neat Cyclohexanone

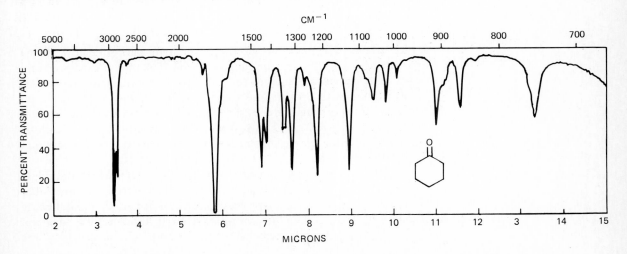

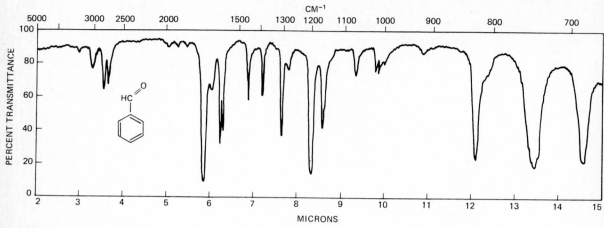

FIGURE 10.11 *Infrared Spectrum of Neat Benzaldehyde*

conjugation shifts the peaks to lower frequencies: α,β-unsaturation to 1680–1705 cm^{-1}, and aryl to 1695–1715 cm^{-1}. Aldehydes show two additional weak bands at 2700–2775 and 2820–2900 cm^{-1}. These features can be seen in the infrared spectrum of benzaldehyde (Figure 10.11).

Amines. Primary amines, which have *two* N—H bonds, show *two* stretching bands in the range of 3300–3500 cm^{-1}; secondary amines show *one* band. Tertiary amines have no N—H stretching absorption. Primary amines show a NH$_2$ scissoring mode in the range of 1560–1640 cm^{-1}. Secondary amines absorb near 1500 cm^{-1}. Aromatic amines, for example, aniline (Figure 10.12), show strong absorption in the range of 1250–1370 cm^{-1}.

Hydrocarbons. The infrared spectra of all hydrocarbons show strong C—H stretching vibrations in the range of 3300–2850 cm^{-1}. If a spectrometer with high resolution is used, the subclasses of hydrocarbons can be distinguished, but with ordinary instruments other regions of the spectrum are usually more revealing. Alkenes show a medium intensity C=C stretching absorption near 1670 cm^{-1}. Alkynes have a distinctive C≡C stretching absorption near 2200 cm^{-1}. Aromatic hydrocarbons do not show truly distinctive absorptions.

Figure 10.13 shows the spectrum of neat cyclohexene in which the C—H and C=C bands are visible. It shows also, as is normal, many other uncharacteristic longer wavelength absorptions related to C—C stretching vibrations and various bending vibrations. Even though these are not readily interpreted, they do provide a "fingerprint" characteristic of the cyclohexene molecule.

Sec. 10.2] INFRARED SPECTROSCOPY

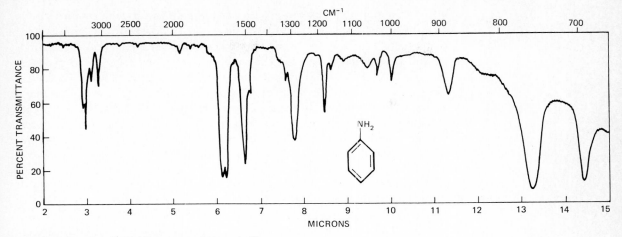

FIGURE 10.12 *Infrared Spectrum of Neat Aniline*

Sulfonic acids and derivatives. Sulfonic acids show a medium-intensity band in the range of 3100–3400 cm^{-1} from OH stretching. The anhydrous acids give strong S—O absorptions at 1340–1350 and 1150–1165 cm^{-1}; the hydrated acids show an additional band at 1120–1230 cm^{-1}, which has been attributed to S—O double bond stretch of the hydronium sulfonate salt, $RSO_3^- H_3O^+$.

Sulfones (R_2SO_2), sulfonamides (RSO_2NR_2), and covalent sulfonates (RSO_3R) show a similar pair of strong S—O absorptions at 1300–1375 and 1120–1195 cm^{-1}.

FIGURE 10.13 *Infrared Spectrum of Neat Cyclohexene*

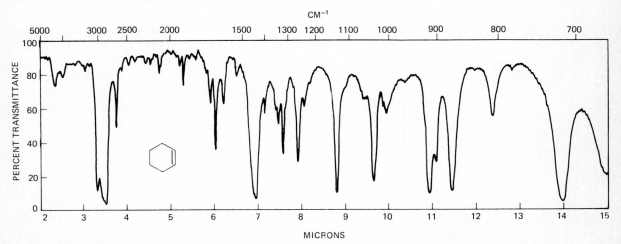

Infrared Sampling Techniques. With pure liquid samples, the simplest technique is to prepare a thin film between a pair of polished NaCl salt plates. A drop of the sample is placed in the center of one plate, and the second plate is placed gently on top. The sandwich is pressed together to produce an *even* film of liquid. Salt plates are soft and easily scratched. They also cleave readily if uneven pressure is applied. Another problem is that they dissolve or etch when exposed to water or hydroxylic solvents. Even the moisture on your fingers is sufficient to damage them; the polished faces should never be touched. The best solvents for cleaning salt plates are carbon tetrachloride and chloroform. Because of their toxicity with prolonged exposure, these solvents should be handled only in the hood or other well-ventilated area. The plates are best handled with rubber-tipped tweezers, and they should be stored in a desiccator.

Another common device used for solutions and liquids that are not too viscous is the solution cell, made from salt plates separated by a thin spacer and held together in a metal frame. Solution cells, because of their precision construction, are relatively expensive. A typical cell has a path length of 0.2 mm and holds about 0.5 mL of solution. With such a cell 5% solutions yield good spectra. If infrared spectra are determined in solution, it is normal practice to place a solution cell filled with pure solvent in the reference beam to cancel (approximately) any absorptions of the solvent.

With solid samples, two other techniques are used. The *mull* method requires that a 5- to 10-mg sample be ground vigorously in an agate or mullite mortar until the sample is glossy and cakes to the side of the mortar (an indication that the particle size is in the micrometer range). A drop of Nujol (commercial mineral oil) is added, and the sample is reground to disperse the solid in the Nujol. The mixture should resemble Vaseline in viscosity. If it is too thick, another drop of Nujol is added and the sample ground again. If the solid particles are too coarse, the sample will scatter the infrared light at short wavelengths and yield a distorted spectrum. The mull is transferred to a salt plate with a microspatula; a second plate is put on top and rotated until all air bubbles are squeezed from the window area, the two plates are held together by the viscous sample and supported in the IR beam by a V-shaped sample holder.

The spectra of solids can also be determined by the potassium bromide *pellet* method. Approximately 1–2 mg of solid is ground *briefly* with about 200 mg of spectral-grade potassium bromide (too much grinding produces a powder that rapidly absorbs water from the atmosphere). The powder is transferred to a pellet press,[3] and pressure is applied to produce a fused translucent wafer that is then mounted in a special holder.

---

[3] The KBr Mini Press, available from Wilks Scientific Corporation, Norwalk, CT, is a simple, effective, and inexpensive press. The Mini Press also serves as a holder for the wafer. The press should be *cleaned thoroughly* and *dried carefully* after each use.

10.3 Nuclear Magnetic Resonance

In the previous section it was shown how functional groups can be identified readily by infrared absorption spectroscopy. This section will describe the complementary technique of nuclear magnetic resonance (NMR), which is useful for identifying the hydrocarbon skeleton to which the functional group is attached. Proton NMR allows the qualitative and quantitative characterization of the different types of hydrogen in the molecule (aromatic, aliphatic, etc.). The newer ^{13}C NMR techniques identify the carbons of the skeleton. Although NMR by itself is a valuable structural tool, the combination of NMR and IR techniques is extremely powerful and sufficient to identify all but the most complex molecules.

Theory of ^{1}H and ^{13}C NMR.[4] The nucleus of a hydrogen atom is surrounded by a weak magnetic field associated with the nuclear spin. When hydrogen atoms or materials containing hydrogen atoms are placed in a strong external magnetic field, the weak nuclear magnetic moments line up either parallel or antiparallel to the direction of the magnetic lines of force. The two possible orientations differ in energy by an amount proportional to the strength of the applied magnetic field; if the nuclei are then exposed to electromagnetic radiation of the correct frequency (energy $= h\nu$) the parallel moments absorb energy, flip over, and become antiparallel moments. It is this flipping process, called *nuclear magnetic resonance*, that gives rise to the absorption spectrum.

The condition for proton resonance absorption is that $\nu = 4.2577 \times 10^3 \mathscr{H}$, where ν is the radiating frequency and $\mathscr{H}$ is the magnetic field strength at the nucleus measured in gauss.[5] If all of the protons were exposed to the same magnetic field, they would absorb at the same frequency. Actually, when a molecule is subjected to a magnetic field, the bonding electrons circulate in such a way as to generate additional weak magnetic fields of their own. In general, these new magnetic fields oppose the large, fixed magnetic field, and the underlying proton nuclei are partially shielded. The greater the shielding, the lower the frequency required to produce resonance. To a first approximation, the amount of shielding is proportional to the electron density around the proton (there are secondary effects, which will be discussed later), and thus protons adjacent to electron-donating groups will absorb at lower frequencies. The relationship of varied shielding to NMR absorption peak shift is illustrated in Figure 10.14.

---

[4] This section describes the origin of proton NMR signals. It applies also to ^{13}C NMR, except that the condition for resonance is different.

[5] For ^{13}C nuclei the condition is $\nu = 1.0705 \times 10^3 \mathscr{H}$.

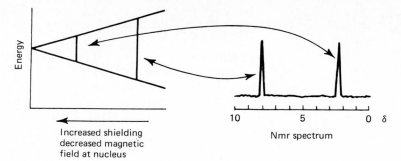

FIGURE 10.14
Relationship of Shielding to Chemical Shift

Chemical Shift. Because of the correlation of electron density with structure, NMR is a powerful tool for determining the structural environment of hydrogen atoms in organic molecules. By common usage the reference standard for practical proton NMR measurements is the highly symmetrical tetramethylsilane (TMS), $(CH_3)_4Si$, a volatile liquid that is easily separated from the sample after the measurement. The protons of TMS are more highly shielded than those found in most other organic compounds. In recording NMR spectra, the most widely used measure is the chemical shift, δ, defined as

$$\delta = \frac{\Delta v}{v_0} \times 10^6 \text{ ppm}$$

where Δv is the difference in absorption frequencies of TMS and the proton being examined, and v_0 is the radiation frequency of the NMR machine.[6] The chart paper used in a calibrated NMR instrument allows both δ and Δv to be read directly.[7]

The chemical shifts observed in several representative proton environments are listed in Table 10.2. The NMR spectrum of methyl *p*-bromobenzoate is reproduced in Figure 10.15, which shows the downfield aromatic absorptions and (relative to the aromatic protons) the upfield methyl absorptions. The reference peak of TMS, which was added to the sample, appears at $\delta = 0.0$

The major determinant of chemical shifts is the inductive electron withdrawal by electronegative groups and is cumulative and approximately additive so that, for example, the chemical shift (in parts per million) for the

---

[6] An older unit, now disappearing from the literature, is the tau unit defined as $\tau = 10.0 - \delta$.

[7] In practice NMR machines are designed to operate at a constant frequency, and it is the magnetic field that is varied. However, because of the proportional relationship between absorption frequency and magnetic field, the change in magnetic field can be expressed as an equivalent change in frequency. The protons of molecules less shielded than those found in TMS are referred to as deshielded, as occurring downfield from TMS, or as having positive chemical shifts.

TABLE 10.2 Nuclear Magnetic Resonance Chemical Shifts

Structural environment	Chemical shift from TMS, ppm
$-\underset{\vert}{\overset{\vert}{C}}-CH_3$ (saturated)	0.8–1.1
$-CH_2-$ (saturated)	1.2–1.3
$-\underset{\vert}{\overset{\vert}{C}H}-$ (saturated)	1.4–1.6
$X-\underset{\vert}{\overset{\vert}{C}}-CH_3$ (X = halogen, $-O$, $>N$)	1.0–2.0
$\overset{\diagdown}{\underset{\diagup}{C}}=\overset{\diagup}{\underset{\diagdown}{C}}-CH_3$	1.6–1.9
$ArCH_3$	2.1–2.5
$O=\underset{\vert}{\overset{\vert}{C}}-CH_3$	2.1–2.6
$-C\equiv C-H$	2.4–3.1
$-O-CH_3$	3.5–3.8
$\overset{\diagdown}{\underset{\diagup}{C}}=CH_2$ (nonconjugated)	4.6–5.0
$\overset{\diagdown}{\underset{\diagup}{C}}=\overset{\overset{H}{\diagup}}{\underset{\diagdown}{C}}$	5.2–5.7
$Ar-H$	6.6–8.0
$-C\overset{\overset{O}{\diagup\!\diagup}}{\underset{\diagdown}{H}}$	9.8–10.8

protons in methane is 0.8; in methyl chloride, 3.0; in methylene chloride, 5.3; and in chloroform, 7.3. The other important factor, magnetic anisotropy of neighboring bonds, is deeply interwoven and difficult to separate from the inductively caused chemical shifts. However, two situations reveal its presence clearly. In aromatic compounds the ring protons typically absorb at $\delta = 6.6$–8.0, a region far below typical alkene absorption even though both employ sp^2-hybridized carbon atoms to bond to hydrogen and must therefore have similar inductive effects and electron densities at the protons. In acetylenes, the carbon atoms are sp-hybridized and are more electronegative than the sp^2-hybridized carbon atoms of aromatics or alkenes. Nevertheless, the protons of acetylenes occur upfield (less shielded) at $\delta = 2.4$–3.1. These magnetic anisotropy effects arise whenever the electrons of a bond can circulate with particular ease in one direction under the influence of the applied magnetic field. In aromatic compounds this easy circulation occurs in the plane of the benzene ring (hence the name "ring

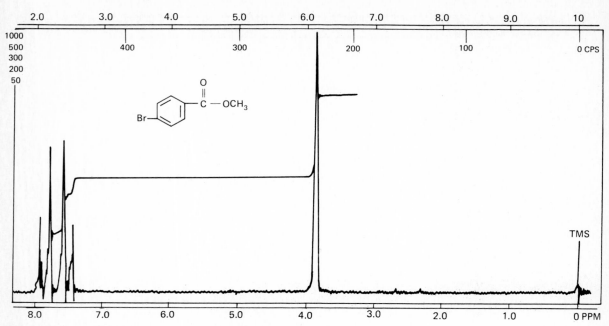

FIGURE 10.15 *NMR Spectrum of Methyl* p-*Bromobenzoate*

current"); in acetylenes it occurs around the axis of the cylindrically symmetric triple bond. These extra currents act as little electromagnets, which produce extra magnetic fields that alter the magnetic field strengths at nearby protons. Figure 10.16 depicts the magnetic fields induced in aromatics and acetylenes and shows how the local fields at the hydrogen nuclei are altered.

If the NMR machine is operated at low radiation levels, the areas under the peaks are proportional to the numbers of protons giving rise to the absorptions. The machine can be operated in an integral mode such that the pen displacement on the chart is proportional to the accumulated area

FIGURE 10.16 *Alteration of Magnetic Field at the Proton Due to Induced Local Field*

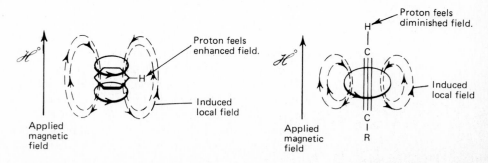

Sec. 10.3] NUCLEAR MAGNETIC RESONANCE

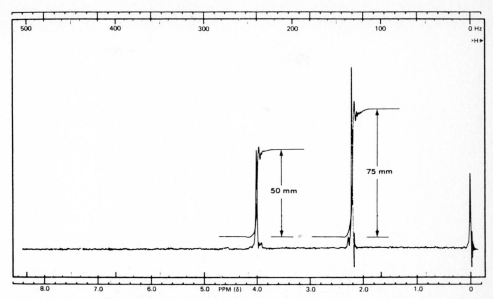

FIGURE 10.17 *NMR Spectrum of 1,2,2-Trichloropropane with Superposed Intensity Integrals*

under each peak. Figure 10.17 reproduces the NMR absorption spectrum of 1,2,2-trichloropropane on which is superposed the integral for each peak. As expected, the areas are in the ratio of 3 : 2 with the methylene hydrogens appearing further downfield than the methyl hydrogens.

Exchange. Table 10.2 does not contain an entry for hydroxyl protons of carboxylic acids or alcohols because the chemical shifts are highly dependent on the solvent. One of the novel features of NMR spectroscopy is that when protons with different chemical shifts exchange positions, the NMR response depends on the rate of exchange relative to one cycle of the applied radiation (at 100 mHz one cycle is 10^{-8} s). If exchange is rapid on this "NMR time scale," only a single sharp peak will be observed at an averaged position proportional to the concentrations of each kind of exchanging proton. If the exchange is slow, two separate sharp peaks appear, each at its characteristic chemical shift position. If the exchange rate is close to the NMR cycle time, the peaks of the exchanging protons are broadened and span the entire gap between the characteristic chemical shift positions. If that gap happens to be large, the absorption peaks can be so broad that they become lost in the background noise. The exchange phenomenon, one of the few directly observable consequences of the uncertainty principle in quantum mechanics, can be analyzed rigorously and used to calculate rates of rapid proton exchanges. In principle the same phenomenon applies to protons that exchange positions by rotation around bonds;

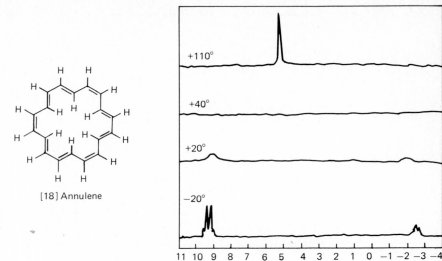

FIGURE 10.18
NMR Spectra of [18]Annulene at Different Temperatures

if the rotation rate happens to approximate the NMR cycle time, broadening is observed. With some NMR machines, the temperature of the sample and consequently the rate of exchange can be controlled to make the broadening phenomenon appear or become more distinct.

A striking example of a temperature-dependent exchange process is shown by the NMR spectrum of [18]annulene (Figure 10.18). At $-20°$ two sets of peaks are present centered at 8.94 δ (12 outer protons) and -2.50 δ (6 inner protons).[8] At a high temperature ($+110°$) the ring rapidly flips inside out and the entire set of 18 protons appears as a single sharp peak at 5.45 δ. At intermediate temperatures the peaks are broad; at 40° the peaks are so broad as to be indistinguishable from the baseline noise.

Spin–Spin Interactions. In the preceding paragraphs we have described how the application of a magnetic field to isolated protons cause their magnetic moments (spins) to align themselves either with the field (α spins) or against it (β spins). The difference in energy between the two states is proportional to the strength of the magnetic field.[5] If two protons are within a few angstroms of each other, their weak magnetic moments (spins) can interact to change the energies of the spin states, and this results in a change in the chemical shifts of the two protons. Since a sample of a substance contains many molecules (6×10^{23}/mole), which can have all of the possible different combinations of proton spins, spin–spin interaction can lead to a complex pattern of overlapping absorption spectra. Fortunately, many of the widely

---

[8] The unusual chemical shift of the inner protons results from the large ring current acting in [18]annulene (see Figure 10.16 and the related discussion).

occurring patterns are simple and can be readily used to help identify an unknown.

For example, 1,1,1,2,3,3-hexabromopropane, $Br_2CHCHBrCBr_3$, has two different kinds of protons and might be expected to give rise to two peaks in the area ratio of 1:1. In fact, four peaks are observed (Figure 10.19) as a consequence of the interaction between the spins and different ways the neighboring spins can be aligned. The pair of peaks associated with the proton on carbon-2 have an *average* value close to the chemical shift that would be expected in the absence of neighboring spins. The same is true for the pair of peaks associated with the proton of carbon-3. The difference in chemical shifts (splitting) between either pair of peaks is called the coupling constant, conventionally designated as J. The splitting is the same for both pairs of peaks because it arises from the same interaction. It is common practice to add subscripts to identify the protons that are coupled (e.g., J_{23} for the hexabromopropane).

The simple picture of spin–spin splitting applies as long as the coupling constant is small compared to the difference in chemical shifts between the centers of the doublets. When this condition is not met, additional quantum-mechanical effects enter; the result is that for two proton systems the inner peaks in the spectrum become stronger at the expense of the outer peaks and the observed spacing between the peaks is no longer simply J. In the limit of no chemical shift difference (e.g., $Br_2CH-CHBr_2$), the outer peaks have zero intensity. This is why the protons of a CH_2 group generally show only a single sharp peak even though the protons are strongly coupled to each other. The important determinant is the ratio of the coupling constant to the chemical shift difference. The spectra of molecules having small ratios of coupling constant to chemical shift difference are referred to as first-order spectra; those displaying the additional quantum-mechanical effects are designated as second-order spectra.

Molecules containing two magnetically nonequivalent protons are known as AB systems. Molecules with three nonequivalent protons are

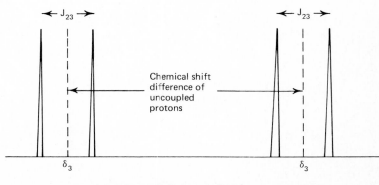

FIGURE 10.19
Spin–Spin Splitting in 1,1,1,2,3,3-Hexabromopropane

called ABC systems if each proton has a different chemical shift. In ABC systems, three coupling constants J_{AB}, J_{AC}, and J_{BC} must be considered. If all three are small compared to the chemical shift differences between protons A, B, and C, a first-order spectrum results, consisting of 12 peaks. The A protons are split into a doublet by the B protons, and each member of this doublet is split into another doublet by coupling with the C protons. The same splitting pattern applies to the B and C proton absorptions.

If two of the three protons have the same chemical shift (e.g., Br_2CH-CH_2Br) and are coupled identically to the third proton, the system is called an A_2B. Because J_{AA} is large compared to the zero difference in chemical shift of the two A protons, the two magnetically equivalent nuclei give no observable splitting of each other. If J_{AB} is large compared to the difference in the chemical shifts of the A and B protons, the spectrum appears as a triplet for the B protons (total area of 1 unit with components in the ratio of 1:2:1 separated by an amount J_{AB}) and a doublet for the A protons (total area of 2 units with two components in the ratio of 1:1 separated by an amount J_{AB}).

For the general case of A_nB the B protons will occur as $n+1$ peaks with a total area of 1 unit separated by J_{AB} Hz. The relative areas of the components within the multiplet are the coefficients of the different powers of x and y obtained by expanding the expression $(x+y)^n$. The B protons occur as a doublet of equal height with a total area of n units separated by J_{AB} Hz.

Although second-order spectra of ABC and larger spin systems are quite complex, the quantum-mechanical equations describing them can still be solved exactly. If the chemical shifts and coupling constants are known or can be estimated, computer programs are available that will produce the spectrum for systems containing up to seven interacting spins. The reverse process of extracting chemical shifts and coupling constants from an experimental spectrum is much more difficult. Table 10.3 gives some characteristic coupling constants.

TABLE 10.3
Spin–Spin Coupling Constants

Type	J, Hz	Type	J, Hz
H–C–H (geminal)	12–15	H,C=C,H (cis/trans)	13–18
H–C–C–H	2–9	H,C=C,H	7–12
C=C–H	0.5–3	aromatic H (ortho, meta, para)	o 6–9 m 1–3 p 0–1

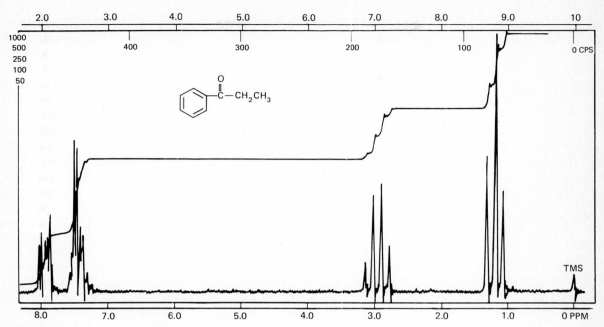

FIGURE 10.20 *NMR Spectrum of Propiophenone*

As an example of the use of spin–spin coupling in the interpretation of NMR spectra, consider the spectrum shown in Figure 10.20, obtained from a compound with an empirical formula $C_9H_{10}O$. Three groups of peaks are apparent. The relative peak areas, obtained by determining the spectrum in the "integral mode," are in the ratio of 5:2:3. From the chemical shift data in Table 10.2 the complex group of peaks of area 5 centered near $\delta = 7.5$ ppm can be assigned to a phenyl group. This complex set of peaks arises from a second-order A_2B_2C spin system. The area ratio of 2:3 for the remaining peaks suggests an ethyl group, which can be confirmed by noting that the absorptions for the CH_2 group are split into a quartet as expected for a set of three equivalent adjacent hydrogens ($n + 1$ rule); the CH_3 absorptions are split into a triplet by the set of two equivalent adjacent hydrogens. The coupling constant is identical for both sets of peaks as it must be if they are mutually coupled, and has a value close to 7 Hz as expected for hydrogens on carbons joined by a freely rotating single bond (see Table 10.3).

Other Nuclei. Nuclear magnetic resonance measurements have been most widely made for protons, but some other nuclei also have magnetic moments and NMR spectra. One of the more interesting for structure elucidation is ^{13}C, since, by measurement of NMR spectra of ^{13}C nuclei, it is possible to gain information directly about the carbon skeleton. Until recently a major limitation on ^{13}C spectroscopy was the low natural abundance, about 1%, of ^{13}C atoms. The dominant ^{12}C isotope does not have a magnetic moment and gives no NMR signal. It was only with the development of advanced digital electronics that it became practical to measure the weak signals from the ^{13}C nuclei. Happily, there is a benefit of the low natural abundance of ^{13}C nucleus—the probability is low that a given ^{13}C nucleus will be adjacent to another ^{13}C nucleus, thus ^{13}C—^{13}C spin–spin coupling is unimportant and the spectra are much simpler than proton spectra. Spin–spin coupling does occur between ^{1}H and ^{13}C nuclei, which makes it possible to identify the number of protons on each carbon atom. By suitable control of the instrument's electronics the ^{1}H—^{13}C spin–spin coupling can be turned off to give a simple indication of the number of protons attached to a given carbon. At this time the cost of such an elaborate instrument is still quite high, but it is low enough to be routine in research situations.

The ^{13}C chemical shifts are quite large, roughly 20 times those found for the proton attached to the same carbon. Table 10.4 lists ^{13}C chemical shift ranges for some of the more important classes of carbon atom found in organic molecules. Elaborate rules and correlation charts have been developed that allow prediction of ^{13}C shifts to within a few parts per million. Figure 10.21 is the ^{13}C NMR spectrum of propiophenone.

TABLE 10.4
^{13}C *Chemical Shifts*

Carbon class	Chemical shift range (TMS = 0), ppm
Alkane	
—CH_3	10–30
—CH_2	15–55
—CH	25–60
—C	30–40
Alkene	
=CH_2	100–120
=CHR and =CR_2	110–150
Alkynes	
—C≡C—H	60–70
—C≡C—R	70–90
Aromatic	
Hydrocarbons	120–150
Substituted	90–160
Carbonyl	
Aldehydes	190–220
Ketones	120–220
Carboxylic acids	165–190
Esters	160–180
Amides	150–175

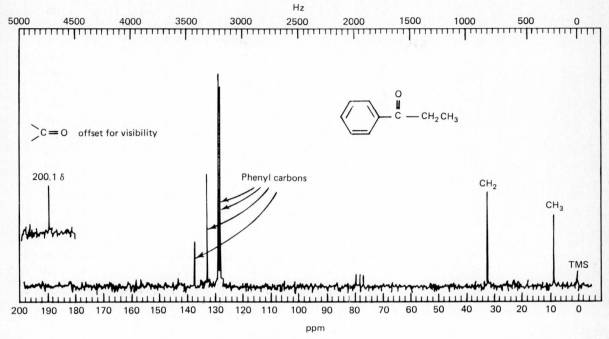

FIGURE 10.21 ^{13}C *NMR Spectrum of Propiophenone, Noise Decoupled*

Representative Applications of NMR in Structure Elucidation. The NMR spectra of hydrocarbons are quite distinctive. With saturated hydrocarbons, the primary protons absorb near $\delta = 1.0$ ppm (measured downfield from TMS), the secondary protons near 1.25 ppm, and the tertiary protons near 1.5 ppm. Substantial shifts to lower field (large δ) occur when electronegative groups are present or when the protons fall in the deshielding cone of a nearby aromatic ring. The NMR spectrum of ethyl benzene in Figure 10.22 shows CH_3 absorptions centered near $\delta = 1.25$ ppm and CH_2 absorptions centered near $\delta = 2.7$ ppm. The CH_2 protons are shifted downfield by the aromatic ring. The spin–spin splitting pattern (the CH_3 protons are split into a triplet by the adjacent CH_2 group; the CH_2 protons are split into a quartet by the adjacent CH_3 group) confirms this assignment.

Alkene protons absorb near $\delta = 5.0$ ppm. In Figure 10.23, the NMR spectrum of α-methylstyrene shows a group of peaks for each of the two alkene protons. The fine structure of these peaks comes from the coupling of the alkene protons to each other and to the methyl protons, which absorb at $\delta = 2.1$ ppm.

Alkyne protons absorb in the range $\delta = 2-3$ ppm. The NMR spectrum of phenylacetylene, shown in Figure 10.24, contains this distinctive absorption.

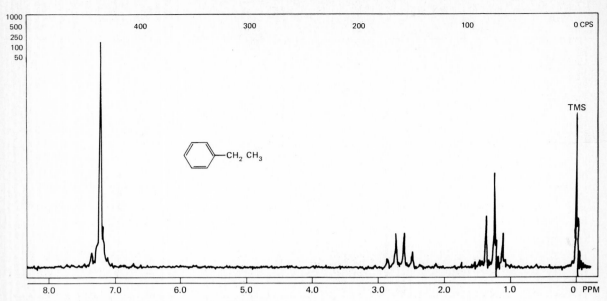

FIGURE 10.22 NMR Spectrum of Neat Ethylbenzene

FIGURE 10.23 NMR Spectrum of Neat α-Methylstyrene

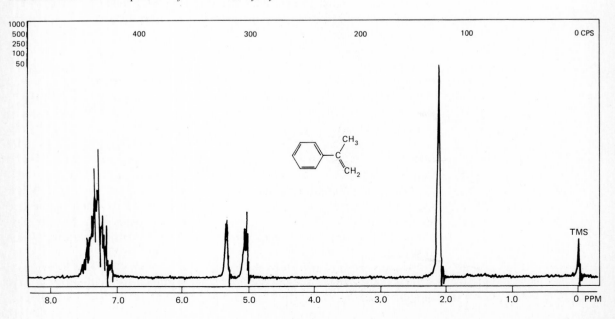

Sec. 10.3] NUCLEAR MAGNETIC RESONANCE

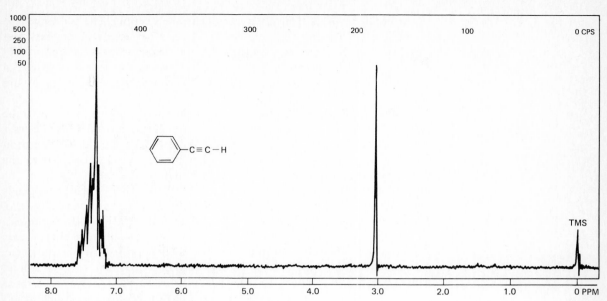

FIGURE 10.24 NMR Spectrum of Neat Phenylacetylene

FIGURE 10.25 NMR Spectrum of Neat Acetaldehyde

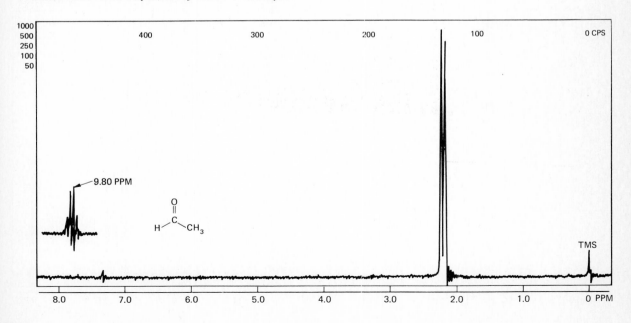

Aromatic protons typically absorb near $\delta = 7.5$ ppm. The protons of an aromatic ring are strongly coupled to each other and will appear as a complex set of peaks when they have different chemical shifts (Figures 10.23 and 10.24). When the aromatic protons have similar chemical shifts (i.e., a large J/δ ratio), this complexity is reduced to a single peak, as in Figure 10.23.

In the NMR spectrum the protons on the carbon atoms adjacent to the carbonyl group of aldehydes and ketones are shifted downfield by about 1 ppm (see the spectrum of acetaldehyde, Figure 10.25). The aldehydic proton occurs at a very low field, near $\delta = 10$ ppm; it is weakly coupled to the protons on the carbon adjacent to the carbonyl.

In the NMR spectrum the position of the hydroxyl proton depends critically on the degree of hydrogen bonding and hence on the choice of solvent. The common range is $\delta = 3.0$–5.5 ppm. If a trace of acid is present, the hydroxyl protons exchange rapidly and appear as a single sharp peak, although they may be spin–spin coupled to other protons in the molecule. Assignment of an NMR peak to OH can be confirmed by addition of trifluoroacetic acid, which absorbs beyond $\delta = 10$ ppm. Because of rapid exchange with the acid protons, the OH proton absorption either disappears or is displaced markedly, whereas the other C—H peaks do not shift significantly.

The OH group exerts a strong substituent effect on protons attached to the carbon atom bearing the OH group, which causes them to be shifted

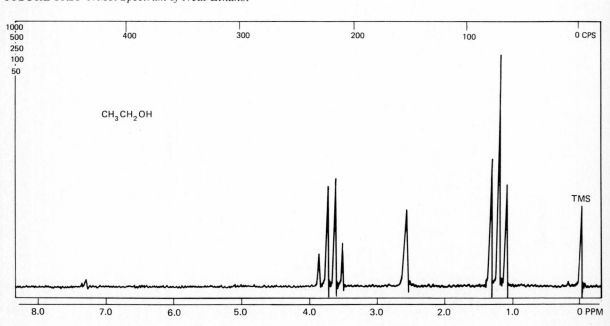

FIGURE 10.26 NMR Spectrum of Neat Ethanol

downfield by about 2–3 ppm. The NMR spectrum of ethanol in $CDCl_3$ containing a trace of acid to sharpen the OH peak shows these features (Figure 10.26). This spectrum shows the quartet and triplet peaks characteristic of an ethyl group (compare with Figures 10.20 and 10.22).

NMR Sampling Techniques. Proton NMR spectra are normally determined on 10–20% solutions. After solubility, the most important factor in solvent choice is the presence of interfering solvent absorptions. When it will dissolve the sample, the best solvent is carbon tetrachloride. Another widely used solvent is deuterated chloroform, $CDCl_3$. The deuterium nucleus gives NMR peaks but not in the proton region; most $CDCl_3$ solvent batches contain some $CHCl_3$, which gives a weak peak at $\delta = 7.27$. It is essential that the NMR sample contain a reference such as 1% TMS. This can be added when the sample is prepared, but it is more convenient to use solvents that have been spiked with TMS. If TMS must be added, remember that it is extremely volatile (bp 27°) and will evaporate rapidly.

About 0.5 mL of solution is pipetted into a 5-mm NMR tube (TMS is added with a microdropper if it is not already present in the solvent), and the tube is capped immediately to prevent evaporation. In drawing the sample into the pipet in preparation for adding it to the NMR tube, it is good practice to twist a small wad of clean surgical cotton around the pipet tip. The cotton filter removes solid impurities from the sample as the pipet is filled; the cotton filter is discarded before the sample is transferred to the NMR tube.

In order to improve the homogeneity of the magnetic field felt by the sample in the NMR machine, the tube is spun about its long axis. This is accomplished by slipping the tube into a tightly fitting plastic "spinner," which spins the tube when the instrument directs a stream of air against it. The positioning of the spinner on the tube is critical, both for even spinning and for proper location of the sample within the instrument. Most installations provide some kind of gauge for proper positioning. One word of caution here: NMR tubes are fragile and the tube must be aligned perfectly as the tube is placed in the instrument.

The chances are that your laboratory instructor will run your sample; the operation of the instrument will therefore not be described here. The principal steps are to finely tune the machine to give the best resolved spectrum and to adjust the instrument setting so that the TMS peak is properly aligned on the chart paper.

10.4 Ultraviolet and Visible Spectroscopy

The electrons of a molecule can be excited to higher energy levels by application of electromagnetic radiation with a frequency corresponding to the energy gap between the ground state and the excited state. Although the processes are identical, a distinction is made for practical reasons between visible spectroscopy (400–800 nm) and ultraviolet spectroscopy (400 nm and below). Ultraviolet and visible absorption spectra are displayed as plots of *absorbance* versus wavelength, where absorbance is defined as

$$\text{Absorbance} = \log_{10} \frac{\text{light intensity entering sample}}{\text{light intensity leaving sample}}$$

It can be shown that the absorbance is proportional to the concentration of the sample and to the pathlength of the sample through which the light travels (*Beer–Lambert law*). A quantity commonly used to characterize electronic spectra is the *molar extinction coefficient*, ε, defined as

$$\varepsilon = \frac{\text{absorbance}}{\text{concentration (moles/L)} \times \text{pathlength (cm)}}$$

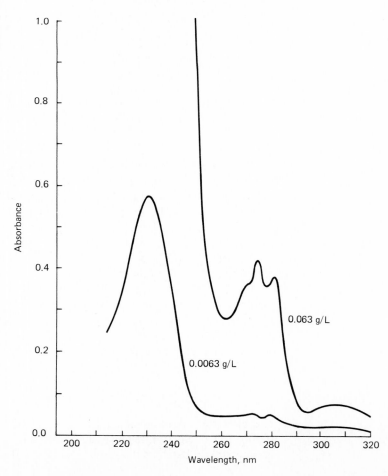

FIGURE 10.27
Ultraviolet Absorption Spectrum of Methyl Benzoate

TABLE 10.5
Ultraviolet Absorption Spectra

Structural unit	Wavelength at maximum absorption, nm	Typical ε
Isolated double bond	180–195	10,000
Isolated carbonyl group	270–290	25
Conjugated diene (*cis*)	near 240	5,000
Conjugated diene (*trans*)	near 220	16,000
Alkylbenzenes	260–280	200
	and near 210	8,000
Naphthalene	314	315
	275	5,625
	220	112,200

The ultraviolet spectrum of methyl benzoate measured in a 1-cm cell is reproduced in Figure 10.27. The value of ε at the absorption maximum at 230 nm is 12,000; the ε of the smaller peak at 275 nm is 850. Table 10.5 lists several structural units that have characteristic absorptions.

As Table 10.5 suggests, UV–visible absorption is primarily used to detect and identify conjugated π-electron networks. As a general rule, the longer the conjugated chain, the longer the wavelength of the absorption maximum. For example, an isolated double bond absorbs at ~ 190 nm, a diene at 220 nm, and the chain of 11 double bonds in β-carotene all the way out in the visible region at 484 nm (in hexane solvent).

β-Carotene
(orange)

Ultraviolet and Visible Spectral Sampling Techniques. Unlike infrared spectroscopy, in ultraviolet and visible spectroscopy it is customary to measure the extinction coefficients of the absorption peaks as well as their positions. Dilute solutions are required, typically 10^{-4} mole/L, which must be prepared by *quantitative* dilutions using appropriate combinations of transfer pipets and volumetric flasks. Because of the sensitivity of the measurement, considerable care must be exercised to avoid contamination. The most common solvents employed are water, ethanol, acetonitrile, and cyclohexane.

Questions

1. What molecular weights of the parent ion would be observed in a high-resolution mass spectrum for the following molecules?
   (a) C_5H_{12} (b) C_4H_8O (c) $C_3H_8N_2$

2. High-resolution mass-spectrometric analysis of compound A gave a molecular formula $C_9H_{10}O_2$. The infrared spectrum showed strong absorption at 1745 cm^{-1} as well as many other medium-intensity bands. The ultraviolet spectrum showed a maximum at 257 nm, $\varepsilon = 195$. The NMR spectrum consisted of three sharp peaks at $\delta = 7.22$ ppm (area 5), $\delta = 5.00$ ppm (area 2), and $\delta = 1.96$ ppm (area 3). What is the structure of compound A?

3. From a high-resolution mass spectrum of compound B, a molecular formula of $C_6H_{14}O$ could be assigned. The molecule did not absorb in the ultraviolet or visible regions. In the infrared, the strongest absorption above 1400 cm^{-1} occurred at 2900 cm^{-1}. In the NMR compound B showed a septet at $\delta = 3.62$ ppm (area 1), $J = 7$ Hz, and a doublet at $\delta = 1.10$ (area 6), $J = 7$ Hz. What is the structure of compound B?

III PREPARATIONS AND REACTIONS OF TYPICAL ORGANIC COMPOUNDS

11 General Remarks

This chapter describes how to calculate the percent yield of a reaction, how to prepare your notebook in advance of the laboratory, the information to be written in the notebook during the laboratory, and how to submit samples of products. The procedures described here follow closely the current practice at Cornell, and represent our view of how to maximize laboratory learning experience. However, since "there are many roads to Rome," your instructor may well modify some of the procedures.

The preparation of typical organic compounds affords a marvelous opportunity to compare your theoretical understanding of organic chemistry obtained from lectures and reading with reality. Those abstract symbols and formulas represent real substances—glistening solids and odoriferous liquids. Also keep in mind that although you are working with particular compounds, the preparations and reactions usually represent general methods that can be applied to entire classes of molecules.

11.1 Preparation Before the Laboratory

For experiments dealing with syntheses and reactions, the advanced preparation of your notebook involves several steps in addition to those listed in Section 1.6 for the exercises on separation and purification of organic compounds. Nevertheless, you should review that material now.

The most important addition is a table of physical constants of the substances to be manipulated (see step 5 below). By collecting this information and having it available you will be able to understand more readily the reasons for the particular procedure and will often be able to overcome independently any small difficulties that may arise in the course of the laboratory work. You should proceed in the following manner.

1. Read the descriptive pages concerning the laboratory operation to be carried out (these are found immediately preceding each experiment). In the notebook, write a title and general statement of the process to be studied.
2. Read the laboratory directions *for the entire procedure* and note particularly any cautions for handling materials.

    To aid in understanding the reasons for the procedure it is helpful to consult the textbook or lecture notes for a discussion of the particular class of compounds that is to be studied. Consideration should be given to important general principles, such as the law of mass action and the influence of solvents and catalysts on rate of reaction.
3. In your notebook, give a concise statement of the type of reaction that is to be carried out, such as "Conversion of an alcohol to an alkene" or "Oxidation of a secondary alcohol to a ketone." Write *balanced* equations, using condensed structural formulas, for the *main reaction* or sequence of reactions involved in converting the starting materials to the final products. Along the reaction arrow indicate the conditions used—temperature, solvent, catalyst (if any), and so forth.
4. Write balanced equations for *significant side reactions* that may divert an appreciable amount of the starting materials and lead to formation of by-products or accessory products that must be removed in the purification of the main product.

    Write balanced equations for any *test reactions* that are used to test for completion of the reaction, to detect the presence of an impurity, to confirm the identity of the product by conversion to a derivative, and so on.
5. Prepare a table of the physical constants of all organic and inorganic substances that enter into the main and side reactions and are produced in these reactions. The form shown in the Sample Notebook Page may be used. The physical constants of common organic and inorganic compounds may be found in the chemical handbooks.

    Include in the table the weight (in grams) of each reactant and the number of moles, or fraction of a mole, actually used. From these data and the balanced equation for the main reaction, determine which starting material is the limiting factor and calculate the theoretical yield (in grams) based on this reactant. This calculation is described in detail in Section 11.2.
6. It is helpful also to indicate schematically the successive steps involved in the *purification procedure*, starting with the substances

Experiment 13.3 (A) n-Butyl Bromide
Conversion of a Primary Alcohol to an Alkyl Bromide

Main Reactions:

$$CH_3CH_2CH_2CH_2-OH + HBr \xrightarrow[reflux]{H+} CH_3CH_2CH_2CH_2-Br + H_2O$$

$$NaBr + H_2SO_4 \longrightarrow HBr + NaHSO_4$$

Side Reactions:

No important organic side reactions

$$2NaBr + 3H_2SO_4 \text{ (concd)} \longrightarrow Br_2 + SO_2 + 2H_2O + 2NaHSO_4$$

Test Reactions: None

Purification:

$$\left.\begin{array}{l} C_4H_9-Br, C_4H_9-OH \\ NaHSO_4, H_2SO_4 \\ (Br_2), H_2O \end{array}\right\} \xrightarrow{\text{steam distillation}} \left.\begin{array}{l} C_4H_9-Br \\ C_4H_9-OH \\ (Br_2), H_2O \end{array}\right\} \xrightarrow{80\% H_2SO_4} \left.\begin{array}{l} C_4H_9-Br \\ (H_2O) \\ (H_2SO_4) \end{array}\right\}$$

$$\begin{array}{l}(1) NaHCO_3 \\ (2) H_2O\end{array} \longrightarrow \left.\begin{array}{l}C_4H_9-Br \\ (H_2O)\end{array}\right\} \xrightarrow{CaCl_2} C_4H_9-Br \text{ purified by distillation}$$

Physical Constants:

Substance	Mol. wt.	Grams used	Moles used	Sp. gr. (20°)	Mp	Bp	Solubility (g/100 mL)		
							Water	Ethanol	Ether
n-C_4H_9-OH	74	37	0.50	0.810	-80°	117°	9	∞	∞
H_2SO_4	98		1.25	1.83	11°	d 340°	∞	reacts	sol.
Na Br·2H_2O	139	87	0.62	—	51°	—	80 cold	slight	insol.
n-C_4H_9-Br	137	(68.5)	(0.50)	1.277	-112°	102°	insol.	∞	∞

Quantities Used:

$NaBr \cdot 2H_2O \quad \dfrac{87g}{139} = 0.62 \text{ mole}$

$H_2SO_4 \text{ (96\%)} \quad \dfrac{129g \times 0.96}{98} = 1.25 \text{ mole}$

n-C_4H_9-OH $\quad \dfrac{37g}{74} = 0.50$ mole = limiting factor

Theoretical Yield: based on n-butyl alcohol (0.5 mole) = $0.5 \times 137g$ = 68.5 g n-butyl bromide

19 March 1984: Reaction mixture refluxed 2 hr. organic layer had slight brown color. After distillation the organic layer in distillate was colorless. Product was washed and allowed to stand over $CaCl_2$ in corked flask.

21 March 1984: Product filtered into dry 100-mL distillation apparatus and distilled: atmospheric pressure = 745 mm
 Tare of receivers (without corks): A = 39.5 g. B = 42.0 g.
 Fractions collected: A, up to 99°; B, from 99 to 102°.
 Receiver A, 40.3 g; net weight (40.3 − 39.5) = 0.8 g.
 Receiver B, 90.5 g; net weight (90.5 − 42.0) = 48.5 g.

Percent Yield: $\dfrac{48.5}{68.5} \times 100\% = 70.8\%$ of theoretical

Other Methods of Preparation:

$$3 \text{ n-}C_4H_9-OH + PBr_3 \longrightarrow 3 \text{ n-}C_4H_9-Br + H_3PO_3$$

Sample Notebook Page

present in significant amounts in the reaction mixture at the completion of the reaction and showing what substances are removed at each step. The solubility relationships discussed in Sections 6.3 and 9.6 and the table of physical constants (see step 5) are useful in this connection.

7. After you have prepared your notebook in accordance with the foregoing instruction, your instructor may ask to see it for preliminary approval *before you start to perform the experiment*. In certain experiments, you will be asked to arrange the apparatus for the experiment and have it approved by the instructor.

8. In laboratory work it is essential to make efficient use of the time assigned; you are expected to plan your laboratory schedule and to make preliminary preparations before coming to the laboratory.

    Since many experiments require that the reactants be refluxed for several hours, you should plan to perform other laboratory work while the operation of refluxing is being carried out.

11.2 Calculation of Yields

The yield (sometimes called the actual yield) is the amount of the purified product actually obtained in the experiment. The theoretical yield (sometimes called the calculated yield) is the amount that could be obtained under theoretically ideal conditions; that is, the main reaction is assumed to proceed to completion without side reactions or mechanical losses, so that the starting materials are entirely converted into the desired product and no material is lost in isolation and purification.

The percentage yield is obtained by comparing the actual yield with the theoretical yield, in the following manner.

$$\text{Percent yield} = \frac{\text{actual yield}}{\text{theoretical yield}} \times 100\%$$

The percent yield is the measure of the overall efficiency of the preparation since many factors, such as incomplete reactions, side reactions, and mechanical losses, affect the actual yield.

If a preparation involves two reacting substances and the amounts actually used are not in the exact proportions demanded by the equation, it is necessary to determine by calculation which of the reactants is the limiting factor (commonly called the limiting reagent) for the calculation of the theoretical yield, as shown by the following example. In this connection the terms *mole* and *moles used* are commonly employed. A *mole* of a compound is equal to the molecular weight expressed in grams. The term *moles used* is employed to express the number of moles or the fraction of a mole of a particular compound actually used in an experiment. The number of moles is equal to the weight of the substance divided by the molecular weight.

Theoretical and Percentage Yields. Suppose that methyl ethyl ether, $CH_3-O-C_2H_5$, has been prepared by the action of methyl iodide upon sodium ethoxide (the Williamson synthesis of ethers). This preparation is carried out by reacting metallic sodium with ethanol and treating the resulting sodium ethoxide with methyl iodide.

$$C_2H_5-OH + Na \longrightarrow C_2H_5-ONa + \tfrac{1}{2}H_2 \qquad (11.1)$$

$$C_2H_5-ONa + CH_3-I \longrightarrow C_2H_5-O-CH_3 + NaI \qquad (11.2)$$

Equations (11.1) and (11.2) may be summarized as follows.

$$\underset{\text{1 mole}}{C_2H_5-OH} + \underset{\text{1 mole}}{Na} + \underset{\text{1 mole}}{CH_3-I} \longrightarrow \underset{\text{1 mole}}{C_2H_5-O-CH_3} + \underset{\text{1 mole}}{NaI} + \underset{\tfrac{1}{2}\text{ mole}}{\tfrac{1}{2}H_2}$$
$$(11.3)$$

From this it is evident that the proportions demanded by the equation are 1 mole of C_2H_5-OH : 1 mole of Na : 1 mole of CH_3-I, and the resulting products would be 1 mole of $C_2H_5-O-CH_3$: 1 mole of NaI : $\tfrac{1}{2}$ mole of H_2.

Suppose that the following quantities of the reagents are actually used in a laboratory preparation.

$$92.0 \text{ g of absolute ethanol} = \frac{92}{46} \text{ moles} = 2.0 \text{ moles } C_2H_5-OH$$

$$5.5 \text{ g of metallic sodium} = \frac{5.5}{23} \text{ mole} = 0.24 \text{ mole Na}$$

$$28.4 \text{ g of methyl iodide} = \frac{28.4}{142} \text{ mole} = 0.20 \text{ mole } CH_3-I$$

The amounts of the reagents are converted from grams into moles (by dividing by the molecular weights) to compare them with the molar proportions expressed in the equation. The *relative* proportions actually used are thus 10 of C_2H_5-OH : 1.2 of Na : 1.0 of CH_3-I. Obviously, ethanol and sodium are used in excess and methyl iodide is the limiting reactant that will determine the theoretical yield.

From equation (11.2) or (11.3), it can be seen that 1 mole of methyl iodide reacting with a sufficient quantity of sodium ethoxide will produce, under ideal conditions, exactly 1 mole of methyl ethyl ether. Consequently, the maximum amount of methyl ethyl ether that could be produced in the above preparation (the theoretical yield) is 0.2 mole. By multiplying this fraction of a mole by the weight of 1 mole of methyl ethyl ether (60 g), the theoretical yield is converted into grams.

$$\text{Theoretical yield} = 0.2 \times 60 \text{ g} = 12.0 \text{ g}$$

If the actual yield of methyl ethyl ether in the above preparation was 8.2 g, the percent yield would be

$$\text{Percent yield} = \frac{8.2}{12.0} \times 100 = 68.3\%$$

Problem. In another preparation of methyl ethyl ether, suppose that the amounts of the reagents actually used were 6.9 g of metallic sodium, 46 g of absolute ethanol, and 49.7 g of methyl iodide. Calculate the theoretical yield, in grams. (Answer = 18.0 g.)

Assuming that the actual yield in the previous preparation was 15.0 g, calculate the percent yield. (Answer = 83.3%.)

11.3 Laboratory Directions

The laboratory directions given in this manual are deliberately detailed. In advanced work and research one frequently must follow directions that assume a general knowledge of manipulative technique and of the chemistry involved. As an example, consider the preparation of *n*-butyl bromide described in detail in Section 13.3(A). Were this preparation to appear in a technical journal the description might be as follows.

A mixture of 60 mL of water, 130 g (1.3 mole) of concentrated H_2SO_4 (cool), 37 g (0.5 mole) of *n*-butyl alcohol, and 87 g (0.65 mole) of NaBr was refluxed for 2 hr, and then distilled until no more product was collected. The crude distillate was washed with water, cold 80% H_2SO_4, saturated $NaHCO_3$, and finally with water. After drying over $CaCl_2$, the product was distilled, yield 45 g, bp 99–103°.

The journal directions assume the worker realizes that the sodium bromide should be pulverized before addition and that good mixing is essential. Typically, experimental sections of professional journals do not specify the amounts of washing reagents or drying agents. In fact, many journal authors might have condensed the experimental description to read

n-Butyl alcohol (0.5 mole, 37 g) was converted to the bromide by heating with NaBr (0.65 mole) and 70% H_2SO_4 (190 g). The crude bromide was washed, dried, and redistilled; yield 45 g, bp 99–103°.

The reader would have to understand the chemistry well enough to isolate the bromide by distillation from the reaction mixture as well as to use the washes specified in the complete directions in the given order. The ability to fill in experimental detail is an important part of being a good laboratory worker and you are advised, as you carry out the experiments in this manual, to ask yourself at each stage why a certain operation or reagent is used.

11.4 In the Laboratory

Carry out the experiment *according to the laboratory directions*, and promptly record your observations, in ink, directly in your notebook. As you work, compare what you see happening with what you anticipated; record all discrepancies. Always observe and record the boiling point of a liquid preparation, and the melting point of a solid organic preparation. The only exception is with boiling points and melting points above 250°. Such high temperatures require special apparatus.

Record the actual yield and calculate the percent yield. Record any general observations and conclusions drawn from your performance of the experiment.

11.5 Samples and Reports

At the conclusion of the experiment you will be required to submit the substance prepared (along with samples of intermediates if a sequence of reactions was involved). All preparations should be placed in an appropriate bottle (wide-mouth bottles for solids and narrow-mouth bottles for liquids) of suitable size, with the experiment number and name of product, the melting or boiling point, as actually observed, your name, and the actual yield (in grams) and tare of the bottle included on the label, as follows.

```
Expt 16(B)     Cyclohexene
         bp 80–85°
        Mike Clements
Yield, 7 g      Tare, 21.5 g
```

Some form of report will also be required. At a minimum this could be your notebook[1] with a final report card summarizing the observed and physical properties of the substance and giving the percent yield. At the other extreme, you could be asked to write a formal report that gives all of these data and discusses the chemistry involved.

Your instructor will examine your notebook (or report) and the product and may ask questions designed to test your knowledge of the fundamental principles involved in the experiment and your ability to make and apply generalizations of the chemistry. Your grade will depend on the quality and quantity of your product, your laboratory technique, and your notebook and reports, as well as your understanding of the chemistry.

---

[1] At Cornell we have found it very convenient to use notebooks with tear-out carbon copies. These are submitted after each period, which simplifies notebook checking by the instructor and leaves the notebook in the possession of the students for preparation of the next period's work.

12 Free-Radical Halogenation

12.1 Mechanism of Free-Radical Chlorination

The preparation of chlorinated hydrocarbons is an extremely important industrial reaction because of their wide use as solvents. When a hydrocarbon is heated with chlorine gas to about 120° or the mixture is exposed to light at a lower temperature, the highly exothermic substitution of a chlorine atom for hydrogen occurs.

$$\text{RH} + \text{Cl}_2 \xrightarrow[\text{or light}]{\text{heat}} \text{R}-\text{C} + \text{HCl} \qquad \Delta H \approx -20 \text{ to } -25 \text{ kcal/mole}$$

The mechanism of this reaction has been studied extensively and shown to proceed by way of a free-radical chain process involving chlorine radicals, Cl·. Several steps are involved, which can be categorized according to whether they create or destroy free radicals (initiation or termination) or simply transfer the radical character from one atom to another (propagation).

$$\begin{array}{rl}
\textit{Initiation:} & \text{Cl}-\text{Cl} \xrightarrow[\text{or light}]{\text{heat}} 2\ \text{Cl}\cdot \\
\textit{Propagation:} & \text{Cl}\cdot + \text{RH} \longrightarrow \text{R}\cdot + \text{HCl} \\
& \text{R}\cdot + \text{Cl}-\text{Cl} \longrightarrow \text{R}-\text{Cl} + \text{Cl}\cdot \\
\textit{Termination:} & \text{Cl}\cdot + \text{Cl}\cdot \longrightarrow \text{Cl}-\text{Cl} \\
& \text{R}\cdot + \text{R}\cdot \longrightarrow \text{R}-\text{R} \\
& \text{R}\cdot + \text{Cl}\cdot \longrightarrow \text{R}-\text{Cl}
\end{array}$$

The reaction starts by the heat- or light-induced cleavage of chlorine to form chlorine free radicals. One of these can react with a hydrocarbon molecule, RH, to produce HCl and a different radical, R·.

The main product-forming step is the reaction of the alkyl radical, R·, with molecular chlorine. In this process a new chlorine radical is produced, which can react with another molecule of hydrocarbon to yield a new alkyl radical, which yields another product molecule and yet another chlorine radical. In principle, the initial formation of a single chlorine atom could lead to the conversion of all of the hydrocarbon and chlorine gas present to chlorocarbon, by an unending cycle of these two propagation steps. In practice this does not happen because one or more of the termination steps intervenes. The recombination of two chlorine radicals, for example, breaks the cyclic chain of reactions in which those atoms are involved. If the chain process is to stay alive, the initiation steps must keep up with the termination steps.

In discussing free-radical chain reactions, it is convenient to talk about the "chain length," which is the average number of cycles of the propagation steps that occur for each initiation step before the cycle is terminated. The chain length clearly depends on the rate constants of the several steps involved and thus will vary with the reaction conditions. Under favorable conditions the chain length can be 100 to 10,000. If free-radical inhibitors are present (species that combine quickly with free radicals) the chain length may approach unity or, with particularly effective free-radical scavengers, even zero.

12.2 Chlorination by Means of Sulfuryl Chloride and AIBN

In the teaching laboratory it is inconvenient to carry out chlorination with chlorine gas because of the elaborate safety procedures required. A more convenient alternative is to use sulfuryl chloride, SO_2Cl_2, which can propagate the chain in a similar fashion.

$$R\cdot + Cl-\underset{\underset{O}{\|}}{\overset{\overset{O}{\|}}{S}}-Cl \longrightarrow R-Cl + SO_2Cl\cdot$$

$$SO_2Cl\cdot \longrightarrow SO_2 \text{ (gas)} + Cl\cdot$$

Sulfuryl chloride is both colorless and more stable than chlorine; therefore, neither heat nor light is suitable for initiating the reaction. A widely used industrial substitute is azobisisobutyronitrile (AIBN), which undergoes radical fragmentation when exposed to ultraviolet light (~340 nm) or heated to about 80° or above.[1] The isobutyronitrile radical fragments then attack the sulfuryl chloride to generate SO_2 and chlorine radicals.

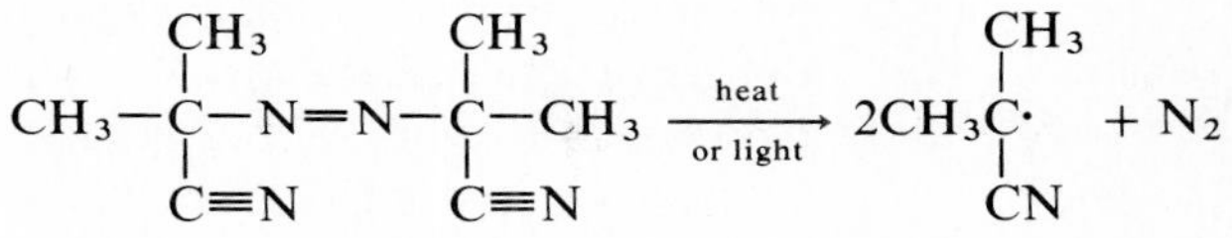

$$CH_3-\underset{\underset{C\equiv N}{|}}{\overset{\overset{CH_3}{|}}{C}}-N=N-\underset{\underset{C\equiv N}{|}}{\overset{\overset{CH_3}{|}}{C}}-CH_3 \xrightarrow[\text{or light}]{\text{heat}} 2CH_3\underset{\underset{CN}{|}}{\overset{\overset{CH_3}{|}}{C}}\cdot \;+\; N_2$$

AIBN

12.3 Energetics of Halogenation

Thus far the discussion has centered on chlorination. It is interesting to compare the energies of reaction for the four common halogens given in Table 12.1. These data show that fluorination is so highly exothermic that even carbon–carbon bonds could be cleaved by the explosive release of energy. At best this makes fluorinations difficult to control and leads to complex mixtures of carbon skeletons. Fluorocarbons are usually made by indirect methods.

Iodination, by contrast, is endothermic, and thus the favored reaction is reduction by *de*iodination.

$$CH_3I + HI \longrightarrow CH_4 + I_2$$

This leaves chlorination and bromination as the two accessible halogenations for general laboratory work.

---

[1] Benzoyl peroxide is also frequently used to initiate free-radical reactions, but great care must be exercised because it can decompose explosively when heated or with initiation of exothermic reactions. AIBN is considered to be much safer, but even with this chemical there have been two reports in the literature of explosions on contact with acetone. The cause of these explosions is unknown.

TABLE 12.1
Enthalpies of Methane Halogenations
$(CH_4 + X_2 \rightarrow CH_3X + HX)$

X	ΔH, kcal/mole
F	−102.8
Cl	−24.7
Br	−7.3
I	+12.7

12.4 Selectivity in Halogenations

If each hydrogen atom in a hydrocarbon reacted with a chlorine free radical at the same rate, the proportion of the different halogenated isomers would follow the number of hydrogen atoms of each kind. For example, in 2,3-dimethylbutane there are 12 primary hydrogens and only two tertiary hydrogens. If each hydrogen atom reacted at the same rate, the ratio of primary to tertiary chlorinated products would be $12/2 = 6$.

In fact, the tertiary hydrogens react typically about three times as fast (the exact factor depends on the experimental conditions). It follows that the ratio of total rates of attack on the 12 primary and the 2 tertiary hydrogens is

$$\frac{12 \times 1}{2 \times 3} = 2.0$$

Thus the fraction of primary product is $12/(12 + 6) = 0.67$, a value much less than might have been anticipated from the sixfold preponderance of primary hydrogens.

The main reason for the greater reactivity (selectivity) of the tertiary hydrogens lies with the greater stability of the tertiary free radical that is formed in the rate-determining step. However, there is an additional feature to consider. When 2,3-dimethylbutane is brominated the only product formed in significant amount is the tertiary bromide, corresponding to a tertiary/primary selectivity in excess of 100. Since the same tertiary hydrocarbon radical is being formed, the different results for chlorination and bromination must lie in the relative positions of the two transition states. Free-radical cleavage of a C—H bond by Cl· is a highly exothermic process, and in the transition state relatively little free-radical character on carbon is developed; although the 3° radical is approximately 7 kcal/mole more stable than the 1° radical, only about 1 kcal/mole of the difference shows up in the transition states (Figure 12.1).

In bromination the same 7 kcal/mole difference exists but, because of the weaker HBr bond strength, the reaction profile shown in Figure 12.1 is distorted by the elevation of the two free radicals on the right in relation to

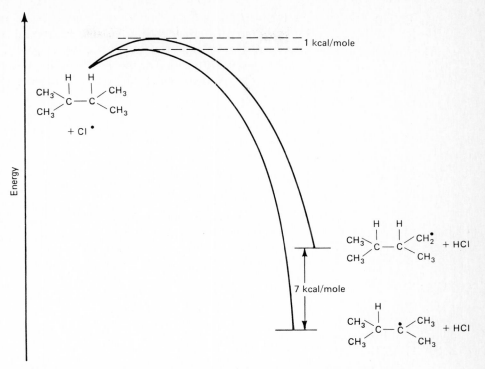

FIGURE 12.1 *Reaction Profiles for the Reaction of Cl· with 2,3-Dimethylbutane*

the starting materials on the left. As a consequence, the transition states lie much closer to the right-hand side of the diagram and the difference between them more fully expresses the 7-kcal/mole difference in energy of the 1° and 3° radicals. Bromination is therefore much more selective.

12.5 Substituent Effects

In the previous section we have discussed how the different stability of 1°, 2°, and 3° radicals leads to different rates of attack on hydrogen atoms attached to these positions. We have also seen that the selectivity (ratio of rates) depends on whether the transition states lie close to the free radicals being formed (high selectivity as in bromination) or close to the starting hydrocarbon (low selectivity as in chlorination). These differences depend on the differences in exothermicity of the abstraction step.

Another factor to be considered in interpreting the rates of free-radical reactions is the effect of substituents on the stability of transition states.

Substituents have little effect on the stability of the starting hydrocarbon or the free radicals formed from it because these species have little polarity. By contrast, the transition state, with a polar partial H—X bond, is quite sensitive to substituents. The magnitude of the substituent effect depends on the polarity of the partial H—X bond, and this depends on the nature of X·.

$$\underset{\text{Starting material}}{\underset{\text{Sub}-\text{CH}_2}{\overset{\diagdown}{\text{C}}}-\text{H} + \text{X}\cdot} \longrightarrow \underset{\text{Transition state}}{\left[\underset{\text{Sub}-\underset{\text{polar}}{\underbrace{\text{CH}_2}}}{\overset{\diagdown}{\text{C}}\cdots\overset{\delta+}{\text{H}}\cdots\overset{\delta-}{\text{X}}}\right]^{\ddagger}} \longrightarrow \underset{\text{Free radical}}{\underset{\text{Sub}\diagup\text{CH}_2}{\overset{\diagdown}{\text{C}}\cdot} + \text{HX}}$$

For a chlorine substituent attached to the carbon next to the C—H bond being attached, the rate of H-atom abstraction is typically diminished by a factor of 2 or 3. When the chlorine substituent is attached directly to the carbon bearing the hydrogen atom, the rate is diminished by a factor of 5 or more.

12.6 Preparations and Reactions

(A) Photochemical Chlorination of 2,3-Dimethylbutane[2]

In a 100-mL Erlenmeyer flask place 5 mL (3.4 g, 0.04 mole) of 2,4-dimethylpentane, 1 mL (1.7 g, 0.012 mole) of sulfuryl chloride, SO_2Cl_2, and about 20 mg (0.1 mmole) of azoisobutyronitrile.[3]

▶ *CAUTION* Sulfuryl chloride is extremely corrosive; if any is spilled or spattered on the skin it should be removed immediately by washing thoroughly with water. Even the vapors of sulfuryl chloride are dangerous. All transfers of the material should be carried out in the hood.

Irradiate the flask for 30 min using a long-wavelength ultraviolet lamp.[4] At the end of the irradiation period, cautiously pour the reaction mixture into a separatory funnel containing 10 mL of methylene chloride and 25 mL of water. Separate the layers and wash the organic layer with 5% sodium

---

[2] This experiment is designed to measure the relative rates of attack on 1° and 3° hydrogen atoms for which it is desirable to chlorinate only a small fraction of the hydrocarbon.
[3] Avoid contact of AIBN with acetone or acetone vapors. See footnote 1.
[4] The azoisobutyronitrile has an ultraviolet absorption maximum near 340 nm. Sufficient light is emitted by a "grow-lamp" used for plants if the flask is held close to it. Germicidal ultraviolet lamps should *not* be used without a Pyrex filter unless special precautions are taken to protect the eyes from the shorter wavelength radiation.

bicarbonate. Wrap a wad of cotton around the tip of a Pasteur pipet and transfer the methylene chloride solution to a dry container for analysis. If the solution is cloudy, it can be dried with magnesium sulfate.

Analyze the mixture by gas–liquid chromatography using a 5-ft 5% DC-200 (silicon oil) column at 50–60°. The approximate relative retention times of the major components are 2,3-dimethylbutane (bp 58°; time = 1.0) and methylene chloride (bp 40°; time = 1.0); 1-chloro-2,3-dimethylbutane (bp 124°; time = 3.5); 2-chloro-2,3-dimethylbutane (bp 112°; time = 4.8).

From the areas of the peaks calculate the relative amounts of the 1° and 3° chloro derivatives (assume that the detector response is the same for each isomer). Calculate the relative reactivities of the 1° and 3° hydrogens (take the 1° hydrogen to be 1.00), taking into account the different numbers of each kind present in the starting material.

(B) Substituent Effects in Free-Radical Chlorination[5]

In a 100-mL round-bottom flask, place 15 mL (13.3 g, 0.14 mole) of 1-chlorobutane, 5 mL (8.4 g, 0.06 mole) of sulfuryl chloride (SO_2Cl_2), and 0.1 g (0.6 mmole) of azoisobutyronitrile.[3]

▶ *CAUTION*   Sulfuryl chloride is extremely corrosive; if any is spilled or spattered on the skin, it should be removed immediately by washing thoroughly with water. Even the vapors of sulfuryl chloride are dangerous. All transfers of the material should be carried out in the hood.

Attach a water-cooled condenser and an efficient gas trap to remove the HCl and SO_2 gases that are evolved (see Section 8.4). Heat the mixture under gentle reflux on a steam bath for 20 min. Remove the steam bath and after the reaction mixture has cooled somewhat, add another 0.1 g of azoisobutyronitrile.[3] Resume heating for an additional 10 min to complete the reaction.

Cool the reaction mixture to room temperature, pour it into 50 mL of water contained in a separatory funnel, and separate the layers. Wash the organic layer[6] with 5% sodium bicarbonate and then dry it over calcium chloride.

Analyze the mixture by gas–liquid chromatography using a 5-ft 5% carbowax 400 column at about 55–60°. The approximate relative retention times of the various components are 1-chlorobutane (bp 77–78°; time = 1.00); 1,1-dichlorobutane (bp 114–115°; time = 3.0); 1,2-dichlorobutane (bp 121–123°; time = 4.7); 1,3-dichlorobutane (bp 131–133°; time = 6.3); 1,4-dichlorobutane (bp 161–163°; time = 14.7).

---

[5] This experiment is patterned after that described by Reeves, *J. Chem. Ed.*, **48**, 636 (1971).
[6] The densities of the organic layer and the aqueous washes are similar. The identity of the aqueous layer should be verified at each stage by testing a small portion for water solubility.

From the areas of the peaks calculate the relative amounts of each of the dichlorobutanes (assume that the detector response is the same for each isomer). Calculate the relative reactivities of each hydrogen in 1-chlorobutane (take the 4-hydrogen reactivity to be 1.00) and compare these to the relative hydrogen reactivities of *n*-butane (primary = 1.0; secondary = 3.6 at 80°).

Questions

1. It is observed that hydroquinone (*p*-dihydroxybenzene) and catechol (*o*-dihydroxybenzene) inhibit free-radical reactions. Explain.
2. A widely used free-radical initiator is benzoyl peroxide (in spite of its extreme explosive hazard), which decomposes vigorously around 80° to give benzoic acid, benzene, and carbon dioxide in varying amounts depending on the solvent. Explain the origin of these products and the catalytic effect of the peroxide when used to initiate halogenation. What halogenated hydrocarbon will contaminate the product?
3. In the gas phase the relative reactivities of bromine radicals for hydrogen atom abstraction is tertiary : primary = 20,000 : 1. If 2,3-dimethylbutane were exposed to bromine vapor and light, what amount of primary bromide would be expected to form?
4. Explain in qualitative terms why free-radical bromination is so much more selective than chlorination.

13 Conversion of Alcohols to Alkyl Halides

13.1 Preparation of Alkyl Halides

Alkyl halides are prepared by many different methods and by the use of a variety of reagents. Direct halogenation of alkanes to form alkyl chlorides and bromides often leads to mixtures of isomers that are difficult to separate with ordinary laboratory fractionating equipment. Likewise, addition of halogen acids to alkenes frequently produces mixtures. The most useful laboratory methods involve the conversion of alcohols to alkyl halides. Fortunately, a large variety of alcohols are available commercially.

Alcohols may be converted to alkyl halides by means of the halogen acids (HCl, HBr, and HI), phosphorous halides (PCl_3, PBr_3, or $P + I_2$), or thionyl chloride ($SOCl_2$).

There are two different mechanisms for nucleophilic substitution, designated as S_N1 (unimolecular) and S_N2 (bimolecular). The S_N1 process involves conversion of the alcohol, via an oxonium-type intermediate (I), to the corresponding carbocation (II), which reacts rapidly with the nucleophile (Cl^-, Br^-, etc.).

$$(CH_3)_3C-OH \xrightleftharpoons{H^+} (CH_3)_3C-\overset{+}{O}H_2 \xrightleftharpoons{-H_2O} CH_3-\overset{\overset{\displaystyle CH_3}{|}}{\underset{\underset{\displaystyle CH_3}{|}}{C^+}}$$

$$\qquad\qquad\qquad\qquad\qquad\quad\text{I}\qquad\qquad\qquad\quad\text{II}$$

The rate depends only on the oxonium ion concentration and is independent of the halide ion concentration (S_N1). This behavior is typical for tertiary alcohols; a side reaction can occur involving elimination of H^+ from the carbocation to form an alkene (E1 elimination).[1]

$$(CH_3)_3C-Cl \underset{(S_N1)}{\xleftarrow{Cl^-}} CH_3-\underset{CH_3}{\overset{CH_3}{\underset{|}{\overset{|}{C^+}}}}-\xrightarrow[(E1)]{-H^+} CH_3-\underset{CH_3}{\overset{}{\underset{|}{C}}}=CH_2$$

II

In the S_N2 mechanism the protonated alcohol molecule is approached by the nucleophile from a position directly behind the carbon bearing the hydroxyl group. In terms of the tetrahedral disposition of the groups around the central carbon this is called a *backside* attack. In the intermediate transition state (III) the substituents have a planar distribution.

$$\bar{N}: \leadsto \overset{}{\underset{}{C}}-\overset{+}{O}H_2 \rightleftharpoons \left[N\cdots\overset{}{\underset{}{C}}\cdots\overset{+}{O}H_2\right]^{\ddagger} \rightleftharpoons N-\overset{}{\underset{}{C}} + OH_2$$

III

At a fixed acidity, the rate is second order (S_N2), and depends on the concentration of the attacking nucleophile *and* the alcohol. This mode is typical for primary alcohols. Here also a side reaction leading to an alkene can occur.

The ease of reaction of alcohols toward halogen acids follows the sequence tertiary > secondary > primary, and the tendency to form alkenes, or to undergo rearrangement, follows the same order. Reactivity of the halogen acids toward alcohols declines in the order HI > HBr > HCl. Thus, conditions suitable for the conversion of a particular alcohol to a halide are influenced by the structure of the alcohol *and* the specific halogen acid to be used.

Tertiary alcohols react readily with any of the three halogen acids, even at 25° and in the absence of catalysts. Secondary alcohols react more slowly; moderate heating and acidic catalysts (50% sulfuric acid, zinc chloride) are used to promote the conversion to bromides and chlorides. Primary alcohols require more vigorous conditions and more active catalysts (65–70% sulfuric acid for conversion to bromides). The Lucas test for differentiating the classes of alcohols (Chapter 9) is based on relative rates of conversion to alkyl chlorides.

Primary and secondary alkyl bromides containing two to five carbon atoms are prepared by heating the alcohol with concentrated hydrobromic acid or a hydrobromic–sulfuric acid mixture. A convenient procedure is to use sodium bromide and an excess of strong sulfuric acid, but this method is not suitable for higher bromides because the high concentration of salts

---

[1] See Chapters 16 and 17 which deal with formation of alkenes from alcohols.

present greatly reduces the solubility of the alcohol in the reaction medium. For secondary alcohols, a lower concentration of sulfuric acid (50%) is used; stronger acid is unnecessary and promotes a side reaction (alkene formation). For higher molecular weight alcohols the action of anhydrous hydrogen bromide at 100–120° in the absence of a solvent, or use of phosphorus tribromide, is a satisfactory procedure. Water-soluble tertiary alcohols are converted rapidly to bromides merely by shaking with concentrated aqueous hydrobromic acid.

Hydrochloric acid containing dissolved zinc chloride (a Lewis acid) is used for the conversion of primary and secondary alcohols to alkyl chlorides. Primary alcohols require heating with a saturated solution of zinc chloride in concentrated hydrochloric acid. Thionyl chloride is a useful reagent for the preparation of primary alkyl chlorides.

Alkyl iodides are obtained readily from alcohols by means of strong aqueous hydriodic acid, but a more economical method is to treat the alcohol with iodine and red phosphorus.

$$6\ R-CH_2OH + 3\ I_2 + 2\ P \longrightarrow 6\ R-CH_2I + 2\ H_3PO_3$$

Alkyl iodides may be prepared also by reaction of alkyl chlorides or bromides with sodium iodide in acetone solution (Finkelstein reaction; see Chapter 14).

13.2 Reactions of Alkyl Halides

Alkyl halides are relatively reactive molecules and are widely used in laboratory and industrial syntheses. Their reactivity and the mechanism of their reactions vary over a wide range. In some features, they show a resemblance to the corresponding alcohols.

Alkyl bromides are particularly useful in laboratory operations. They undergo substitution (replacement) reactions with nucleophilic reagents (alkoxides, phenoxides, cyanide anion, and ammonia) to produce ethers, nitriles, and amines. They alkylate sodium derivatives of malonic and acetoacetic esters, and react with certain metals to form organometallic compounds (such as Grignard reagents).

Reactions of the primary alkyl halides generally follow an S_N2 mechanism and the relative reactivity of the halides follows the sequence methyl ≫ primary > secondary. In many typical reactions the rate for the methyl halide may be 10 to 20 times as fast as the ethyl analog, which will be about 2 times that of the higher primary homologs of straight-chain structures. For primary halides bearing a single group in the β or γ position (isobutyl, isopentyl) the rate is about half that of the straight-chain isomer.

Greatly enhanced reactivity occurs in the allyl (R—CH=CH—CH$_2$—X) and benzyl-type (aryl—CH$_2$—X) halides.

Ionic substitution reactions of tertiary halides follow an S$_N$1 mechanism: the rate-determining step is an ionic cleavage of the carbon–halogen bond to form the carbocation. Secondary halides may react by an S$_N$1 or an S$_N$2 mechanism or a hybrid of both. The S$_N$2 process is favored by powerful nucleophilic reagents (see Section 14.3), high concentration of the reagent, and a weakly polar solvent. The S$_N$1 mechanism is favored by weakly nucleophilic reagents and low concentrations, and especially by reaction media of high solvating power. Molecular rearrangements often occur in S$_N$1 replacement reactions and in E1 elimination reactions leading to alkenes but are infrequent in S$_N$2 reactions and E2 (bimolecular) elimination reactions.

Many solvents have been used for S$_N$2 reactions. Methanol and ethanol are readily available and inexpensive, and water may be added to increase the solubility of inorganic salts if necessary. Glacial or aqueous acetic acid is also useful. Acetone is a *polar aprotic solvent*: one that has a high dielectric constant but lacks hydroxyl groups. Usually replacement reactions are very much faster in this kind of medium. Acetone, dimethyl sulfoxide, (CH$_3$)$_2$SO (DMSO), dimethylformamide, (CH$_3$)$_2$N—CHO (DMF), and acetonitrile, CH$_3$—CN are good examples of effective aprotic solvents.

There is a suspicion that many alkyl halides, particularly those that are effective alkylating agents, can cause cancer on prolonged exposure.

13.3 Preparations

(A) *n*-Butyl Bromide

In a 250-mL round-bottom flask place 30 mL of water and add in small portions, with cooling, 35 mL (65 g, 0.65 mole) of concentrated sulfuric acid. To the cold diluted acid add 25 mL (18.5 g, 0.25 mole) of *n*-butyl alcohol,[2] while mixing thoroughly and cooling. While shaking the flask to prevent formation of lumps, add 31 g (0.3 mole) of finely pulverized sodium bromide crystals (NaBr)[3] in small portions. Good mixing is essential.

---

[2] *n*-Propyl bromide can be made by the same procedure, using an equivalent amount of *n*-propyl alcohol (18.5 mL, 15 g) instead of butyl alcohol. For the preparation of *n*-pentyl bromide, using 27 mL (22 g) of *n*-pentyl alcohol, the amount of concentrated sulfuric acid should be increased to 38 mL (70 g). This method is unsuitable for the preparation of secondary bromides.

[3] Instead of sodium bromide and sulfuric acid, a hydrobromic-sulfuric acid solution may be used. This is prepared by adding 17 mL (30 g) of concentrated sulfuric acid, with cooling, to 37 mL (55 g, 0.33 mole) of 48% hydrobromic acid in a 500-mL flask. *n*-Butyl alcohol, 0.5 mole (37 g, 40 mL), is added and the reaction carried out as described previously.

Introduce a few boiling chips into the flask, attach an upright condenser, and connect the top of the condenser to a gas absorption trap to absorb hydrogen bromide that evolves during the reaction (see traps shown in Figure 8.2). Heat the flask *gently* while swirling the contents constantly, until most of the sodium bromide has dissolved. Adjust the heating so that the mixture boils *gently*, and swirl the contents occasionally, for 30 min.

Allow the flask to cool slightly and disconnect the condenser. Arrange the apparatus for distillation, add 15 mL of water and a few boiling chips, and distill the mixture until a test portion of the distillate contains little or no water-insoluble material (20–30 min).

Transfer the distillate to a separatory funnel, add 20–25 mL of water, and shake well. Draw off *carefully* the butyl bromide layer[4] into another separatory funnel and discard the aqueous layer. To remove unchanged butyl alcohol, wash the crude alkyl bromide by shaking it thoroughly with about 13 mL of *ice-cold* 80% sulfuric acid (prepared by adding 10 mL of concentrated acid to 2.5 mL of water). Separate the sulfuric acid carefully and completely (Which layer?).[5] Wash the butyl bromide layer with 15 mL of saturated sodium bicarbonate solution (**Caution**—*foaming!*); shake the mixture *gently* first before inserting the stopper. Insert the stopper, shake more vigorously, invert the separatory funnel, and release the internal pressure by opening the stopcock. Finally, wash the product with water and draw it off carefully.

Collect the butyl bromide in a small dry Erlenmeyer flask, add 3–4 g of anhydrous calcium chloride, cork the flask firmly, and shake vigorously. It is advantageous to allow the material to dry overnight or longer. Through a funnel, decant (or filter) the dried product into a distilling flask of proper size, add a few small boiling chips, and distill. Collect in a weighed bottle the portion boiling at 99–103°. If an appreciable low-boiling fraction is obtained, dry this portion with 1–2 g of anhydrous calcium chloride and redistill. The yield is 22–24 g.

The NMR spectrum of neat *n*-butyl bromide is shown in Figure 13.1.

(B) *sec*-Butyl Chloride

In a 250-mL round-bottom flask, place 45 g (0.33 mole) of anhydrous zinc chloride and 28 mL (32 g, 0.33 mole) of concentrated hydrochloric acid that has been cooled to 5°. Shake the mixture and keep it cool until the zinc chloride has dissolved, then add 15 mL (12 g, 0.16 mole) of *sec*-butyl

---

[4] Careful separations are necessary. The aqueous layer removes sodium sulfate or sodium carbonate, which would produce troublesome precipitates when calcium chloride is added as a drying agent.

[5] A general procedure to determine which layer is the organic material is to separate the layers carefully and add a few drops of each layer separately to 5 mL of water in a test tube. The organic layer will be insoluble in the water, but the usual washing liquids (sulfuric acid, sodium bicarbonate solution, etc.) will be soluble in water.

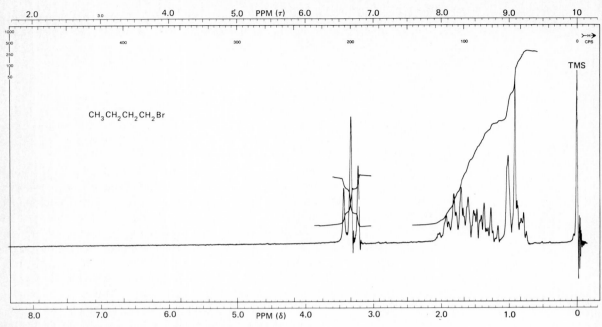

FIGURE 13.1 *NMR Spectrum of Neat* n-*Butyl Bromide*

alcohol. Attach a reflux condenser, add two small boiling chips, and boil the mixture briskly for 1 hr. Cool the reaction mixture and transfer to a separatory funnel. Draw off and discard the lower, aqueous layer.

Transfer the crude butyl chloride to a 250-mL round-bottom flask and add an equal volume of concentrated sulfuric acid. Attach a reflux condenser, heat until the butyl chloide boils *gently*, and continue the boiling for 15 min. This treatment eliminates impurities that would not be removed by ordinary distillation. Cool the flask slightly, remove the condenser, set it downward for distillation, and connect it to the flask. Distill off the *sec*-butyl chloride (avoiding excessive heating of the residual sulfuric acid) and transfer the distillate to a separatory funnel. Wash the product with water, decant into a small dry flask, and shake it with 2–3 g of anhydrous calcium chloride. Decant the dried liquid through a funnel into a small dry distilling flask, add a few small boiling chips, and redistill. Collect the fraction boiling at 67–69° in a weighed receiver. The yield is 9–12 g. If an appreciable low-boiling fraction is obtained, dry this again and redistill it.

(C) *t*-Butyl Chloride

In a 125-mL separatory funnel, place 70 mL (82 g, 0.8 mole) of concentrated hydrochloric acid that has been cooled to 5°. Add 24 mL (18.5 g, 0.25 mole) of *t*-butyl alcohol and shake the mixture occasionally during

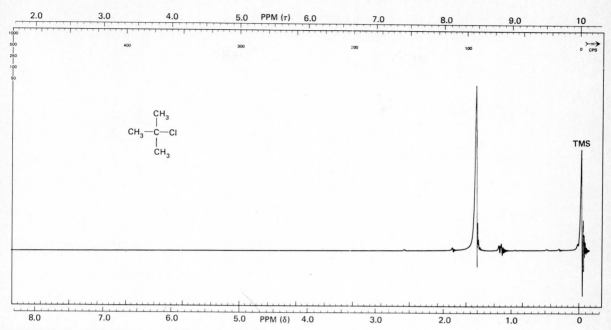

FIGURE 13.2 *NMR Spectrum of Neat* t-*Butyl Chloride*

20 min. From time to time, relieve internal pressure by inverting the funnel and slowly opening the stopcock.

Allow the mixture to stand undisturbed until the layers have separated sharply, then separate and discard the spent hydrochloric acid (which layer?). Wash the product with 10–15 mL of water, then with 5% sodium bicarbonate solution (**Caution**—*foaming*!) and again with water. Transfer the chloride to a small dry flask and dry it with 5–6 g of anhydrous calcium chloride. Decant the dried liquid through a funnel into a dry distilling flask, add a few small boiling chips, and distill. Collect the fraction boiling at 49–52° in a weighed bottle. The yield is 15–18 g. If an appreciable low-boiling fraction is obtained, dry this again with calcium chloride, and redistill it.

The NMR spectrum of neat *t*-butyl chloride is shown in Figure 13.2.

Questions

1. Why is strong sulfuric acid more effective than water in removing a small amount of an alcohol from an alkyl halide?
2. What alkyl halide is formed by ionic addition of hydrobromic acid to 1-butene? Would the same bromide be formed in the presence of peroxides (free-radical catalysts)?
3. How may *n*-butyl alcohol be converted to each halide?
   (a) *n*-butyl chloride

(b) *n*-butyl iodide
(c) *n*-butyl bromide (without using hydrobromic acid).

4. Show how *n*-butyl bromide could be converted to
   (a) *n*-octane
   (b) *n*-butylmagnesium bromide
   (c) *n*-butyl methyl ether
   (d) *n*-butane

5. What experimental conditions would be suitable for conversion of *sec*-butyl bromide to α-methylbutyronitrile by reaction with sodium cyanide (S_N2 mechanism)? Explain.

6. Could *n*- and *sec*-butyl alcohols be converted to the chlorides merely by shaking with concentrated hydrochloric acid, the way *t*-butyl is? (see Lucas' test, Chapter 8).

7. What is the relative reactivity of the four isomeric butyl chlorides in (a) S_N2 and (b) S_N1 reactions?

8. In what type of reactions of isobutyl chloride might you expect to observe molecular rearrangements? Explain.

9. In washing the *t*-butyl chloride product (Expt. 13(C)) with 5% bicarbonate solution, the solution never stops bubbling completely. Explain.

14 Second-Order Nucleophilic Substitution

14.1 Replacement Reactions

In acetone solution, alkyl bromides react with sodium iodide to produce the alkyl iodide and a precipitate of sodium bromide. This particular reaction, known as the Finkelstein reaction,[1] is but one example of an extremely broad class of organic reactions. Some are named after their discoverers (the Williamson ether synthesis, the Menshutkin reaction, etc.), but today most are known simply by their mechanistic class name—second-order nucleophilic substitutions, or S_N2 for short.

Finkelstein reaction:

$$Na^+I^- + R-Br \xrightarrow{acetone} I-R + NaBr\downarrow$$

Williamson ether synthesis:

$$CH_3O^- + C_4H_9-I \longrightarrow CH_3O-C_4H_9 + I^-$$

Menshutkin reaction:

$$\begin{array}{c} CH_3 \\ \diagdown \\ N: \\ \diagup | \\ CH_3 \ CH_3 \end{array} + CH_3OSO_3H \longrightarrow \begin{array}{c} CH_3 \\ \diagdown \\ N^+-CH_3 \ SO_4H^- \\ \diagup | \\ CH_3 \ CH_3 \end{array}$$

[1] Finkelstein, *Ber*, **43**, 1528 (1910).

14.2 Stereochemistry and Kinetics

In a classic series of papers, Hughes and Ingold[2] worked out the details of nucleophilic displacement reactions. They showed that the entering nucleophile attacks the carbon at the rear of the bond to the leaving group.

$$I^- + \underset{\text{Nucleophile}}{\underset{CH_3}{\overset{C_2H_5}{H^{\prime\prime\prime\prime}}}C-Br} \longrightarrow \underset{\text{Substrate}}{} \left[\underset{\text{Transition state}}{\overset{CH_3}{\underset{H\ \ \ CH_3}{I^{\delta-}\text{---}C\text{---}Br^{\delta-}}}} \right]^{\ddagger} \longrightarrow I-\underset{CH_3}{\overset{C_2H_5}{C^{\prime\prime\prime}H}} + Br^-$$

The configuration of the groups attached to the central carbon is inverted and, if the starting material is chiral, there is a corresponding change in optical activity. Inversion is a characteristic feature of S_N2 processes that distinguishes them from S_N1 reactions.

Hughes and Ingold also showed that the rate of the reaction was proportional to the concentration of both the nucleophile and the substrate (the species being attached).

14.3 Nucleophilicity

In acid–base chemistry *basicity* is a measure of the affinity of an electron pair of a base for a *proton*. In the S_N2 reaction the nucleophile shares an electron pair with the carbon atom being attacked. By analogy, *nucleophilicity* is a measure of the affinity of a base for a *carbon* atom. Just as there are strong bases (e.g., CH_3O^-) and weak bases (NO_3^-), there are good nucleophiles (I^-) and poor ones (CH_3OH). As long as one compares only first-row elements, there is a rough correlation between nucleophilicity and basicity. However, in protic solvents second- and third-row elements are considerably more nucleophilic than predicted by such a correlation. Thus iodide, although much less basic than fluoride, undergoes S_N2 reactions almost 10^5 times faster. The lesser reactivity of first-row elements appears to be related to their greater solvation in protic solvents.

14.4 Substrate Structure

In S_N2 reactions there is considerable increase in steric congestion of the backside of the molecule as the nucleophile attacks to form the pentacoordinated transition state. The larger the groups attached to the carbon

---

[2] Ingold, *Structure and Mechanism in Organic Chemistry*, 2nd ed. (Ithaca, NY and London: Cornell University Press, 1969).

TABLE 14.1
Effect of Branching at the α-Carbon on S_N2 Reaction Rates

Alkyl group	Relative S_N2 rate
CH_3-X	30
CH_3CH_2-X	1
$(CH_3)_2CH-X$	0.03
$(CH_3)_3C-X$	Immeasurably slow

being attacked, the slower the rate of reaction. The effect of branching at the α-carbon is illustrated in Table 14.1.

14.5 Solvent

The choice of solvent can have a profound effect on the rate of S_N2 reactions. The fundamental principle is that polar solvents stabilize ions and, to a lesser degree, dipoles. Dispersed charges are stabilized less than concentrated charges. For example, in the Finkelstein reaction of iodide ion with *n*-butyl bromide the charge on the iodide ion is concentrated. However, in the transition state the charge is dispersed over both the iodide and the bromide ions. Thus in passing from the starting materials to the transition state there is a net loss of solvation energy, which means that the reaction rate will increase in lower polarity solvents. In practice one must not use a solvent of too low polarity (such as a hydrocarbon) or so little of the sodium iodide will dissolve that there will be no reaction. In the Finkelstein reaction, acetone is used as the solvent. Among polar solvents it lies at the lower end of the polarity scale. Acetone dissolves sodium iodide but not sodium bromide, which precipitates as the reaction proceeds.

14.6 Preparation of *n*-Butyl Iodide

In a 250-mL round-bottom flask place 100 mL of acetone, add 15 g (0.10 mole) of sodium iodide and 10 mL (13 g, 0.09 mole) of *n*-butyl bromide. Add two boiling chips, attach a reflux condenser, and boil the mixture for 30 min.

Cool the mixture to room temperature with an ice bath and pour it into a large separatory funnel containing 100 mL of water. Drain off the lower layer of *n*-butyl iodide, wash it with water, and dry it over 1–2 g of anhydrous calcium chloride.

▻ *CAUTION* *n*-Butyl iodide, like most good alkylating agents, is suspected of causing cancer on prolonged exposure. This reaction should be carried out in a hood.

Transfer the dried *n*-butyl iodide to a dry distillation apparatus and distill, collecting the fraction 129–131°. The yield is about 13 g.

Questions

1. Arrange the compounds of each set in order of reactivity toward an S_N2 displacement reaction.
   (a) 2-chloro-2-methylbutane, 1-chloropropane, 1-chloro-2,2-dimethylpropane, 1-chloro-2-methylpropane
   (b) 2-bromopropane, 2-chloropropane, 2-iodopropane
2. Compare S_N1 and S_N2 reactivities of alkyl halides with regard to
   (a) stereochemical outcomes
   (b) an increase in the nucleophile concentration
   (c) an increase in the alkyl halide concentration
   (b) an increase in the polarity of the solvent
   (e) occurrence of rearrangements
3. Arrange the following nucleophiles in order of increasing reactivity toward ethyl iodide in ethanol as a solvent: $^{131}I^-$ (radioactive iodide); OH^-; H_2O; $CH_3CO_2^-$.

15 Chemical Kinetics: Solvolysis of *t*-Butyl Chloride

15.1 First-Order Kinetics

The measurement of reaction rates (*reaction kinetics*) at different concentrations of reactants provides important clues to the reaction mechanism. For example, the rate of substitution of hydroxide for chloride in the S_N1 reaction of *t*-butyl chloride with potassium hydroxide in aqueous ethanol depends on the concentration of the halide but is essentially independent of the hydroxide concentration. The molecular interpretation is that the slow, rate-determining step involves the ionization of the halide followed by the fast collapse of the intermediate with either water or hydroxide ion to give *t*-butanol as the product. Collapse with an ethanol molecule leads to *t*-butyl ethyl ether. Such reactions are said to follow first-order kinetics (the "1" of S_N1).

$$CH_3-\underset{\underset{CH_3}{|}}{\overset{\overset{CH_3}{|}}{C}}-Cl \xrightarrow{\text{slow}} \left[CH_3-\overset{CH_3}{\underset{CH_3}{C^+}} \right] \xrightarrow{\text{fast}} CH_3-\underset{\underset{CH_3}{|}}{\overset{\overset{CH_3}{|}}{C}}-OH$$

The S_N1 reaction may be contrasted with S_N2 reactions, which involve simultaneous formation of the new carbon–nucleophile bond as the chloride ion departs and thus follow second-order kinetics.

First-order kinetics can be expressed mathematically as

$$\text{Loss of halide per unit time} = \frac{-d[\text{RX}]}{dt} = k[\text{RX}] \qquad (15.1)$$

where $[RX]$ is the concentration of the halide and k is a proportionality constant known as the *rate constant*. A more practical expression can be obtained by integrating equation (15.1) to obtain

$$\ln \frac{[RX]_0}{[RX]_t} = kt \tag{15.2}$$

where $[RX]_0$ is the concentration of RX at the start of the reaction, and $[RX]_t$ is the concentration at some later time, t. Note particularly that ln is the *natural* logarithm (base e). After some mathematical manipulation, equation (15.2) can be rewritten as

$$\log_{10} [RX]_t = -\frac{kt}{2.303} + \log_{10} [RX]_0 \tag{15.3}$$

where now the more familiar base 10 logarithm is used. The form of equation (15.3) is of a straight line

$$y = mx + b$$

If the reaction rate is first-order, a plot of each $\log_{10} [RX]_t$ value (calculated from measurement of RX at various times) against the time of measurement will give a straight line of slope $-k/2.303$.

In practice any means of measuring the halide concentration will do and might be by spectroscopy, chromatography, or classical analytical chemistry. Because of the properties of logarithms, the actual concentration of RX is not required and any quantity proportional to it is acceptable. Thus if gas chromatography is used, the heights of the RX peaks at different times can be substituted in equation (15.3) and the same value of k will result. Under some circumstances it is more convenient to measure the concentration of a reaction product than a reactant. From the stoichiometry the two can be related and the integrated rate equation expressed in terms of the product concentrations.

$$RX + H_2O \longrightarrow ROH + HX$$

$$[HX]_t = [RX]_0 - [RX]_t$$

At very large times (approximated as t equal to infinity, ∞), $[HX]_\infty = [RX]_0$, since there is no RX remaining. On this basis equation (15.3) becomes

$$\log_{10}([HX]_\infty - [HX]_t) = -\frac{kt}{2.303} + \log_{10} [HX]_\infty \tag{15.4}$$

In equation (15.4) it is permissible to use any quantity proportional to the HX concentration.

The units of k are time^{-1}. A quantity frequently used to characterize a first-order rate process is the "half-life," the time required for half of the reactant to react. From equation (15.2), this is seen to be

$$\text{Half-life} = t_{1/2} = \frac{\ln 2}{k} = \frac{0.69}{k}$$

Conversely, if the time for half of the reaction to occur can be estimated, the value of k can be estimated without making a graph.

It is interesting to note that $t_{1/2}$ does not depend on the concentration of any reactant. If a reaction follows first-order kinetics, as do S_N1 reactions, the time required for completion does not change if solvent is removed or added. This lack of concentration dependence is *not* true of reactions following other kinetic orders.

15.2 Laboratory Practice

A successful kinetics experiment requires both an understanding of the operations to be performed and *advance* preparation of many solutions and reagents, as well as collection of essential apparatus. Kinetics experiments once started must be completed without interruption. To avoid mistakes and "lost points" a table should be prepared in which the laboratory data can be recorded as the experiment progresses.

In the following experiment the S_N1 reaction of t-butyl chloride with a 50 : 50 mixture of water and 2-propanol will be examined. In this *solvolysis* of the chloride it is convenient to follow the reaction progress by titrating small samples of the reaction mixture to determine the amount of hydrochloric acid developed. Since it takes some time to perform a titration, it is necessary to stop (*quench*) the reaction before the sample is titrated. S_N1 reaction rates are very sensitive to the amount of water present in the solvent, going slower the smaller the amount of water. A convenient technique for quenching the t-butyl chloride solvolysis is to add the aliquot to an equal volume of acetone.

Reaction rates are also quite dependent on temperature, typically doubling for each rise of 10°. In precise kinetic work the reacting solution is placed in a constant temperature bath that holds the temperature fluctuation to $\pm 0.05°$. In this experiment the solution is kept on the bench top, and if it is not in a drafty location, it will probably hold its temperature to $\pm 1°$ during the experiment, which corresponds to about $\pm 10\%$ variation in the rate constant. If greater temperature control is desired the reaction flask can be placed in a large beaker of water.

The time at which each sample is quenched must be recorded, so that the times from the start of the reaction can be calculated for use in Equation (15.4). It is desirable to titrate several samples from each of the first two half-lives. There is no point in measuring time intervals with greater precision than is obtained with the titrations. In the following example the half-life is about 50 min; so that recording the times to the nearest half-minute (about 1%) is adequate.

If the course of the reaction is being followed by measuring one of the products, equation (15.4) is to be used. This requires knowing the product concentration at "infinite" time. Since the starting material diminishes by 50% for each half-life, the "infinity" value can be approximated with adequate precision by titration of a sample after 8–10 half-lives, which corresponds to 99.5–99.9% reaction. Another strategy for obtaining the required "infinity" is to withdraw a sample and accelerate the rate of reaction either by raising the temperature or changing the solvent. With $S_N 1$ reactions, which are very sensitive to the polarity of the solvent, a common technique is to place a sample in an equal volume of water. This causes about a 10-fold increase in rate and permits the "infinity" to be measured after one half-life.

15.3 Measurement of the $S_N 1$ Reaction Rate of t-Butyl Chloride

In a cork-stoppered 250-mL Erlenmeyer flask prepare 100 mL of a 1 : 1 (by volume) solution of 2-propanol and water; after mixing it well, allow it to equilibrate to the laboratory temperature while the next step is carried out.[1]

Obtain 150 mL of approximately 0.04 N standardized aqueous sodium hydroxide in a 250-mL Erlenmeyer flask fitted with a cork or rubber stopper. Record the concentration of the base in your notebook. Set up a 25- or 50-mL buret, and fill it with base. Obtain a dropping bottle of bromothymol blue solution and a white background card to enhance the visibility of the green end point. Secure a stopwatch or clock that will record up to 2 hr of elapsed time with a resolution of at least 0.5 min.

In a 100-mL volumetric flask weigh *accurately* a sample (about 1 g) of t-butyl chloride and add the water–2-propanol solvent to the mark. Stopper the flask, mix the contents well, and note the time and laboratory temperature. The flask should remain stoppered at all times except when a sample is being withdrawn.

The first sample should be taken about 10 min after the reaction is started. Subsequent samples should be withdrawn at about 20, 35, 50, 75,

---

[1] Another suitable solvent is 1 : 1 (by volume) acetone and water. The samples should be withdrawn at 10-min intervals.

and 100 min after the start of the reaction. For each sample a 10-mL aliquot is drawn into a 10-mL pipet using a pipet bulb (*not by mouth!*) and transferred to a 125-mL Erlenmeyer flask containing about 10–15 mL of ordinary acetone as a reaction quench. Record the time of addition in the notebook. Three drops of bromothymol blue solution are added and the solution titrated to a green end point that persists for about 20 s. Before the next sample is withdrawn, the pipet must be cleaned by rinsing it with a *little* acetone and dried by drawing air through it by connecting it to a rubber hose leading to the water pump.

To determine the "infinity" titer, a 10-mL sample is added to 10 mL of water. The additional water causes about a 10-fold increase in reaction rate so that the sample is ready to titrate after about 1 hr. If sufficient sample is available it is desirable to have duplicate or triplicate "infinity" values so that the error can be reduced by taking an average value.

Prepare a plot of $\log_{10}$(titer at infinity − titer at time t) versus the time of sampling. Draw the best straight line through the points and from the slope calculate the value of the rate constant. Compare the measured infinity value with the quantity calculated from the concentrations of the chloride, the base, and the volume of the pipet.

If a calculator or a computer with a program for fitting a straight-line function is available, evaluate the slope, k, and compare the result with the slope obtained from the graph.

Questions

1. Show how equation (15.2) can be transformed into equation (15.3).
2. Derive equation (15.4).
3. In the solvolysis of *t*-butyl chloride some di-*t*-butyl ether is formed. Does this affect the application of equation (15.4)?
4. Explain why addition of water to the solvent in the solvolysis of *t*-butyl chloride causes an increase in reaction rate.

16 Alkenes

16.1 Sources of Alkenes

Alkenes are often obtained from alcohols or alkyl halides by elimination reactions. Alcohols undergo elimination of water by heating with sulfuric or phosphoric acid[1] or by passing the alcohol vapor over alumina or silica catalysts at high temperatures. Alkyl halides undergo loss of halogen acid (dehydrohalogenation) by heating with a solution of potassium hydroxide in ethanol.

Alkenes are produced industrially in enormous quantities by the pyrolysis (*cracking*) of alkanes at 400–600°, by passage over metal oxide catalysts. The large alkane molecules undergo rupture of carbon–carbon bonds to form a mixture of smaller alkanes and alkenes. Catalytic dehydrogenation of alkanes is also an important industrial method for producing alkenes. Alkanes may also be converted to aromatic hydrocarbons (arenes); thus, *n*-heptane is changed stepwise to methylcyclohexane and finally to toluene, and *n*-hexane furnishes benzene.

$$C_{12}H_{26} \xrightarrow{500°} C_6H_{12} + C_6H_{14}, \quad C_5H_{10} + C_7H_{16}, \quad \text{etc.}$$

The ease of dehydration of alcohols follows the sequence: tertiary > secondary ≫ primary. *t*-Butyl alcohol is converted rapidly to isobutylene (2-methylpropene) by 40–50% sulfuric acid at 85°; *sec*-butyl alcohol requires 60–65% acid at 100° for alkene formation and *n*-butyl

---

[1] Other dehydrating agents (iodine, oxalic acid, potassium bisulfate) are illustrated in Chapter 17.

alcohol 75–80% acid at 135–140°. Phosphoric acid causes less oxidative degradation than strong sulfuric acid, but the rate of reaction is slower and higher temperatures are required. In the present experiment, the dehydration of secondary alcohols by 65% sulfuric acid at 95–110° is illustrated.

16.2 Carbocation Rearrangements

The mechanism of dehydration of alcohols varies in detail with the reagent and the structural type of the alcohol. For primary alcohols the mechanism is complex: at the high temperatures required for alkene formation with strong acids (170° for ethanol), the alcohol is in equilibrium with the corresponding dialkyl ether. Attack of the protonated alcohol or ether (oxonium ions) by the acid anion can occur by way of a one-step concerted process (E2 mechanism) in which the proton and water molecule are lost simultaneously. Primary carbocations are relatively unstable, and their possible intervention in alkene formation remains controversial.

$$\begin{array}{c} H \\ \overset{+}{O}H \\ \diagdown C-C\diagdown \\ H \end{array} \xrightarrow{-OSO_3H} \begin{array}{c} H \\ OH \\ \diagdown C=C\diagdown \\ H \quad OSO_3H \end{array} \longrightarrow \diagdown C=C\diagdown + H_2SO_4 + H_2O$$

With tertiary and most secondary alcohols, protonation occurs more readily and the intermediate oxonium structure leads to a carbocation that loses a proton to form the alkene (E1 mechanism).

$$(CH_3)_3C-OH \xrightarrow{H^+} (CH_3)_3C-\overset{+}{O}H_2 \longrightarrow (CH_3)_3\overset{+}{C} \longrightarrow (CH_3)_2C=CH_2$$

If two different alkenes can be formed, there is usually a selectivity in the mode of elimination and one of the isomers predominates. For example, 2-pentanol yields mainly 2-pentene and very little 1-pentene. Likewise 4-methyl-2-pentanol in the present experiment furnishes only about 5% of 4-methyl-1-pentene.

$$CH_3CH_2CH_2-\underset{\underset{\displaystyle OH}{|}}{CH}-CH_3 \longrightarrow CH_3CH_2CH_2-\overset{+}{CH}-CH_3 \longrightarrow CH_3CH_2CH=CH-CH_3$$

Sometimes the double bond is formed at a position removed from the carbon atom that was bonded to the hydroxyl group and occasionally the carbon skeleton itself is altered during the reaction. The formation of 2-pentene from 1-pentanol may occur through rapid protonation of 1-pentene as it is formed, under the reaction conditions, to give the more stable secondary carbocation. Alternatively, the secondary carbocation might be formed directly from the protonated alcohol, as the water molecule departs, by rapid migration of hydrogen *with its bonding electrons* (hydride shift).

Similar migration of a methyl group with its bonding electrons (methide shift) is observed in the dehydration of neopentyl alcohol, $(CH_3)_3C-CH_2-OH$. Shift of the methyl group converts the primary neopentyl carbocation into the more stable tertiary pentyl carbocation, which leads to the formation of 2-methyl-2-butene.

$$(CH_3)_3C-\overset{+}{C}H_2 \longrightarrow (CH_3)_2\overset{+}{C}-CH_2CH_3 \longrightarrow (CH_3)_2C=CH-CH_3$$

Addition of a carbocation to an alkene furnishes a new carbocation and leads stepwise to alkene polymers of increasing molecular weight (dimers, trimers, and higher polymers). Usually small amounts of such dimers and trimers are formed in laboratory preparations of alkenes from alcohols.

16.3 Dimerization of Isobutylene (2-Methylpropene)

In the presence of 60–65% sulfuric acid under mild conditions, *t*-butyl alcohol undergoes dehydration to isobutylene, which is converted mainly to octenes rather than higher polymers (see Section 16.5(C)).

$$CH_3-\underset{CH_3}{\overset{CH_3}{\underset{|}{C}}}-\overset{+}{O}H_2 \longrightarrow CH_3-\underset{CH_3}{\overset{CH_3}{\underset{|}{\overset{|}{C}}^+}} \xrightarrow{CH_2=C(CH_3)_2} CH_3-\underset{CH_3}{\overset{CH_3}{\underset{|}{C}}}-CH_2-\overset{+}{\underset{CH_3}{\underset{|}{C}}}-CH_3 \longrightarrow$$

$$CH_3-\underset{CH_3}{\overset{CH_3}{\underset{|}{C}}}-CH_2-\underset{CH_3}{\overset{|}{C}}=CH_2 \;+\; CH_3-\underset{CH_3}{\overset{CH_3}{\underset{|}{C}}}-CH=\underset{CH_3}{\overset{|}{C}}-CH_3$$

2,4,4-Trimethyl-1-pentene (~80%)   2,4,4-Trimethyl-2-pentene (~20%)

Industrially, isobutylene is obtained by thermal *cracking* of petroleum distillates. The mixture of octenes can be hydrogenated to give 2,2,4-trimethylpentane (known industrially as isooctane), which is used to establish the octane number of motor fuels. On this scale *n*-heptane, a poor fuel, is assigned a *zero* rating, and isooctane, a very good fuel, is assigned a 100 rating. The octane number of a particular fuel is the percentage of isooctane in a mixture of heptane and isooctane that duplicates the knocking characteristics, in a standard internal combustion engine, of the fuel being examined.

Tetraethyllead, $Pb(C_2H_5)_4$, has been used to improve the octane number of motor fuels, but is considered to be environmentally objectionable. A compound considered more suitable is methylcyclopentadienylmanganese tricarbonyl.[2]

***Cis-trans* Isomerism.** Alkenes in which two different groups are attached to *each* of the carbon atoms of the double bond, as in 2-butene or 2-pentene, are capable of existing in two stereoisomeric forms arising from differences in the spatial distribution of the groups (*cis* and *trans* configurations). In these molecules the bonding of the atoms immediatlely surrounding the double bond is planar.

$$\underset{cis}{\underset{H}{\overset{CH_3}{>}}C=C\underset{H}{\overset{C_2H_5}{<}}} \qquad \underset{trans}{\underset{H}{\overset{CH_3}{>}}C=C\underset{C_2H_5}{\overset{H}{<}}}$$

The groups are held in relatively fixed positions: an energy barrier of 35–40 kcal restricts rotation of the groups about the double bond. The *cis* and *trans* stereoisomers exhibit discernible differences in physical properties and also in their rates of reaction with a given reagent.

16.4 Reactions of Alkenes

Alkenes are used in the laboratory and in industry as starting materials or intermediates for the synthesis of many important compounds. These transformations generally involve the characteristic addition reactions of the carbon–carbon double bond (ionic mechanism), with electrophilic reagents such as sulfuric acid or halogen acids, hypochlorous acid, chlorine or bromine, and oxidizing agents.

With unsymmetrical alkenes such as propylene and isobutylene, the mode of addition is highly selective: the electron-deficient fragment of the reagent adds to the carbon atom bearing the larger number of hydrogen atoms (the Markovnikov rule).

$$CH_3-CH=CH_2 + HBr \longrightarrow CH_3-CHBr-CH_3$$

This orientation of addition is attributed to electron release by the alkyl group(s), which increases electron density at the less highly substituted carbon atom. An exception to the rule is the addition of hydrogen bromide

---

[2] For examples of metal derivatives of cyclopentadiene see Section 43.2(B).

to terminal alkenes in the presence of peroxides. Under these conditions (free-radical mechanism), propylene gives 1-bromopropane, since the reaction involves attack of the CH_2 group by an electrophilic bromine *atom*.

Substitution reactions of alkenes occur under special conditions and at high temperatures. Propylene and isobutylene react with chlorine at 500° to give allyl chloride, $CH_2=CH-CH_2Cl$, and methallyl chloride, respectively. These reactive alkenyl chlorides are useful synthetic intermediates.

Alkenes are oxidized readily by palladium or platinum salts to carbonyl compounds. Conversion of ethylene to acetaldehyde by this means recently has become an important industrial process; the palladium is recovered and is converted easily to palladous chloride for reuse. Propylene gives acetone, and 1- and 2-butenes give methyl ethyl ketone.

$$CH_2=CH_2 \xrightarrow[H_2O]{PdCl_2} H^+[C_2H_4 \cdot PdCl_2OH]^- \longrightarrow CH_3CH=O + Pd^0 + 2\ HCl$$

An interesting and useful reaction is the addition of diborane, $(BH_3)_2$, to an alkene in ether solution (hydroboration),[3] to form a trialkylborane. The electron-deficient boron atom adds almost entirely to the carbon atom having the larger number of hydrogens and the hydride anion adds to the adjacent atom. Since the alkylboranes are oxidized readily to alcohols by alkaline hydrogen peroxide, this sequence effects hydration of the alkene.

$$6\ CH_2=CH-R + (BH_3)_2 \xrightarrow{ether} 2\ B(CH_2CH_2-R)_3 \xrightarrow[NaOH]{H_2O_2} 6\ HO-CH_2CH_2-R$$

The result is counter to the Markovnikov mode of addition of water to alkenes in the presence of acids. By the hydroboration route isobutylene can be converted to the primary alcohol, isobutyl alcohol.

16.5 Preparations

(A) Methylpentenes

Arrange a distillation assembly similar to that shown in Figure 2.10 using a 100-mL round-bottom flask and fractionating column. Fit the lower end of the condenser with an adapter that protrudes into a 50-mL receiver cooled in an ice-water bath.

Place 20 mL of water in a 125-mL Erlenmeyer flask and add *carefully*, while swirling 21.5 mL (39.5 g, 0.4 mole) of concentrated sulfuric acid. Cool the diluted acid (65% H_2SO_4) to 20–25° and add slowly 19 mL (15.3 g,

---

[3] Zweifel and Brown, *Organic Reactions*, **13**, 1 (1963); also Brown and Tierney, *J. Am. Chem. Soc.*, **80**, 1552 (1958); Brown, *Boranes in Organic Chemistry* (Ithaca, NY: Cornell University Press, 1972).

0.15 mole) of 4-methyl-2-pentanol, with good mixing. Transfer the solution to the reaction flask through a funnel and add two small boiling chips.

Heat the reaction mixture on a steam bath until distillation of the volatile product ceases (do not allow the distillation temperature to exceed 90°). Transfer the distillate to a separatory funnel, shake well with 5–6 mL of 10% sodium hydroxide solution and allow the layers to settle. Draw off the lower aqueous layer, wash the hydrocarbon layer with 5–6 mL of water, and finally pour the hydrocarbon through the mouth of the funnel (Why?) into a dry 25-mL Erlenmeyer flask. Add 2 g of anhydrous calcium chloride and allow the flask to stand, occasionally swirling it, for 15–20 min; longer standing does no harm.

▶ *CAUTION* The methylpentenes are volatile and flammable. Take care to avoid fire hazards and to minimize loss by evaporation.

Carefully decant the dried product into a distilling flask of the proper size, add a few small boiling chips, and distill from a water bath. Collect in a dry weighed bottle the portion boiling at 54–68°. The yield is 9–10 g. The product consists of three structural isomers: 4-methyl-2-pentene, 4-methyl-1-pentene (bp 53.9°), and 2-methyl-2-pentene (bp 67.3°). The first of these exists in two stereoisometic forms (*cis*-, bp 56.4°; *trans*-, bp 58.6°). 2-Methyl-2-pentene results from molecular rearrangement (hydride shift) within the intermediate secondary carbonium ion to form a more stable tertiary carbonium ion before the transformation into an alkene occurs. Elaborate fractional distillation techniques are required to separate the mixture into the individual components.

The composition of the product can be ascertained by vapor phase chromatography (see Section 7.4).

Tests for Unsaturation.

Bromine test. Add a little of the hydrocarbon *dropwise* to 3 mL of a 2% solution of bromine in carbon tetrachloride. Note the amount required to discharge the bromine color. Do a similar test with cyclohexane.

Permanganate test (Baeyer's test). To a solution of 2 drops of the hydrocarbon in 2 mL of pure acetone add dropwise, while shaking, a 0.5% aqueous solution of potassium permanganate. Note the number of drops added before the permanganate color appears. Compare the result with that from cyclohexane.

(B) Cyclohexene

Arrange a distilling assembly as described in Section 16.5(A), with fractionating column and 50-mL cooled receiver. Prepare the cooled 65% sulfuric acid[4] as directed there and to it add slowly 16 mL (15 g, 0.15 mole) of cyclohexanol (practical or technical grade) instead of the methylpentanol.

Add two boiling chips and heat the reaction flask *gently* so that cyclohexene and water distill through the column. Continue the distillation until about 20 mL of liquid remains in the distilling flask.

CAUTION Cyclohexene is a volatile flammable liquid. Take care to minimize fire hazards and loss by evaporation.

To the distillate add 2–3 mL of 10% sodium carbonate solution to neutralize traces of acid (test with litmus paper) and swirl to mix the two layers. Transfer the liquids to a separatory funnel, allow the layers to separate, and draw off the lower, aqueous layer. Wash the cyclohexane layer with a little water, draw this off, and pour the hydrocarbon layer *through the mouth of the funnel* (Why?) into a small dry flask. Add 2 g of anhydrous calcium chloride and allow the material to stand for 10–20 min with occasional swirling.

Decant the dried liquid into a small distilling flask, add 2 or 3 boiling chips, and distill carefully. Collect the material boiling at 80–85° in a weighed flask or bottle. If there is an appreciable low-boiling fraction, dry this again, and redistill it. The yield is 6–8 g.

Carry out the tests for unsaturation given in 16.5(A).

(C) 2,4,4-Trimethyl-1- and -2-pentenes (Diisobutylenes)

In a 250-mL round-bottom flask, place 20 mL water and add cautiously, with swirling, 20 mL (37 g, 0.35 mole) of concentrated sulfuric acid. Cool the diluted acid to about 50° and add slowly 19 mL (15 g, 0.2 mole) of *t*-butyl alcohol. Attach at once a reflux condenser and boil the material gently for 30 min.

Cool the reaction mixture to room temperature, transfer it to a separatory funnel and carefully draw off the aqueous acid layer. Wash the hydrocarbon layer with water to remove traces of acid and dry it with 1–2 g of anhydrous calcium chloride. Decant the dry liquid into a small flask arranged for distillation and distill, collecting the fraction boiling at 100–108°. The yield is 8–9 g (12 mL). The recorded boiling points of the octenes are 2,4,4-trimethyl-1-pentene, 101–102°; 2,4,4-trimethyl-2-pentene, 104°.

Carry out the tests for unsaturation given in 16.5(A). Vapor-phase chromatography affords an excellent method for examination of the mixed octenes. For details of this method see Chapter 7.

---

[4] In place of sulfuric acid, 6 mL of 85% phosphoric acid may be used, but the reaction is slower. There is less discoloration.

Questions

1. Write structural formulas and systematic names for
   (a) all of isomeric pentenes (C_5H_{10})
   (b) all of the isomeric methylcyclohexenes
2. What alkene will be the main product when each of the following alcohols is dehydrated (cf. carbocation rearrangements)?
   (a) 3-methyl-1-butanol
   (b) 3-methyl-2-butanol
   (c) 3,3-dimethyl-2-butanol
3. What isomeric hexenes (propylene dimers) can be formed by addition of the 2-propyl carbocation to propylene and subsequent loss of a proton? How could the structure of these hexenes be established (by chemical methods)?
4. What product(s) would you expect to be formed from 2,4,4-trimethyl-1-pentene by each of the following?
   (a) oxidation with aqueous palladous chloride
   (b) conversion to an alcohol by the hydroboration route

17 A Multiple-Step Synthesis

17.1 From n-Butyl Alcohol to 2-Methylhexenes

The sequence of reactions undertaken in this experiment involves four steps and illustrates a typical situation in synthetic organic chemistry. Chemical transformations and isolation procedures always take place with some loss of material, and this reduces, sometimes quite drastically, the yield of product.

$$C_4H_9-OH + HBr \longrightarrow C_4H_9-Br \tag{17.1}$$

$$C_4H_9-Br + Mg \longrightarrow C_4H_9-MgBr \tag{17.2}$$

$$CH_4H_9MgBr + CH_3-\overset{O}{\overset{\|}{C}}CH_3 \longrightarrow C_4H_9-\underset{OH}{\overset{|}{C}(CH_3)_2} \tag{17.3}$$

$$C_4H_9-\underset{OH}{\overset{|}{C}(CH_3)_2} \xrightarrow{acid} C_7H_{14} \tag{17.4}$$

(alkene mixture)

With four steps and an average yield of 80% 1 mole of starting material will furnish 0.4 mold of finished product. This figure falls rapidly as more steps are involved: with the same average yield of 80% six steps will give 0.26 mole and eight steps, 0.17 mole. Lower average yields lead to more drastic losses; an average yield of 70% after eight steps, gives only 0.057 mole of the end product. This emphasizes the need for good yields of the intermediate products.

In the present sequence, the first step is the conversion of *n*-butyl alcohol to *n*-butyl bromide, as described in Section 13.3(A). The following step, formation of an organomagnesium halide (Grignard reagent), requires strictly anhydrous conditions and necessitates careful attention to drying the alkyl halide, the ether used as solvent, and the reactant ketone (acetone). Conversion of the resulting tertiary alcohol, 2-methyl-2-hexanol (I; R = C_3H_7), to the 2-methylhexenes affords an example of the partitioning of a tertiary carbocation between isomeric alkenes during an E1-type elimination of water.[1]

For the final dehydration step, one of several different reagents may be selected to afford an opportunity to see how a specific reagent can affect the distribution of isomers in the resulting alkenes.

$$R-CH_2-\underset{\underset{CH_3}{|}}{\overset{\overset{CH_3}{|}}{C}}-OH \underset{}{\overset{H^+}{\rightleftharpoons}} R-CH_2-\underset{\underset{CH_3}{|}}{\overset{\overset{CH_3}{|}}{\overset{+}{C}}}-OH_2 \underset{}{\overset{-H_2O}{\rightleftharpoons}} R-CH_2-\overset{+}{C}\underset{CH_3}{\overset{CH_3}{<}}$$

$$\quad\quad\quad\quad\text{I} \quad\quad\quad\quad\quad\quad\quad\quad\quad \text{II} \quad\quad\quad\quad\quad\quad\quad\quad \text{III}$$

The carbocation intermediate (III) can lose one of the six hydrogens from the adjacent methyl groups to give the alkene IV (2-methyl-1-hexene) or lose a hydrogen from the adjacent CH_2-group to give the alkene V (2-methyl-2-hexene)

$$\text{III} \xrightarrow[\text{from CH}_3]{-H^+} R-CH_2-C\underset{CH_3}{\overset{\overset{CH_2}{\|}}{<}} \quad\text{or}\quad \xrightarrow[\text{from CH}_2]{-H^+} \underset{H}{\overset{R}{>}}C=C\underset{CH_3}{\overset{CH_3}{<}}$$

$$\quad\quad\quad\quad\quad\quad\quad\quad \text{IV} \quad\quad\quad\quad\quad\quad\quad\quad\quad\quad\quad \text{V}$$

If there were no selectivity one would expect IV and V to be formed in the ratio 3 of IV to 1 of V. The more highly substituted alkene, V, is the more stable product and at equilibrium would be present in the ratio of about 1 of IV to 10 of V. It will be of interest to compare the results with different dehydrating agents on the distribution of isomers IV and V and the yields. If a *free* carbocation is involved in the process, all of the catalysts should give the same result. But if the catalyst is intimately associated with the elimination step, then it may influence the relative amounts of IV and V.

---

[1] See also Section 16.3.

17.2 Grignard Synthesis of an Alcohol[2]

Organomagnesium halides, Grignard reagents, are among the most versatile synthetic intermediates for laboratory work. They are formed by simple, direct reaction of magnesium metal with alkyl or aryl halides (usually bromides) in the presence of a solvent such as ether or tetrahydrofuran. There is some uncertainty about the structure and mechanism of reaction of Grignard reagents. They seem to exist as coordination compounds in a complex equilibrium involving R_2Mg, $MgBr_2$, and $RMgBr$. For convenience, they are simply designated as $RMgBr$.

Unless the reactants, solvent, and apparatus are dried carefully and the magnesium is pure and relatively free of oxide coating, the reaction does not start readily. Addition of a small crystal of iodine aids in inducing reaction, probably by exposing a small fresh surface of the metal. A few drops of 1,2-dibromoethane also may be used to start the reaction. Alkyl and aryl bromides are the preferred halides in most cases. With chlorides (except relatively reactive ones) the reaction is more difficult to start; with iodides there is greater tendency to favor a side reaction—coupling at the metal surface to form the hydrocarbon R—R (the Wurtz reaction).

The most important uses of the Grignard reagent involve two types of reaction, in both of which the alkyl or aryl group of R—MgX is transferred *with its bonding electrons* to a carbon atom of the reactant.

1. Addition to the carbonyl function of an aldehyde, ketone, ester, amide, acid halide, or carbon dioxide (or the cyano group of a nitrile).[3]

$$(CH_3)_2C=O \xrightarrow{RMgX} (CH_3)_2\underset{R}{C}-OMgX \xrightarrow[H^+]{H_2O} (CH_3)_2\underset{R}{C}-OH$$

$$C_6H_5-C\equiv N \xrightarrow{RMgX} C_6H_5-\underset{R}{C}=N-MgX \xrightarrow[H^+]{H_2O} C_6H_5-\underset{R}{C}=O$$

2. Replacement of the alkoxyl groups of esters and acetals and of the halogen atom of a reactive organic halide (also ring opening of alkylene oxides).

---

[2] Section 35.2 gives another example of the Grignard reaction in which an ester, methyl benzoate, is allowed to react with 2 moles of phenylmagnesium bromide to form triphenylmethanol.

[3] The relative reactivity of various functional groups toward phenylmagnesium bromide follows roughly the order: —CH=O > —CO—CH$_3$ > —N=C=O > —CO—F > —CO—C$_6$H$_5$, —CO—Cl, —CO—Br > —CO$_2$Et > —C≡N. See Entemann and Johnson, *J. Amer. Chem. Soc.*, **55**, 2900 (1933).

$$C_6H_5\text{—CO—OCH}_3 \xrightarrow{2 \text{ RMgX}} C_6H_5\text{—}\underset{R}{\overset{R}{\text{C}}}\text{—OMgX} \xrightarrow[H^+]{H_2O} C_6H_5\text{—}\underset{R}{\overset{R}{\text{C}}}\text{—OH}$$

$$\underset{\underset{O}{\diagdown\diagup}}{CH_2\text{—}CH_2} \xrightarrow{RMgX} R\text{—}CH_2CH_2\text{—OMgX} \xrightarrow[H^+]{H_2O} R\text{—}CH_2CH_2\text{—OH}$$

The halomagnesium complex produced in the reaction is usually hydrolyzed by cold dilute mineral acid, to free the organic product. For acid-sensitive compounds strong aqueous ammonium choride solution may be used.

Compounds containing active hydrogen (water, alcohols, ammonia and amines, acetylenes, phenols, acids) convert a Grignard reagent to the parent hydrocarbon, R—H. Halogens, oxygen, and atmospheric carbon dioxide also react with R—MgX. Halides of boron, tin, mercury, and many metals react with R—MgX to furnish access to organic derivatives of the less reactive elements.

17.3 Preparation of 2-Methyl-1-hexene and 2-Methyl-2-hexene

n-**Butyl Bromide.** Follow the procedure given in Experiment 13(A), for conversion of 0.5 mole (37 g, 46 mL) of *n*-butyl alcohol to the bromide, using the sodium bromide–sulfuric acid method. Dry and distill the product *carefully*, and protect it from atmospheric moisture. Record the weight and yield. Reserve a small sample (0.5 mL), and use the remainder in the next step. If necessary, make adjustments in the quantities of materials used to correspond to your supply of butyl bromide.

n-**Butylmagnesium Bromide.** Assemble an apparatus like that shown in Figure 8.1b with reflux condenser, separatory funnel for addition, and a 500-mL reaction flask. Prepare a bath of ice and water to permit rapid cooling if the reaction should become too vigorous. During all of this operation make certain that there is no flame *anywhere near* to ignite the ether vapor.

▶ *CAUTION* Ether is extremely volatile and highly flammable. Extreme care must be taken to avoid fire hazards.

In the reaction flask place 0.33 mole (8 g) of clean magnesium turnings and add a *small* crystal of iodine. In a clean dry flask place 0.33 mole

(45–46 g, 36–37 mL) of pure *n*-butyl bromide and add 100 mL of *pure anhydrous* ether.[4] Pour a portion of the bromide solution into the separatory funnel and allow 10–15 mL to flow onto the magnesium in the flask. Under favorable conditions, the reaction will start within a few minutes, accompanied by vigorous boiling of the ether. As soon as this occurs, introduce 25 mL of dry ether directly through the top of the condenser to moderate the vigor of the reaction.

If the reaction does not start promptly, warm the flask gently in a bath of tepid water, and be prepared to moderate the reaction quickly by cooling and adding dry ether if the reaction starts suddenly. If necessary, add another small crystal of iodine (or a small quantity of a previously prepared Grignard reagent) to initiate the reaction. For the success of the experiment it is *absolutely essential* that the reaction begin before the remainder of the solution of butyl bromide is added to the magnesium.

When the initial vigorous reaction has moderated, allow the remainder of the butyl bromide solution to flow dropwise into the flask at a rate such that the ether refluxes gently without external heating. Swirl the flask frequently. After all of the bromide has been introduced, reflux the mixture gently for 20–30 min on a steam bath. Do not heat so vigorously that ether vapor escapes through the condenser. At this point almost all of the magnesium will have dissolved. The volume of the solution should not be less than 100 mL; if necessary add more dry ether to bring the volume to 100–125 mL.

2-Methyl-2-hexanol. Before adding acetone to the Grignard solution, cool the reaction flask in an ice–salt mixture to as low a temperature as possible. The yield is increased by carrying out the next step at low temperature, mixing thoroughly and slowly adding the acetone.

In a small dry flask mix 0.35 mole (20 g, 26 mL) of dry acetone (dried at least overnight over anhydrous potassium carbonate or magnesium sulfate) with 40 mL of anhydrous ether. Transfer the solution to the separatory funnel and allow it to drop *very slowly* into the cooled Grignard solution, while swirling and shaking the flask to insure good mixing and effective cooling. Each drop of the solution reacts vigorously, producing a hissing sound and forming a white precipitate that usually redissolves when the flask is shaken. After all of the acetone has been added, remove the cooling bath and allow the mixture to stand at room temperature, occasionally shaking it, for 20 min or longer.

Pour the reaction mixture slowly and carefully, while stirring, onto a mixture of chipped ice and dilute sulfuric acid (prepared by adding 0.4 equivalent (20 g, 11 mL) of concentrated sulfuric acid to 100 mL of water, and adding about 100 g of chipped ice). Rinse the reaction flask with a little of the dilute sulfuric acid and a little ordinary ether, and add these washings

---

[4] Ether used as the solvent must be of the anhydrous grade.

to the main product. Transfer the mixture to a separatory funnel and separate the two layers; *save both layers.* Extract the aqueous layer with two 50-mL portions of ordinary ether and combine the ether extracts with the ether layer from the first separation. The aqueous layer may now be discarded.

Wash the ether layer with 25 mL of water to which 2–3 mL of strong sodium bisulfite solution has been added,[5] then with 25 mL of cold water to which 5 mL of saturated sodium bicarbonate solution has been added, and separate the layers carefully. Dry the ethereal solution overnight or longer, over anhydrous magnesium sulfate or potassium carbonate, filter from the drying agent, and remove the solvent by distilling the dried solution from a flask fitted with a fractionating column, using a water or steam bath.

Transfer the residue to a small distilling apparatus and collect the material boiling above 75° and below 135°, in 15° fractions. Collect the product over the range 135–140°. If an appreciable quantity of low-boiling material is obtained in the fraction above 120°, dry this again and redistill it, collecting additional product in the 135–140° range. Additional product may also be obtained by redistilling the higher-boiling fraction collected over the range 142–145°. The yield is 12–20 g. The reported boiling point of 2-methyl-2-hexanol is 141–142°.

Reserve 10 mL (8 g) of the product for conversion to alkenes, and place the remainder in a small bottle with a label.

2-Methylhexenes. Arrange a 100-mL round-bottom flask for simple distillation (Figure 2.10), using as receiver a small pear-shaped flask that can be cooled in an ice bath. In the reaction flask, place a dehydration catalyst chosen from the following list.[6]

1. Iodine: Use several small crystals; only a little is needed. The distillate is likely to have a red color, which will fade in several hours; it can be removed quickly with a little aqueous sodium bisulfite.
2. Oxalic acid: Use 5 g finely ground; the solid should dissolve on heating.
3. Potassium bisulfate ($KHSO_4$): Use 5 g finely ground; the solid may not dissolve on heating.
4. Phosphoric acid: Use 5 mL of 85% H_3PO_4.

Place 0.07 mole (8 g, 10 mL) of 2-methyl-2-hexanol in the 100-mL flask, mix it well with the catalyst, and heat the material *slowly.* Distill the resulting alkenes *carefully,* not allowing the distillate temperature to rise above 90°.

---

[5] Sodium bisulfite removes traces of free iodine that can catalyze alkene formation when the product is distilled. Any acid remaining in the product has a similar effect.

[6] Arrange to have fellow students use different promotors for the reaction so that results can be compared and discussed.

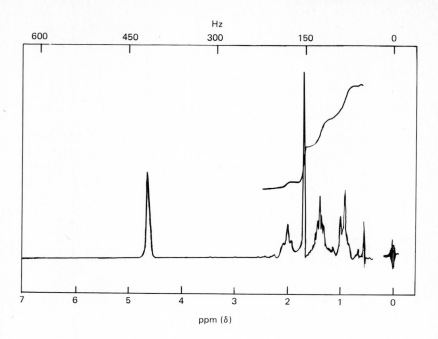

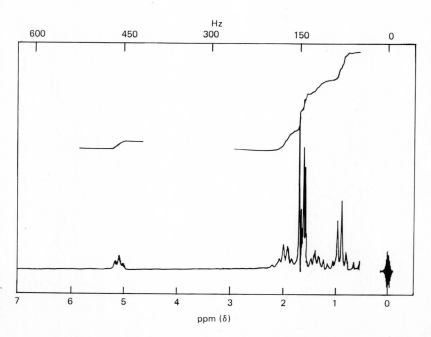

FIGURE 17.1
NMR Spectra of
2-Methyl-1-hexene and
2-Methyl-2-hexene

Rapid distillation or overheating will result in incomplete reaction by forcing unreacted alcohol into the distillate.

▶ *CAUTION*   The 2-methylhexenes are volatile and flammable.

With a small pipet remove carefully the lower aqueous layer from the distillate. Dry the upper alkene layer with a little anhydrous magnesium sulfate. Pipet off the alkenes into a tared vial and store the sample, tightly capped, until it can be examined by gas chromatography or NMR spectroscopy. Record the weight and yield of the alkenes and the distilling range of your sample; recorded boiling point of pure 2-methyl-1-hexene is 91°; of 2-methyl-2-hexene, 94.5°

Chromatographic Analysis of the 2-Methylhexenes. Subject your sample of these alkenes to gas-chromatographic analysis using a polar silicon oil column such as DC-710 (see Chapter 7). To confirm the identity of the components, it is highly desirable to have pure samples of 2-methyl-2-hexene and 2-methyl-1-hexene available for comparing their individual retention times. For the ratio of isomers in your mixture, assume a molar response factor of 1 : 1.

NMR Analysis of the 2-Methylhexenes. If an NMR instrument is available, the ratio of isomers can be determined by comparing the mixture NMR spectrum with the spectra of the pure components shown in Figure 17.1. The protons bound to sp^3-hybridized carbons give rise to different, but complex, patterns between 0.7 and 2.3 ppm. In a mixture of the two alkenes, these peaks will be superposed, which prevents accurate differentiation. However, the protons bound to the sp^2-hybridized carbons give rise to distinct absorptions: a singlet at 4.6 ppm for the 1-alkene, and a triplet at 5.1 ppm for the 2-alkene. With a mixture of the two, these peaks are easily distinguished and can be integrated, after correction for the relative number of protons giving rise to the peaks, to give the ratio of 2-methyl-1-hexene and 2-methyl-2-hexene in the sample.

Questions
1. Indicate two methods of preparing 2-methyl-2-hexanol using methylmagnesium bromide as the Grignard reagent.
2. Write reactions for the action of *n*-butylmagnesium bromide with the following.
    (a) carbon dioxide
    (b) *n*-butyraldehyde
    (c) ethanol
    (d) propyne
    (e) ethyl formate (H—CO—OEt)
    (f) formaldehyde, anhydrous

3. What product would you expect to be formed by addition of H—Br to each of your isomeric 2-methylhexenes?
4. What do you suggest for chemical methods to use in establishing the structure of 2-methyl-1-hexene and 2-methyl-2-hexene?
5. Have you discerned a specific effect of any one of the dehydration catalysts used in the E1 elimination reaction on 2-methyl-2-hexanol?
6. From your actual yield of the 2-methylhexenes, calculate what the average yield was overall for the four steps of this synthesis.

18 Hydration of Alkenes and Alkynes

18.1 Hydration of Double Bonds

An important industrial reaction is the hydration of alkenes to produce alcohols. For example, t-butyl alcohol, a valuable antiknock fuel additive, is prepared by the hydration of isobutylene with 60–65% aqueous sulfuric acid.

$$(CH_3)_2C=CH_2 + H^+ \rightleftharpoons (CH_3)_3C^+ \xrightarrow{H_2O} (CH_3)_3C\overset{+}{O}H_2 \rightleftharpoons (CH_3)_3COH + H^+$$

The reaction proceeds through a t-butyl cation and is the reverse of the acid-catalyzed dehydration of alcohols discussed in Chapter 16. The direction the reaction takes is determined by the reaction conditions. Hydration is favored by low temperatures and an aqueous medium.

In the isobutylene hydration, two carbocation intermediates might conceivably be formed, giving rise to two different alcohols.

In practice, the tertiary cation is so much more stable than the primary cation that only the tertiary alcohol is observed. In general, the position of the hydroxyl group is the site of the more stable carbocation, that is, the reaction follows *Markovnikov's rule*.

One consequence of the carbocation mechanism for hydration is that in some circumstances the intermediate carbocation can rearrange before it reacts with water to give rise to alcohols with altered carbon skeletons. Such rearrangements are more likely when secondary cations are being formed. For example, when 3,3-dimethyl-1-butene is hydrated with sulfuric acid and water, the major product is the rearranged alcohol 2,3-dimethyl-2-butanol. Apparently the rate of rearrangement of the methyl group to the first formed adjacent carbocation center is faster than the rate of attack by water.

$$\text{CH}_3-\underset{\underset{\text{CH}_3}{|}}{\overset{\overset{\text{CH}_3}{|}}{\text{C}}}-\text{C}\overset{\text{H}}{\underset{\text{CH}_2}{\diagdown\!\!\!\!\!\diagup}} \xrightarrow{\text{H}^+} \text{CH}_3-\underset{\underset{\text{CH}_3}{|}}{\overset{\overset{\text{CH}_3}{|}}{\text{C}}}-\overset{\text{H}}{\underset{\text{CH}_3}{\text{C}^+\diagdown}} \xrightarrow[\text{slow}]{\text{H}_2\text{O}} \text{CH}_3-\underset{\underset{\text{CH}_3}{|}}{\overset{\overset{\text{CH}_3}{|}}{\text{C}}}-\overset{\text{H}}{\underset{\text{CH}_3}{\text{C}-\text{OH}}}$$

minor

fast ↓

$$\text{CH}_3-\underset{\underset{\text{CH}_3\text{CH}_3}{|}}{\overset{+}{\text{C}}}-\overset{\text{H}}{\underset{}{\text{C}}}-\text{CH}_3 \xrightarrow{\text{H}_2\text{O}} \text{CH}_3-\underset{\underset{\text{CH}_3\text{CH}_3}{|}}{\overset{\overset{\text{OH}}{|}}{\text{C}}}-\overset{\text{H}}{\underset{}{\text{C}}}-\text{CH}_3$$

major

18.2 Oxymercuration–Demercuration of Alkenes

A major improvement on the acid-catalyzed hydration of alkenes is provided by the oxymercuration–demercuration procedure developed by Brown and Geohegan.[1] The standard procedure utilizes a 1:1 water–tetrahydrofuran solution of mercuric acetate, which reacts with most alkenes to give a mercury-containing intermediate that yields an alcohol on reduction with sodium borohydride.

$$\text{R}-\text{CH}=\text{CH}_2 \xrightarrow[\text{H}_2\text{O/THF}]{\text{Hg(OAc)}_2} \underset{\text{H}}{\overset{\text{R}\;\;\;\text{OH}}{\diagdown\!\!\text{C}\!\!\diagup}}-\text{CH}_2-\text{HgOAc} \xrightarrow[\text{NaOH}]{\text{NaBH}_4} \underset{\text{H}}{\overset{\text{R}\;\;\;\text{OH}}{\diagdown\!\!\text{C}\!\!\diagup}}-\text{CH}_3 + \text{Hg}$$

---

[1] Brown and Geohegan, Jr., *J. Org. Chem.*, **35**, 1844 (1970).

The alcohol is formed in excellent yields (typically 90–100%) and follows the Markovnikov substitution pattern. The advantage of the Brown procedure over acid-catalyzed hydration is that it is much faster and generally gives the unrearranged alcohol.[2]

The oxymercuration–demercuration of 1-hexene to give 2-hexanol is described in detail in Section 18.4(A). In this standard procedure, tetrahydrofuran is used as the solvent. All chemists should be aware that *tetrahydrofuran tends to form dangerous peroxides on exposure to air*, and these have led to a number of laboratory explosions. As supplied by the manufacturers, tetrahydrofuran is stabilized by adding small amounts of antioxidant, which prevents buildup of peroxides from short air exposure. However, on purification the antioxidant may be removed or destroyed and the tetrahydrofuran becomes susceptible to peroxide formation with even brief exposure to air. As a general rule only freshly purchased, stabilized tetrahydrofuran should be used.

18.3 Hydration of Alkynes

The triple bond of alkynes can also be hydrated with aqueous acid. The intermediate vinyl alcohols are unstable and rearrange immediately to give a carbonyl compound.

$$R-C\equiv CH \xrightarrow[\text{Hg}^{2+}]{\substack{H_2SO_4 \\ H_2O}} R-\underset{\substack{| \\ \text{vinyl alcohol} \\ \text{Enol}}}{\overset{OH}{C}}=CH_2 \rightleftharpoons R-\underset{\text{Keto}}{\overset{\overset{O}{\|}}{C}}-CH_3$$

The initial addition follows Markovnikov's rule so that with terminal acetylenes a methyl ketone rather than an aldehyde is formed. The only exception is acetylene, HC≡CH, which as a consequence of its unique structure must yield acetaldehyde. In practice, better yields are obtained if a mercuric ion catalyst is added. In this respect the hydration of alkynes differs sharply from that of alkenes, which require that a full equivalent of mercuric ion be used.

---

[2] The preparation of 2-hexanol by acid-catalyzed hydration of 1-hexene with 85% sulfuric acid at 100° has been described by McKee and Kauffman, *J. Chem. Educ.*, **59**, 695 (1982).

18.4 Reactions and Preparations

(A) Oxymercuration–Demercuration of 1-Hexene

In a 50-mL round-bottom flask, dissolve 1.1 g (0.0033 mole) of mercuric acetate, $Hg(OCOCH_3)_2$, in 5 mL of water.

▶ *CAUTION* Mercuric acetate is very poisonous. It is advisable to wear disposable gloves during this experiment and to wash your hands thoroughly after you are finished.

Add 5 mL of tetrahydrofuran and 0.5 mL (0.34 g, 0.0040 mole) of 1-hexene.

▶ *CAUTION* Tetrahydrofuran is volatile and extremely flammable. Only stabilized material should be used.

A yellow precipitate forms immediately. Swirl the flask to dissolve the precipitate. Attach a water-cooled condenser, add two boiling chips, and add through the condenser 2 mL of 6 M aqueous sodium hydroxide followed by 0.5 mL of a 0.3-M solution of sodium borohydride in 3 M aqueous sodium hydroxide.[3] Boil the mixture for 15 min on a steam bath.

Cool the mixture to room temperature and allow the mercury droplets to coalesce and settle (this may require waiting until the next laboratory period). Carefully decant the upper liquid layers from the mercury into a small separatory funnel. Discard the mercury and any residual solution clinging to it into the waste mercury container situated in the hood. (*Do not pour the mercury into the sink.*)

Add 10 mL of methylene chloride to the water–tetrahydrofuran mixture contained in the separatory funnel and separate the lower organic layer. Extract the aqueous layer with two more 10-mL portions of methylene chloride. Dry the combined organic layers over magnesium sulfate.[4]

Remove the drying agent by filtration and analyze the organic solution by gas chromatography (silicon oil column) for the relative concentrations of the 2-hexanol (Markovnikov) and 1-hexanol (anti-Markovnikov) products. A reference solution of the alcohols in methylene chloride can be used to determine their relative retention times.

(B) 2-Heptanone by Hydration of 1-Heptyne

In a 500-mL round-bottom flask place 20 mL of water and add to it *slowly and cautiously* 10 mL of concentrated sulfuric acid. To the warm acid solution add 0.5 g (0.0023 mole) of mercuric oxide. Add two boiling chips,

---

[3] Sodium borohydride hydrolyzes slowly in aqueous sodium hydroxide to give sodium borate and hydrogen gas. The solution should be freshly prepared.

[4] A quicker way to dry the solution is to first remove any dispersed water droplets with anhydrous sodium sulfate and then, after filtration or careful decantation, distill off about half of the methylene chloride.

attach a reflux condenser, and add through the condenser 50 mL of methanol followed by 15 mL (11.0 g, 0.11 mole) of 1-heptyne.[5]

CAUTION Mercuric oxide is very poisonous. It is advisable to wear disposable gloves during this experiment and to wash your hands thoroughly after you are finished.

The mercury complex of the alkyne will precipitate, and an exothermic reaction will begin. After the initial reaction subsides, heat the reaction mixture until it boils and continue heating for 30 min.

Cool the reaction mixture to room temperature and add 100 mL of water. Rearrange the apparatus for simple distillation and distill the mixture until about 120 mL of distillate has been collected. Add approximately 15 g of sodium chloride to salt out the product.[6] Separate the layers, extract the lower aqueous layer twice with 20-mL portions of dichloromethane, and combine the original organic layer with the extracts.

The residue from the distillation, which contains the mercuric salts, should be poured carefully into the mercury waste container in the hood. (*Do not* pour this *severely poisonous solution* into the sink. Wash your hands carefully after this operation.)

Dry the combined organic layers over anhydrous magnesium sulfate and distill off the solvent and any unreacted 1-heptyne (bp range 35–100°). Transfer the residue to a small distillation flask and distill. The yield of 2-heptanone bp 148–150° is about 10 g.

2-Heptanone has a penetrating fruity odor and is said to be responsible for the "peppery" odor of Roquefort cheese. It is used commercially as a constituent of artificial carnation essence.

Small samples of 2-heptanone can be converted to the semicarbazone and 2,4-dinitrophenylhydrazone as described in Section 9.

Questions

1. Explain why hydration of double bonds is favored by low temperature and an aqueous medium.
2. What is Markovnikov's rule and what is its mechanistic basis?
3. Why does the oxymercuration–demercuration reaction on 1-hexene yield 2-hexanol rather than 1-hexanol? Why does 1-hexyne yield 2-hexanone rather than hexanaldehyde?

---

[5] 1-Heptyne is available inexpensively from Heico Division, Whittaker Corporation, Delaware Water Gap, PA.

[6] See Section 6.3 for a discussion of "salting out."

19 Glaser–Eglinton–Hayes Acetylene Coupling

19.1 Introduction

A major challenge in the design of an organic synthesis is the construction of the carbon skeleton. For this reason any method for creating new C—C bonds is of interest, particularly if it leaves reactive sites that can subsequently be converted into the desired functional groups.

The usual methods for forming new C—C bonds add one or two carbon atoms at a time. When appropriate, it is far more efficient to couple the two halves of the molecule in a single step. More than a century ago, Glaser discovered that it was possible to couple oxidatively two acetylene molecules using air and a basic solution of cuprous ion as oxidant.

$$C_6H_5-C\equiv C-H \xrightarrow[\text{air}]{Cu^+, NH_4OH} C_6H_5-C\equiv C-C\equiv C-C_6H_5$$

In 1956 Eglinton and Galbraith discovered that catalytic quantities of cupric ion, Cu^{2+}, could be used but reported that the reaction was slow. In 1962 Hayes found that superior catalysis was obtained with tertiary amine complexes of cuprous, Cu^+, salts. Under these conditions, even at room temperature, the reaction proceeded almost as rapidly as the oxygen could be added. For example, the acetylene 1-ethynylcyclohexanol had a reaction half-life (the time for 50% reaction) of 13 min and gave a 93% yield of the coupled acetylene product.

An interesting application of the acetylene coupling reaction was the synthesis of cyclic hydrocarbons containing very large rings.

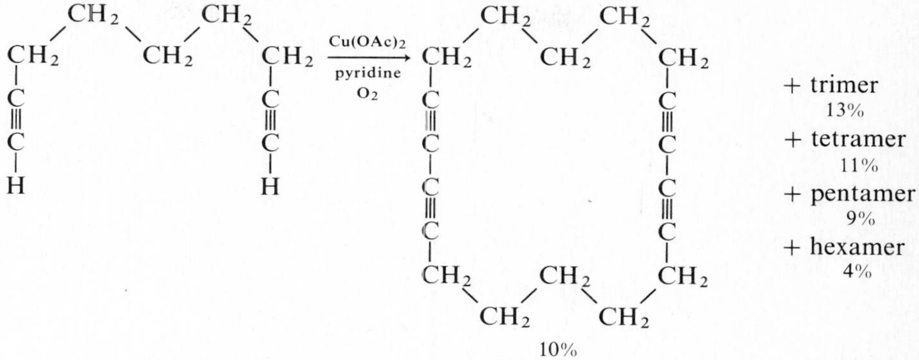

$$\text{2 HC} \equiv \text{C-CH}_2\text{-CH}_2\text{-CH}_2\text{-CH}_2\text{-C} \equiv \text{CH} \xrightarrow[\text{O}_2]{\text{Cu(OAc)}_2 \atop \text{pyridine}} \text{cyclic dimer (10\%)}$$

+ trimer 13%
+ tetramer 11%
+ pentamer 9%
+ hexamer 4%

Catalytic reduction gave the saturated macrocyclic rings including the hexamer, which has a ring of 54 CH_2 groups.

19.2 Mechanism of Acetylene Coupling

The mechanism of the Glaser–Eglinton–Hayes acetylene coupling reaction has been examined and, although the details are still obscure, at least the broad features can be understood. Both the Cu^+ and Cu^{2+} ions appear to be required. The Cu^{2+} is the actual oxidizing agent, but Cu^+ appears to be required to form the copper acetylide, $R-C \equiv C-Cu$. It is proposed that a dimer of the copper salt is formed, which is then oxidized.

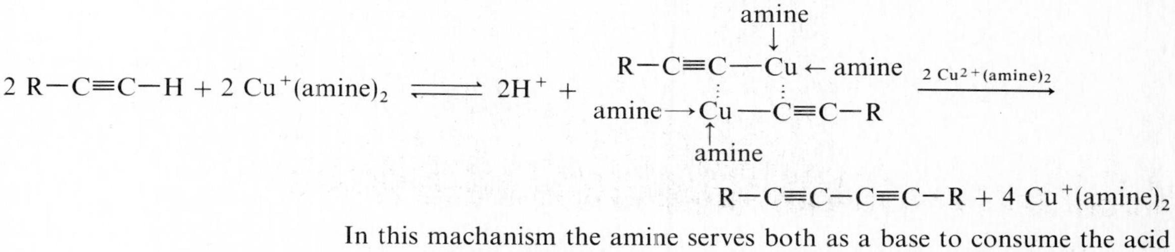

$$2\ R-C\equiv C-H + 2\ Cu^+(\text{amine})_2 \rightleftharpoons 2H^+ + \begin{array}{c} \text{amine} \\ \downarrow \\ R-C\equiv C-Cu \leftarrow \text{amine} \\ \vdots\quad\vdots \\ \text{amine} \rightarrow Cu-C\equiv C-R \\ \uparrow \\ \text{amine} \end{array} \xrightarrow{2\ Cu^{2+}(\text{amine})_2}$$

$$R-C\equiv C-C\equiv C-R + 4\ Cu^+(\text{amine})_2$$

In this machanism the amine serves both as a base to consume the acid generated in the formation of the complex and as a solubilizing agent to prevent the precipitation of the acetylide salt.

19.3 Preparation

Oxidative Coupling of 1-Ethynylcyclohexanol

Assemble the apparatus shown in Figure 19.1 using a 250-mL filter flask and a 100-mL round-bottom flask. The purpose of the filter-flask bubbler is to saturate the air with acetone to minimize evaporation of the reaction

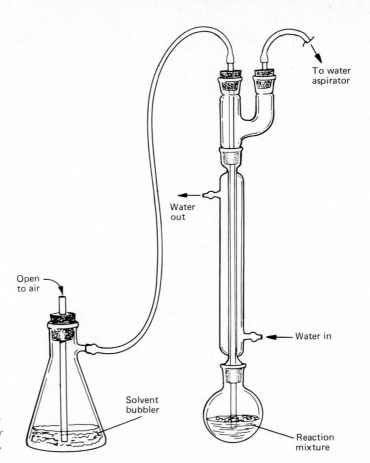

FIGURE 19.1
Apparatus for Introduction of Air into Reaction Mixture

solvent in the round-bottom flask. Place 30 mL of acetone in each of the flasks and to the round-bottom flask add 0.25 g (0.002 mole) of tetramethylethylenediamine and 0.20 g (0.002 mole) of cuprous chloride (the preparation is described below). After these reagents have dissolved, add 5 g (0.040 mole) of 1-ethynylcyclohexanol. Immerse the two flasks in pans of water heated to about 40°, turn on the water aspirator, and for 20 min draw air through the bubbler and through the reaction mixture.

At the end of the reaction period, disconnect the filter-flask bubbler, turn off the water flow to the condenser, and continue to draw air through the round-bottom flask until the acetone has evaporated. Add 20 mL of water and 1 mL of concentrated hydrochloric acid to the residue. The cupric salts will go into solution and leave the product as a solid. Collect the solid on a Büchner funnel and air dry it. The yield is about 4.5 g (93% of theory), mp 177°.

Preparation of Cuprous Chloride Solution.[1] In a 100-mL round-bottom flask, prepare a solution of 6 g (0.02 mole) of powdered copper sulfate crystals ($CuSO_4 \cdot 5H_2O$) and 1.8 g of sodium chloride in 20 mL of hot water. In a beaker prepare a solution of 1.4 g of sodium bisulfite and 0.9 g of solid sodium hydroxide in about 1.4 mL of water, and add this solution with swirling to the hot copper sulfate solution over a period of 5–10 min. Cool the mixture to room temperature, allow the solid to settle, and decant off the liquid. Wash the precipitated cuprous chloride two or three times with water, by decantation. The cuprous chloride is obtained as a white powder that darkens on exposure to the air. Dissolve the cuprous chloride (as $HCuCl_2$) by adding 7 mL of concentrated hydrochloric acid and 2.5 mL of water. Cork the flask to minimize oxidation and place it in an ice bath.

Questions

1. The starting material for this preparation, 1-ethynylcyclohexanol, is an industrial product. Suggest how it is made.
2. Vinylacetylene is synthesized industrially by the dimerization of acetylene with a mixture of cuprous chloride, ammonium chloride, and HCl (the Nieuwland enyne synthesis). Do you think that the mechanism is the same as the Glaser–Eglinton–Hayes reaction? If not, what might the mechanism be?
3. Give the principal products for the reaction of 1-ethynylcyclohexanol and its Glaser–Eglinton–Hayes coupling product with the following reagents.
   (a) H_2/Pt
   (b) aqueous $AgNO_3$
   (c) H_2(1 equiv.)/Pd–$BaSO_4$–quinoline
   (d) excess Br_2/CCl_4, 0°

---

[1] An alternative procedure is to dissolve 2.0 g of commercial cuprous chloride in 7 mL of concentrated hydrochloric acid and 2.5 mL of water.

20 Preparation of Aldehydes and Ketones by Oxidation

20.1 Chromic Acid Oxidation of Alcohols

Chromic acid mixture (dichromate and 40–50% sulfuric acid) oxidizes a primary alcohol stepwise to the aldehyde and the corresponding carboxylic acid (see Section 9.7(A)). The simpler aldehydes (propionaldehyde, butyraldehyde) can be prepared in moderate yields by introducing the dichromate solution dropwise *into* a hot acidified solution of the alcohol, so that the oxidizing agent is not present in excess, and distilling the volatile aldehyde away from the mixture as rapidly as it is produced. Even so, some of the aldehyde is oxidized to the acid, and this is converted in part to the ester, $R-CO-OCH_2R$. Also, part of the aldehyde may react with the alcohol to form the acetal, $R-CH(OCH_2R)_2$. A simple expression for the oxidation of a primary alcohol is

$$Na_2Cr_2O_7 + 2\ H_2SO_4 \longrightarrow 2\ NaHSO_4 + H_2Cr_2O_7 \xrightarrow{H_2O} 2\ H_2CrO_4$$

$$3\ R-CH_2OH + 2\ \underset{\text{(orange-red)}}{H_2CrO_4} + 3\ H_2SO_4 \longrightarrow 3\ R-CH{=}O + \underset{\text{(green)}}{Cr_2(SO_4)_3} + 8\ H_2O$$

The detailed mechanism of oxidation of primary alcohols is of some interest because it indicates that the reaction can be arrested at the aldehyde stage by using chromic anhydride (CrO_3) and operating under anhydrous conditions. This modification has proved to be successful; Collins' reagent, the chromic anhydride–pyridine complex ($CrO_3 \cdot 2C_5H_5N$) in methylene chloride, converts 1-octanol (1 hr at 25°) into the corresponding aldehyde in 95% yield.

$$R-CH_2OH + CrO_3 \longrightarrow \underset{\text{Chromate ester}}{R-CH_2O-CrO_3H} \longrightarrow H_2CrO_3 + R-CH=O \underset{H_2O}{\overset{CrO_3, \text{slow}}{\rightleftarrows}} \begin{matrix} R-CO_2H \\ \uparrow CrO_3 \text{ fast} \\ R-CH-OH \\ | \\ O \\ | \\ H \end{matrix}$$

Under anhydrous conditions the aldehyde hydrate does not form, so the fast oxidation step to the acid is avoided. Extreme care must be taken in preparing and using Collins' reagent because explosive oxidations have been reported when the reagents are mixed in the wrong order. Collins' reagent is not used in the preparations presented here.

Oxidation of secondary alcohols to ketones by means of chromic acid mixture is generally a satisfactory method of preparing ketones, because the latter do not undergo further oxidation so easily. The oxidation is exothermic and the temperature must be controlled to avoid a violent reaction. For water-insoluble compounds, chromic acid ($CrO_3 + H_2O$) in an acetone–sulfuric acid medium (Jones' reagent) or in glacial acetic acid (Fieser's reagent) may be used. Oxidation with chromic acid in acetone solution is rapid and quite selective; usually a double bond is not attacked. During the oxidation the orange-red chromic acid is converted to the green chromium ($+3$) ion, which is the basis of the chromic acid oxidation test for alcohols (see Section 9.7(A)).

Under ordinary conditions tertiary alcohols are relatively stable to cold chromic acid, but under more vigorous conditions tertiary alcohols, and ketones, may be degraded by cleavage of carbon–carbon bonds.

Chromic acid oxidation of secondary alcohols, as with the primary alcohols, occurs through formation of the chromate ester. The next step is the slowest in the reaction and involves removal of a proton from the adjacent carbon atom, forming the ketone and $H_2CrO_3(Cr^{4+})$. Disproportionation of this acid gives CrO_3 and a salt of $Cr_2O_3(Cr^{3+})$.

$$\begin{matrix} H \\ R \\ \diagdown \\ C \\ R \diagup \diagdown O \\ O-Cr-OH \\ \| \\ O \end{matrix} \longrightarrow \begin{matrix} R \\ \diagdown \\ C=O + H_2CrO_3 \\ R \diagup \end{matrix}$$

$$3\ H_2CrO_3 \xrightarrow{H_2SO_4} CrO_3 + Cr_2(SO_4)_3$$

The rate of proton removal to form the ketone from the chromate ester is increased by addition of effective nucleophilic reagents. Also, replacement of the pertinent hydrogen by deuterium causes the oxidation rate of 2-propanol to decline to about one-sixth the normal rate.

20.2 Other Oxidation Methods

On a larger scale aldehydes and ketones can be prepared in excellent yields by dehydrogenation of the corresponding alcohol at elevated temperatures (250–250°) over a metal catalyst, such as platinum, silver, copper, and copper–zinc alloy. Ketones may be prepared also by several other procedures; passing the vapor of an organic acid over an oxide catalyst (ThO_2 or MnO), or by pyrolysis of the calcium or barium salt of an organic acid (Ruzicka reaction); by addition of a Grignard reagent to a nitrile, followed by hydrolysis; and by the acetoacetic ester ketone synthesis.

Diaryl ketones may be prepared by chromic acid oxidation of diarylmethanes ($Ar-CH_2-Ar$) or diarylcarbinols ($Ar-CHOH-Ar$), preferably in glacial acetic acid solution. Chromic oxidation of methyl side chains can be arrested at the aldehyde stage by operating in the presence of acetic anhydride, which serves to protect the aldehyde from further oxidation by conversion to the diacetate, $Ar-CH(OCOCH_3)_2$. The latter can be hydrolyzed readily to regenerate the aldehyde.[1]

20.3 Preparations

(A) Methyl n-Propyl Ketone (2-Pentanone)

In a 500-mL round-bottom flask place 50 mL of water and add carefully, with cooling, 11 mL (20 g, 0.2 mole) of concentrated sulfuric acid. To the cold diluted acid add 32.5 mL (26.5 g, 0.3 mole) of 2-pentanol,[2] and swirl the flask to obtain good mixing. To prepare the chromic acid oxidizing solution, dissolve 30 g (0.1 mole) of sodium dichromate dihydrate in 50 mL of water, add carefully 11 mL (20 g, 0.2 mole) of concentrated sulfuric acid, and cool the solution to room temperature.

Introduce the oxidizing solution *into* the pentanol solution in small portions, swirling the solution, and observe the temperature. By intermittent cooling in a pan of water, as needed, maintain the internal temperature at 25–30°. If the temperature is kept too low, the oxidizing agent may accumulate in the solution and react suddenly with great vigor. When all of the oxidizing agent has been added, and the temperature no longer rises spontaneously, stopper the flask and allow it to stand at room temperature for 1 hr or longer, occasionally swirling it.

Fit the flask with a short fractionating column (it need not be packed),

---

[1] *Organic Syntheses*, Collective Volume **II**, 441 (1943).

[2] Diethyl ketone (3-pentanone) may be prepared from 3-pentanol by the same method; in the preliminary distillation of the product the azeotrope, containing 14% water, distills at 82.9°. The procedure given here is an adaptation of that described by Yohe, Louder, and Smith, *J. Chem. Educ.*, **10**, 374 (1933).

and attach a water-cooled condenser arranged for distillation. Add 150 mL of water and a boiling chip, and distill the mixture until a test portion of the distillate is essentially free of oily droplets. Do not collect an excessive amount of aqueous distillate; the ketone is appreciably soluble in water and a larger portion will be lost in the solution. The azeotrope of 2-pentanone and water, containing 20% of water, distills at 83.3°; stop the distillation when the temperature of the distilling vapor has risen a few degrees above this point.

To the distillate add 0.5 g of solid sodium carbonate to neutralize any acid, and salt out[3] the dissolved ketone by adding 2 g of clean sodium chloride for each 10 mL of water present. Draw off the aqueous layer and transfer the ketone to a small dry Erlenmeyer flask. Add 4–5 g of anhydrous potassium carbonate and shake well. If an aqueous layer is formed, draw this off and add a fresh portion of the drying agent. Filter (or decant) the dried liquid into a small distilling flask, add two small boiling chips, and distill over a wire gauze. Collect the product boiling at 97–102°. The yield is 12–15 g. The product should be colorless.[4]

If a moist low-boiling fraction of product is obtained it may be dried and redistilled or used for preparation of the semicarbazone, 2,4-dinitrophenylhydrazone, and other derivatives given in Chapter 9.

(B) Cyclohexanone

In a 400-mL beaker, dissolve 21 g (0.07 mole) of sodium dichromate dihydrate in 120 mL of water. Add carefully 17 mL (30 g, 0.3 mole) of concentrated sulfuric acid, stirring the mixture, and cool the deep orange-red solution to 30°. Place 21 mL (20 g, 0.2 mole) of cyclohexanol and 60 mL of water in a 500-mL Erlenmeyer flask and *to it* add the dichromate solution in one portion. Swirl the mixture to insure thorough mixing and observe its temperature. The mixture rapidly becomes warm; when the temperature reaches 55°, cool the flask in a basin of cold water, or under the tap, and regulate the amount of cooling so that the temperature remains between 55 and 60°. Continue external cooling only as long as necessary to maintain this temperature; when the temperature of the mixture no longer rises above 60° on removal of external cooling, allow the flask to stand for 1 hr, occasionally shaking it.

Pour the mixture into a 500-mL round-bottom flask, add an additional 120 mL of water and, by means of a short unpacked fractionating column, attach a condenser set downward for distillation. Add two boiling chips and distill the mixture until about 100 mL of distillate, consisting of water and an upper layer of cyclohexanone, has been collected.

---

[3] For a discussion of salting out, see (Section 6.3).
[4] Yellow discoloration of the product may arise from the presence of a little of the intensely yellow diketone, 2,3-pentanedione. This impurity is alkali-sensitive and can be removed by adding about 0.5 g of crushed sodium hydroxide and redistilling.

Saturate the aqueous layer with salt (20–25 g will be required), separate the cyclohexanone layer, and extract the aqueous layer with 15 mL of methylene chloride, or pentane (*flammable!*). Combine the solvent extract with the cyclohexanone layer and dry it with ~6 g of anhydrous magnesium sulfate. Filter the dried solution into a distilling flask of suitable size, attach a condenser, and distill off the solvent from a water bath. Distill the residual cyclohexanone, using a wire gauze, and collect the fraction boiling at 151–155° (mainly 152–154°). The yield is 12–15 g.

Conversion of cyclohexanone to the oxime and phenylhydrazone is described in Chapter 9.

Questions

1. Compare the behavior of 2-methyl-2-butanol, 3-methyl-2-butanol, and 2,2-dimethylpropanol toward mild oxidation with chromic acid solution. What are the common (trivial) names of these alcohols?
2. Write equations showing the action of the following reagents on butyraldehyde.
   (a) hydrogen cyanide
   (b) ethylmagnesium bromide
   (c) iodine–potassium iodide + alkali (iodoform test)
3. Show how the reagents listed in Question 2 would react with 2-pentanone.
4. What product is formed by self-condensation of propionaldehyde, followed by elimination of water from the aldol?

21 Reactions of Aldehydes and Ketones

21.1 Carbonyl Addition Reactions

In Chapter 9 a number of carbonyl addition reactions were described that are useful in the chemical identification of an unknown aldehyde or ketone. The reactions can be summarized according to the equation

$$R_2C=O + NH_2-X \rightleftharpoons R_2C=N-X + H_2O$$

Here X stands for OH (with oxime formation), for $NH-C_6H_3(NO_2)_2$ (with dinitrophenylhydrazone formation), and for $NH-CO-NH_2$ (with semicarbazone formation). In Chapter 9 the reaction conditions were chosen so that for most unknowns, the equilibrium was achieved and the product was favored. However, in typical laboratory experiments, most organic reactions are performed under conditions that do not achieve equilibrium between the products and the starting materials. This means that the sole or major product is due to kinetic control. Under equilibrium conditions the principal final product may be different from that resulting from kinetic control.

A study of equilibria and rates[1] in the reaction of aldehydes and ketones with semicarbazide has led to examples that illustrate the effect of these factors.

$$R_2C=O + NH_2NH-CO-NH_2 \rightleftharpoons R_2C=NNH-CO-NH_2 + H_2O$$

[1] Conant and Bartlett, *J. Am. Chem. Soc.*, **54**, 2881 (1932).

In this reaction two opposing forces are at work. Increased acidity beyond about pH 4.9 decreases the amount of *free* semicarbazide through salt formation, and this reduces the rate because of the low nucleophilic activity of the cation. But the carbonyl compound is subject to addition of a proton to the carbonyl group, which enhances its its electrophilic activity (and rate of reaction). The combination of these two opposing factors leads to a range of pHs over which the reaction proceeds readily. Above and below this pH range, the rate falls off sharply. (See Questions 1 and 2 at the end of this chapter.)

$$H_2NNH-CO-NH_2 + H^+ \rightleftharpoons {}^+H_3NNH-CO-NH_2$$

$$R_2C=O + H^+ \rightleftharpoons [R_2C=OH^+ \leftrightarrow R_2\overset{+}{C}-OH]$$

The optimum conditions for interaction of carbonyl compounds with reagents to form derivatives such as semicarbazones, oximes, and arylhydrazones involve buffered solutions. Sodium acetate or phosphate buffers are often used.[2]

In the experiment described in Section 21.3(A), the semicarbazones of a mixture of 2-furaldehyde and cyclohexanone will be prepared under both kinetically controlled and equilibrium conditions. One of these carbonyl compounds forms a semicarbazone much faster than the other, but at equilibrium the second semicarbazone is formed in greater amount. Both situations will be examined and the dominant product identified.

2-Furaldehyde   Cyclohexanone

21.2 Reduction of Carbonyl Compounds

An important reaction of aldehydes and ketones is their reduction to an alcohol, and many reagent/reaction condition combinations have been developed to carry out this process cleanly. For small-scale syntheses sodium borohydride, $NaBH_4$, is a particularly convenient reducing agent. It has extremely high reducing capacity—1 mole can reduce 4 moles of a ketone.

$$4(C_6H_5)_2C=O + Na\overset{-}{B}H_4 \longrightarrow Na\overset{-}{B}[OCH(C_6H_5)_2]_4 \xrightarrow{H_2O} 4(C_6H_5)_2CH-OH$$

[2] For an interesting series of experiments in this field, see Roberts, Gilbert, Rodewald, and Wingrove, *An Introduction to Modern Experimental Organic Chemistry*, 2nd ed. (New York: Holt, Rinehart and Winston, 1974), pp. 188–190.

Since sodium borohydride decomposes at an appreciable rate in water[3] or methanol, it is desirable to effect reactions in ethanol or 2-propanol. In these media at room temperature sodium borohydride will reduce aldehydes and ketones to the corresponding primary and secondary alcohols. It is quite selective and does *not* reduce nitriles, nitro compounds, carboxylic acids or esters, or lactones. Lithium aluminum hydride (LiAlH$_4$) is a somewhat stronger reagent and will reduce esters, lactones, and amides, but must be used in aprotic media.[4]

In aprotic solvents such as dioxane and 1,2-dimethoxyethane (glyme), sodium borohydride will reduce acid chlorides to alcohols. Secondary and tertiary halides are reduced at 50° by 4 M solutions of sodium borohydride in a 65 volume percent solution of diglyme and 1 M aqueous sodium hydroxide. Other uses of sodium borohydride are replacement of the diazonium group by hydrogen, and reduction of ozonides to the corresponding alcohols. Thus, the ozonide of oleic acid furnished 1-nonanol and 9-hydroxynonanoic acid.

$$C_8H_{17}-CH=CH-(CH_2)_7-CO_2H \xrightarrow{O_3} \text{ozonide} \xrightarrow{NaBH_4}$$
$$C_8H_{17}CH_2OH + HO-CH_2-(CH_2)_7CO_2H$$

In 21.3(B) the reduction of benzophenone to diphenylmethanol (benzohydrol) by sodium borohydride is described.

21.3 Reactions of Carbonyl Compounds

(A) Equilibria and Rates in Carbonyl Reactions: Formation of 2-Furaldehyde and Cyclohexanone Semicarbazones

In a 125-mL flask dissolve 2 g of semicarbazide hydrochloride and 4 g of potassium monohydrogen phosphate (K$_2$HPO$_4$) in 50 mL of water. Mix together in a small flask 1.9 g (2 mL) of cyclohexanone, 2.0 g (1.7 mL) of 2-furaldehyde, and 10 mL of 95% ethanol. Place half of the solution in each of two small test tubes.

In a 50-mL Erlenmeyer flask place 25 mL of the semicarbazide solution and cool it to 0–5° in an ice bath. In the same cooling bath chill one of the test tubes containing the furaldehyde–cyclohexanone solution. When it is thoroughly chilled, empty the contents of the test tube into the semi-

---

[3] The decomposition is slowed markedly by addition of alkali.
[4] Surveys of the uses of these reducing agents are found in Fieser and Fieser, *Reagents for Organic Synthesis*, Vol. **I** (New York; Wiley, 1967), pp. 1049–1055 (for NaBH$_4$) and 581–600 (for LiAlH$_4$). See also, for LiBH$_4$, W. G. Brown, *Organic Reactions*, **6**, 649 (1951).

carbazide solution and mix them well. Crystals will form quickly; replace the flask in the cooling bath for 5 min, then filter the crystals with suction and wash them with about 5 mL of cold water. After drying them thoroughly, weigh them and determine the melting point. Which semicarbazone is this?

The recorded melting points of the semicarbazones are 2-furaldehyde, 202°; cyclohexanone, 166°.

Place the remaining 25 mL of semicarbazide solution in a 50-mL flask and heat it on a steam bath to 85°. Add the 2-furaldehyde–cyclohexanone solution from the second test tube and swirl the solution. Heat the flask for 15 min longer on the steam bath and then cool the solution to room temperature. Finally, chill the reaction mixture in an ice bath for a few minutes to complete crystallization of the product. Collect the crystals with suction, and wash them with a little cold water. Dry the product thoroughly, record the weight, and take the melting point. Which semicarbazone is this?

Account for the results that you have observed.

(B) Reduction by Sodium Borohydride: Diphenylmethanol

In a 100-mL round-bottom flask place 2.75 g (0.015 mole) of benzophenone and add a slurry of 0.3 g (0.0075 mole, a large excess) of sodium borohydride in 15 mL of 2-propanol. Add two boiling chips and reflux the mixture for 30 min on a steam bath. Allow the solution to cool; no harm is done if it stands overnight or longer.

CAUTION Sodium borohydride is strongly caustic. Handle it carefully and do not permit it to touch the skin.

FIGURE 21.1 Infrared Spectrum of Benzophenome (Nujol Mull)

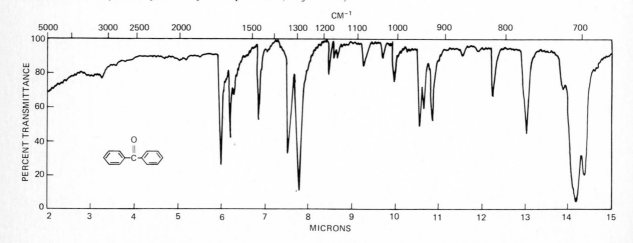

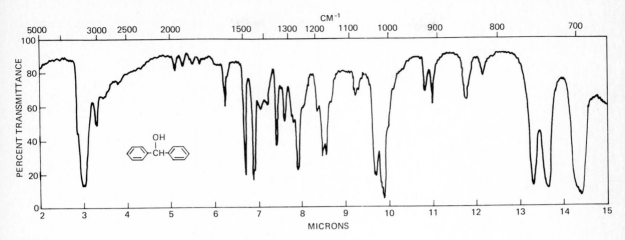

FIGURE 21.2 *Infrared Spectrum of Diphenylmethanol (Nujol Mull)*

To decompose the boric ester complex, add 15 mL of 10% aqueous sodium hydroxide and swirl the reaction mixture vigorously until the precipitate has dissolved completely. Break up any resistant lumps carefully with a stirring rod. Transfer the alkaline solution to a separatory funnel with the aid of 15 mL of water. Extract the diphenylmethanol by shaking it with two successive 20-mL portions of methylene chloride (CH_2Cl_2). Combine the extracts, transfer them to a distilling flask, and carefully distill off the methylene chloride (traces of water in the methylene chloride will steam-distill). On cooling and standing the residue will crystallize to give a nearly quantitative yield of almost pure diphenylmethanol, mp 68–69°. If desired, the product may be recrystallized from 60% water–methanol.

The infrared spectra of benzophenone and diphenylmethanol are shown in Figures 21.1 and 21.2.

Reactions. Diphenylmethanol has a reactive hydroxyl group: resonance involving the two aryl groups of the diphenylmethyl cation facilitates rupture of the C—OH bond in the transition state. It is converted easily into bisdiphenyl methyl ether merely by boiling with dilute mineral acids and reacts readily with hydrogen chloride to give diphenylchloromethane. For characterization diphenylmethanol may be converted to the acetate (mp 41–42°) or benzoate (mp 88°).

Diphenylmethanol reacts directly with some active methylene compounds and 1,4-quinones. With ethyl acetoacetate it gives the α-benzohydryl derivative (I), which produces 4,4-diphenyl-2-butanone on warming with dilute alkali. With 1,4-naphthoquinone in glacial acetic acid and a little sulfuric acid it introduces a diphenylmethyl group at the 2 position (II).

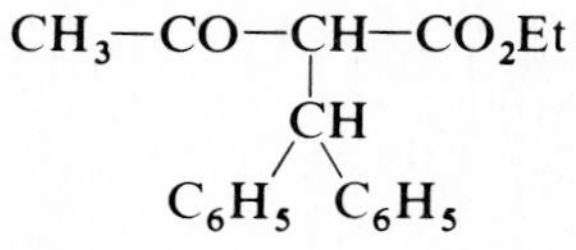

$$CH_3-CO-CH-CO_2Et$$
$$\overset{|}{CH}$$
$$C_6H_5\ C_6H_5$$

I

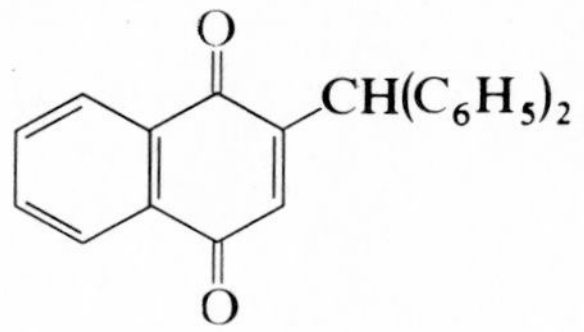

II
(lemon yellow, mp 168°)

Questions

1. In considering the effects of pH on rate, it is convenient to plot log [S] or log [SH$^+$] versus pH, where [S] and [SH$^+$] are the concentrations of species S in its unprotonated and protonated forms. Such plots are called "pH profiles."
   (a) What is the pH profile of a species S that is half-protonated at pH 7?
   (b) What is the pH profile for SH$^+$?

2. (a) If the rate of reaction between A and BH$^+$ is proportional to the product of their concentrations (rate = k[A][BH$^+$]), show that log (rate) = constant + log [A] + log [BH$^+$].
   (b) What is the pH profile (Question 1) of the rate if A is half-protonated at pH 7 and B is half-protonated at pH 3?

3. Explain why the yield of 2-furaldehyde semicarbazone obtained from a mixture of 2-furaldehyde and cyclohexone changes with temperature.

4. Sodium borohydride is a less powerful reducing agent than lithium aluminum hydride. In what other way do these two reagents differ?

5. Devise a synthesis for the following compounds.
   (a) 4-methyldiphenylmethanol, starting from benzene and toluene
   (b) 3,3'-dibromodiphenylmethanol, starting from bromobenzene

22 A Modified Wittig Synthesis

22.1 The Wittig Reaction

In the Wittig reaction,[1] aldehydes and ketones are converted into alkenes by reaction with an alkylidene phosphorane (a phosphorus ylid,[2] II), generally in high yield. The requisite ylids can be obtained by the action of strongly nucleophilic reagents, such as phenyllithium or sodium hydride, upon appropriate quaternary phosphonium halides (I).

$$[(C_6H_5)_3\overset{+}{P}-CH_2R]\overset{-}{X} \xrightarrow{C_6H_5Li} [(C_6H_5)_3P=CH-R \leftrightarrow (C_6H_5)_3\overset{+}{P}-\overset{-}{C}H-R]$$
$$\text{I} \qquad\qquad \text{IIa} \qquad\qquad \text{IIb}$$

$$\text{II} + (C_6H_5)_2C=O \longrightarrow C_6H_5-CH=C(C_6H_5)_2 + (C_6H_5)_3P=O$$

Since the ylids are unstable, they are usually generated in the reaction mixture in the presence of the carbonyl compound. Advantages of the Wittig synthesis are that carbon–carbon bond formation occurs without production of isomeric alkenes, and acid-sensitive alkenes can be prepared because the reaction occurs under mild conditions in an alkaline medium.

If the organic halide ($R-CH_2X$) used for the formation of the quaternary phosphonium halide (I) is a highly reactive one, such as $C_6H_5-CH_2-Cl$, a simpler procedure may be used. An ester of phosphorous acid is converted by the Arbusov reaction[3] to a phosphonic ester (III),

---

[1] Wittig and Geisler, *Ann.*, **580**, 44 (1953); Maercker, *Organic Reactions*, **14**, 270 (1965).
[2] An ylid is a species with formal positive and negative charges on adjacent atoms.
[3] Kosolapoff, *Organic Reactions*, **6**, 276 (1951).

which reacts with carbonyl compounds, in the presence of a base, in the same way as an ylid.

$$(C_2H_5O)_3P + R-CH_2Cl \xrightarrow{heat} (C_2H_5O)_2\overset{+}{P}-CH_2R + C_2H_5Cl$$
$$\underset{O^-}{|}$$
$$III$$

$$III + (C_6H_5)_2C=O \xrightarrow{NaOEt} R-CH=C(C_6H_5)_2 + NaO-\underset{\underset{O}{\parallel}}{P}(OC_2H_5)_2$$

In this reaction a proton is abstracted from the phosphoric ester (III) to give an anion that adds to the aldehyde. After another molecule of base abstracts a proton from the adduct, a transitory four-membered ring (containing the phosphorous and the original aldehyde oxygen) is formed, which then undergoes carbon–oxygen and carbon–phosphorus bond cleavage to give the indicated product plus diethyl phosphate.

From benzyl-type halides and substituted benzaldehydes, unsymmetrical stilbenes can be synthesized,[4] and with benzophenones, triarylethylene derivatives are formed. α,β-Unsaturated aldehydes such as crotonaldehyde and cinnamaldehyde furnish derivatives of 1,4-butadiene.

$$C_6H_5-CH=CH-CH=O + C_6H_5-CH_2-PO(OC_2H_5)_2 \xrightarrow{NaOEt}$$
$$C_6H_5-CH=CH-CH=CH-C_6H_5$$

In the present procedure, *p*-methoxybenzaldehyde (anisaldehyde) is treated with diethyl benzylphosphonate (III) to furnish *p*-methoxystilbene.

$$CH_3O-C_6H_4CH=O + C_6H_5CH_2-PO(OEt)_2 \longrightarrow CH_3O-C_6H_4-CH=CH-C_6H_5$$

α-Halogenated esters also will react with triethyl phosphite to form phosphonic esters that are useful intermediates for modified Wittig syntheses. Ethyl bromoacetate gives triethyl phosphonoacetate (IV), which can be deprotonated with sodium hydride to form the highly active anion (V).

$$(C_2H_5O)_3P + Br-CH_2CO_2Et \longrightarrow (C_2H_5O)_2\overset{+}{P}-CH_2CO_2Et$$
$$\underset{O^-}{|}$$
$$IV$$

$$IV + NaH \longrightarrow (C_2H_5O)_2P=CH-CO_2Et \xrightarrow{R_2CO} \underset{R}{\overset{R}{\diagdown}}C=CH-CO_2Et$$
$$\underset{O^-}{|}$$
$$V$$

---

[4] Seus and Wilson, *J. Org. Chem.*, **26**, 5243 (1961); Wadsworth and Emmons, *J. Amer. Chem. Soc.*, **83**, 1733 (1961).

This ylid-like intermediate reacts with aldehydes and ketones to produce mono- and disubstituted acrylic esters. The Wittig synthesis of such compounds is often more satisfactory than a Reformatsky[5] sequence.

$$R_2C{=}O + Br{-}CH_2CO_2Et \xrightarrow[I_2]{Zn} R_2C(OH){-}CH_2CO_2Et \xrightarrow{Ac_2O} R_2C{=}CH{-}CO_2Et$$

The latter involves reaction of the ketone with ethyl bromoacetate and zinc, with iodine as catalyst, to produce a β-hydroxypropionic ester, which is dehydrated by means of acetic anhydride, potassium acid sulfate, and similar reagents.

22.2 Preparation of p-Methoxystilbene[6]

Diethyl benzylphosphonate. In a 100-mL round-bottom flask place 9 mL (8.3 g, 0.05 mole) of triethyl phosphite and 5.8 mL (6.3 g, 0.05 mole) of benzyl chloride (**Caution**—*lachrymator!*).

▶ CAUTION Avoid contact of phosphorus compounds with the skin. Wash off any spilled material thoroughly with soap and water.

Attach a condenser and heat the mixture gently for 1 hr. When the temperature reaches 130–140° evolution of ethyl chloride (bp +12°) begins. The internal temperature continues to rise and attains about 190° by the end of the hour. Allow the product to cool, and dissolve it in 10 mL of dimethylformamide.

p-Methoxystilbene. In a 125-mL Erlenmeyer flask, place 2.8 g (0.052 mole) of sodium methoxide[7] and the solution of diethyl benzylphosphonate.

---

[5] Shriner, *Organic Reactions*, **1**, 1 and 11 (1942).

[6] The specific example of the modified Wittig synthesis given here may be varied at the second step by using *p*-chlorobenzaldehyde (0.05 mole, 7 g) to give *trans*-4-chlorostilbene, mp 129°. Another example is the use of cinnamaldehyde (0.05 mole, 6.6 g) to produce 1,4-diphenyl-1,3-butadiene (see Fieser and Williamson, *Organic Experiments*, 4th ed. (Lexington, MA: Heath, 1979).

[7] Commercial sodium methoxide gives erratic results unless fresh reagent is available. Sodium methoxide sufficient for five preparations can be prepared by the procedure of Cason described in *Organic Syntheses*, Collective Volume **IV**, 651 (1963). To 130 mL of anhydrous methanol contained in a 250-mL round-bottom flask equipped with an upright condenser add *through* the condenser tube 6.0 g of clean sodium cut in small pieces. To keep the reaction under control one piece of sodium should be allowed to react completely before another is added. After all of the sodium has reacted, the excess methanol is removed by distillation, first at atmospheric pressure and then under an aspirator vacuum using a heating bath maintained at 150°. The resulting free-flowing sodium methoxide can be stored in a desiccator for several weeks.

▶ *CAUTION*   Handle sodium methoxide carefully. Any material spilled on the hands should be washed off promptly with a large quantity of water.

Swirl the mixture, and add drop by drop a solution of 6 mL (6.8 g, 0.05 mole) of *p*-methoxybenzaldehyde in 40 mL of dimethylformamide with intermittent cooling in an ice bath so that the temperature of the reaction mixture is maintained between 30 and 40°. Allow the reaction mixture to stand overnight or longer.

Pour the reaction mixture into about 50 mL of water, while stirring it, and collect the product on a suction filter. After washing thoroughly with water, crystallize the material from ethanol. The recorded melting point of *p*-methoxystilbene is 136°. The yield is 6–7 g.

Questions

1. Give a specific example of a reaction of **(a)** an arsenite (or other As^{3+} compound) and **(b)** a sulfite analogous to the formation of a phosphonate from a phosphite.

2. Indicate an appropriate synthesis for each of the following compounds by two approaches—a Wittig synthesis and one other method (for example, Grignard reaction, Meerwein arylation, Perkin reaction, Reformatsky reaction.
   (a) 4-methoxy-4'-chlorostilbene
   (b) methyl β-methylcinnamate [$C_6H_5-C(CH_3)=CH-CO_2CH_3$]
   (c) 4,4'-distyrylbenzene ($C_6H_5-CH=CH-C_6H_4-CH=CH-C_6H_5$)
   (d) 3,4-dimethoxy-α-methylcinnamic acid

23 The Cannizzaro Reaction

23.1 Reactions of Aromatic Aldehydes

Aromatic aldehydes, like aliphatic aldehydes, undergo addition reactions of the carbonyl group leading to cyanohydrins, acetals, oximes, phenylhydrazones, and similar derivatives (see Chapters 9 and 21). They also undergo reactions such as the Cannizzaro reaction and the benzoin condensation (Chapter 33), which require the absence of acidic hydrogens on the carbon adjacent to the aldehyde group (α-hydrogens).

In the presence of strong alkalis, benzaldehyde (like formaldehyde) undergoes disproportionation to form the corresponding primary alcohol and a salt of the carboxylic acid: the Cannizzaro reaction.[1]

$$C_6H_5-\underset{H}{\overset{O^-}{\underset{|}{C}}}-OH + O=CH-C_6H_5 \longrightarrow C_6H_5-\overset{O}{\underset{\|}{C}}-OH + {}^-O-CH_2-C_6H_5 \longrightarrow$$

$$C_6H_5-\overset{O}{\underset{\|}{C}}-O^- + HO-CH_2-C_6H_5$$

The process involves addition of hydroxyl ion to the carbonyl group of one molecule and transfer of hydride anion from the adduct to a second molecule of benzaldehyde, accompanied by proton interchange to form the benzoate anion and benzyl alcohol. If the reaction is effected under anhydrous

[1] For a discussion of the Cannizzaro reaction see Geissman, *Organic Reactions*, **2**, 94 (1944).

conditions with the sodium derivative of benzyl alcohol ($NaOCH_2C_6H_5$) as catalyst, the product is the ester, benzyl benzoate. Aluminum alkoxides in catalytic amount, under anhydrous conditions, convert aromatic and aliphatic aldehydes to esters (the Tishchenko reaction).

For more efficient conversion of an aromatic aldehyde to the corresponding alcohol, one employs a *crossed* Cannizzaro reaction, with formaldehyde to serve as the donor of hydride ion. An excess of formaldehyde is used, and the aromatic aldehyde is transformed almost entirely to the alcohol; surplus formaldehyde is converted to potassium formate and methanol. Formaldehyde may be used in this manner with 2-furaldehyde and the tertiary aliphatic aldehydes.

$$H-CH=O + C_6H_5-CH=O + KOH \longrightarrow H-CO_2K + C_6H_5-CH_2-OH$$

Benzaldehyde differs from aliphatic aldehydes in its behavior toward ammonia. Three molecules of the aldehyde react with two molecules of ammonia to form a crystalline *hydramide*, hydrobenzamide (mp 101–102°): $C_6H_5-CH=N-CH(C_6H_5)-N=CH-C_6H_5$. Another difference is that aromatic aldehydes do not form cyclic trimers (1,3,5-trioxane derivatives), such as those obtained from formaldehyde and acetaldehyde.

23.2 Preparations and Reactions

(A) Benzyl Alcohol[2]

In a small beaker dissolve 0.27 mole (18 g of 85% pure solid) of solid potassium hydroxide in 18 mL of water and cool the solution to about 25°. Place 0.2 mole (21 g, 20 mL) of benzaldehyde in a 125-mL Erlenmeyer flask (or narrow-mouth bottle) and to it add the potassium hydroxide solution. Cork the flask firmly and shake the mixture thoroughly until an emulsion is formed. Allow the mixture to stand for 24 hr or longer. At the end of this period, the odor of benzaldehyde should no longer be detectable.

To the mixture add just enough distilled water to dissolve the precipitate of potassium benzoate. Shake the mixture thoroughly to facilitate solution of the precipitate. Extract the alkaline solution with three or four 20-mL portions of methylene chloride to remove the benzyl alcohol and traces of any unconverted benzaldehyde. Combine the methylene chloride extracts for isolation of benzyl alcohol and reserve the aqueous solution to obtain the benzoic acid.

Concentrate the methylene chloride solution of benzyl alcohol by distillation from a steam bath, using a water-cooled condenser, until the volume

---

[2] In planning the laboratory schedule, it should be observed that this experiment requires materials to be mixed and allowed to stand for 24 hr or longer.

of the residual liquid has been reduced to 15–20 mL. Cool the liquid, transfer it to a small separatory funnel (using 2–3 mL of methylene chloride to rinse the distilling flask), and shake it thoroughly with two 5-mL portions of 20% aqueous sodium bisulfite to remove any benzaldehyde. Wash the methylene chloride solution finally with two 10-mL portions of water and dry it with 3–4 g of anhydrous magnesium sulfate. Filter the solution into a small dry distilling flask and carefully distill off the methylene chloride. Attach a short air-cooled condenser and distill the benzyl alcohol, by heating the flask directly with a luminous flame kept in motion. Collect the material boiling at 200–206°. The yield is 4–5 g.

Reactions. Benzyl alcohol may be characterized by reaction with phenyl isocyanate to form the *N*-arylcarbamic ester (*phenylurethan*) or by treatment with *p*-nitrobenzoyl chloride, in the presence of pyridine, to obtain the crystalline *p*-nitrobenzoic ester.

$$C_6H_5-N=C=O + C_6H_5-CH_2OH \longrightarrow C_6H_5-NH-CO-OCH_2C_6H_5$$
$$\text{(mp 78°)}$$

$$O_2N-C_6H_4-CO-Cl + C_6H_5-CH_2OH \xrightarrow{C_5H_5N} O_2N-C_6H_4-CO-OCH_2C_6H_5$$
$$\text{(mp 85°)}$$

These reactions are described in Chapter 9.

(B) Benzoic Acid

To free the acid, pour the aqueous solution of potassium benzoate (from which the benzyl alcohol has been extracted) *into* a vigorously stirred mixture of 40 mL of concentrated hydrochloric acid, 40 mL of water, and 40–50 g of chipped ice. Test the mixture with indicator paper to make sure that it is strongly acidic. Collect the benzoic acid with suction and wash it once with cold water. Crystallize the product from hot water, collect the crystals, and allow them to dry thoroughly. The yield is about 8 g.

Aromatic and aliphatic carboxylic acids generally are characterized by conversion to crystalline amides. Benzoic acid may be converted to benzamide (mp 130°) or benzanilide (mp 160°). For this purpose the acid usually is converted by means of thionyl chloride to the acid chloride, which is treated with ammonia or an arylamine to obtain the desired amide (Chapter 9).

A valuable aid in the identification of an unknown organic acid is the determination of its equivalent weight (neutralization equivalent) by titration with a standard base. This method may be used also to check the purity of a sample of a known acid. The procedure is described in Chapter 9.

Questions

1. Write equations for the preparation of benzaldehyde from
   (a) benzene
   (b) toluene
   (c) benzoic acid

2. Benzaldehyde forms two stereoisomeric oximes, mp 35° and 130°. How may their configurations be determined?

3. Write equations for the reaction of benzaldehyde with the following reagents.
   (a) methanol (+hydrogen chloride catalyst)
   (b) semicarbazide
   (c) *p*-tolylmagnesium bromide, followed by dilute acid
   (d) sodium cyanide and ammonium chloride, followed by hydrolysis (the Strecker reaction)
   (e) aluminum isopropoxide (the Meerwein–Pondorff reaction)

4. Compare the aldol condensation and the Cannizzaro reaction from the standpoint of the structure of the aldehyde involved.

5. Acetaldehyde, when treated with an excess of formaldehyde in the presence of a basic catalyst, furnishes pentaerythritol, $C(CH_2-OH)_4$ (mixed aldol + crossed Cannizzaro reaction). Write equations, stepwise, for the reactions involved.

6. When an equimolecular mixture of benzaldehyde and cyclohexanone is treated with semicarbazide hydrochloride and sodium acetate, and the reaction mixture is worked up *within a few minutes,* the product is cyclohexanone semicarbazone. But if the reaction mixture is allowed to stand overnight or longer, the product is benzaldehyde semicarbazone! Can you account for this difference? (Consider rates of reaction *versus* equilibria—Chapter 20.)

24 Esters

24.1 Esterification and Saponification

Esters may be prepared by direct esterification of an acid with an alcohol in the presence of an acid catalyst (sulfuric acid, hydrogen chloride) and by alcoholysis of acid chlorides, acid anhydrides, and nitriles. Occasionally they are prepared by heating the metallic salt of a carboxylic acid with an alkyl halide or alkyl sulfate.

Direct esterification is an acid-catalyzed nucleophilic addition of an alcohol to the carboxyl group of an organic acid. The reaction occurs through the following mechanism, illustrated with acetic acid and ethanol: (1) protonation of the carboxyl group, (2) addition of the alcohol and transfer of a proton to one of the hydroxyl groups, (3) elimination of water and deprotonation. It has been demonstrated that an oxygen atom of the carboxyl group is eliminated as water and the oxygen atom of the alcohol is

$$CH_3C\overset{O}{\underset{OH}{\diagup}} \underset{H^+}{\rightleftharpoons} CH_3C\overset{\overset{+}{O}H}{\underset{OH}{\diagup}} \underset{C_2H_5OH}{\rightleftharpoons} \left[CH_3\underset{OH}{\overset{OH}{\underset{|}{C}}}-\overset{+}{\underset{H}{O}}\diagup^{C_2H_5} \right] \rightleftharpoons$$

$$\left[CH_3-\underset{\overset{+}{O}H}{\overset{OH}{\underset{|}{C}}}-OC_2H_5 \right] \underset{H^+}{\rightleftharpoons} CH_3C\overset{\overset{+}{O}H}{\underset{OC_2H_5}{\diagup}} + H_2O \underset{H^+}{\rightleftharpoons} CH_3C\overset{O}{\underset{OC_2H_5}{\diagup}}$$

retained in the ester. Since the reaction is reversible, the equilibrium must be shifted forward to obtain good conversion to the ester. The use of an excess of one of the initial reactants, removal of one of the products, or a combination of both serves this purpose.

The composition of the equilibrium mixture is given approximately by the mass law, shown in equation (24.1), where K_E is the equilibrium constant for esterification and the symbols [ester], [water], etc., refer to *concentrations* expressed in moles per liter or as mole fractions.

$$K_E = \frac{[\text{ester}][\text{water}]}{[\text{acid}][\text{alcohol}]} \tag{24.1}$$

Starting with 1 mole of acetic acid and 1 mole of ethanol (a total of 2 moles), the equilibrium mixture is found experimentally to contain 0.66 mole of ethyl acetate (and an equimolar amount of water). Thus, the mole fractions of ester and water are 0.66/2, and the mole fractions of unesterified acid and alcohol are 0.34/2. Putting these equilibrium concentrations into the mass law expression gives a K_E value of 3.77 for this particular esterification.

$$K_E = \frac{(0.33)(0.33)}{(0.17)(0.17)} = 3.77 \tag{24.2}$$

Inspection of the equilibrium expression shows that the use of an excess of the alcohol (or an excess of the organic acid) will increase the amount of ester formed. Calculations based upon the K_E value of 3.77 indicate that the use of 2 moles of ethanol to 1 mole of acetic acid will bring about an 80% conversion of the acid to ethyl acetate, and 3 moles of ethanol will effect almost 90% conversion to ester. The choice of reactant to be used in excess will depend upon factors such as availability, cost, and ease of removal of excess reactant from the product.

Under ideal conditions, the composition of an equilibrium mixture is not affected by the presence or absence of a catalyst, but experiments have shown that the observed K_E values may increase as much as twofold if a relatively large amount of the acid catalyst is used. In these situations, the "catalyst" changes the environment within the system and removes through its hydration, the water formed in the reaction.

Driving an esterification to completion by removal of the water formed in the reaction is a common practice, especially in larger scale preparations. One method consists in distilling off a water–alcohol azeotrope, treating the azeotropic mixture with a drying agent, and returning the alcohol to the reaction mixture. Another procedure is to add benzene or a similar hydrocarbon and distill out a ternary azeotropic mixture, benzene–alcohol–water.

The *rate of reaction* is influenced significantly by the structure of the alcohol and the acid, and steric factors play an important part. Increasing the number of bulky substituents in the α- or β-position of the acid brings about a marked reduction in the rate constant for esterification. The reaction rates for two series of acids follow the following sequences.

$$H-CO_2H > CH_3-CO_2H > (CH_3)_2CH-CO_2H \gg (CH_3)_3C-CO_2H$$

$$\underset{(0.51)}{C_2H_5-CH_2-CO_2H} > \underset{(0.037)}{(CH_3)_3C-CO_2H} > \underset{(0.023)}{(CH_3)_3C-CH_2-CO_2H} > \underset{(0.00016)}{(C_2H_5)_3C-CO_2H}$$

Specific rates of esterification with methanol at 40°, relative to acetic acid, are shown in the second series (in parentheses). Esters of sterically hindered acids are prepared by methods other than direct esterification: conversion of the acid to the acid chloride, followed by treatment with an alcohol; or reaction of a salt of the acid with an alkyl halide, in the presence of a secondary amine as catalyst.

Acid-catalyzed esterification is a practical method for the preparation of esters of primary and secondary alcohols with typical organic acids but is not useful for tertiary alcohols. They react very slowly and their equilibrium constants are low; K_E for t-butanol is about 0.005, compared with approximately 2 for 2-propanol and 3-pentanol and 4 for 1-propanol and 1-butanol.

Acid chlorides and anhydrides react rapidly with primary and secondary alcohols to give the corresponding esters. In the absence of a base, acid chlorides convert tertiary alcohols into alkyl chlorides; but in the persence of a tertiary amine (pyridine, triethylamine), tertiary alcohols furnish esters. Acid anhydrides are less reactive than acid chlorides but react with most alcohols upon heating. Acetylations with acetic anhydride are promoted by acid catalysts (sulfuric acid, zinc chloride) and by base catalysts (sodium acetate, tertiary amines).

Hydrolysis of an ester is the reverse of esterification. With an acid catalyst, even in the presence of a large amount of water, an appreciable amount of the ester may be present in the equilibrium mixture. Hydrolysis by strong alkalis, *saponification*, is more rapid, and is more effective because hydroxyl ion reacts with the organic acid and drives the reaction to completion.

$$CH_3-CO_2C_2H_5 + OH^- \longrightarrow [CH_3-CO_2]^- + C_2H_5-OH$$

Saponification affords a means of establishing the structure of an unknown ester, through identification of the resulting alcohol and organic acid.

Since many esters are insoluble in water, solutions of potassium hydroxide or sodium hydroxide in 85–90% aqueous methanol or ethanol are used frequently for saponifications. A high-boiling solvent such as diethylene glycol ($HO-CH_2CH_2-O-CH_2CH_2-OH$, bp 245°) is advantageous for the saponification of esters of high molecular weight, and esters of sterically hindered acids or tertiary alcohols. Compounds of the latter group are very resistant to saponification.

24.2 Glyceryl Esters—Fats and Fatty Oils

Fats and fatty oils represent one of the three main groups of foodstuffs; the others are carbohydrates and proteins. Apart from their use as foods, fats and fatty oils are used in enormous quantities in the manufacture of household and industrial products.

Fats and fatty oils are mixtures of esters of the trifunctional alcohol glycerol, $HOCH_2CHOHCH_2OH$, with the higher fatty (aliphatic) acids, C_6 to C_{24}. The nature of the acyl groups present affects the physical and chemical properties of the glycerides. Fats, which are solid or semisolid at room temperature, are made up largely of the glycerides of long-chain *saturated* acids, chiefly the straight-chain C_{16} and C_{18} acids (palmitic and stearic acids). Typical fatty oils consist mainly of glycerides of *unsaturated* fatty acids, which are chiefly C_{18} acids containing one, two, or three double bonds per molecule (oleic, linoleic, and linolenic acids). A few fatty oils owe their liquid character to the presence of glycerides of the lower saturated fatty acids, C_6–C_{14}; for example, coconut oil contains large amounts of the glycerides of lauric and myristic acids (C_{12} and C_{14}), as well as glycerides of lower acids (C_{10}, C_8, and C_6).

The term saponification came from the ancient art of making soap by heating fats and fatty oils with potassium hydroxide solution obtained by leaching wood ashes (pot ash) with slaked lime. The term is now used to designate the general process of hydrolysis of an ester with caustic alkalies. Soluble soaps are the sodium (or potassium) salts of the higher aliphatic acids.

Vegetable oils that contain a substantial proportion of glycerides of fatty acids having two or more double bonds have the property of drying to a hard durable film on exposure to air. Linseed oil from flaxseed is one of the principal drying oils used in the manufacture of paints and varnishes and materials such as oilcloth and linoleum. Many synthetic substitutes for natural drying oils have been developed, such as modified alkyd, epoxy, and acrylate polymers that become cross-linked on exposure to air to form high-molecular-weight polymers.

In soap manufacture *glycerol* is recovered by evaporation and vacuum distillation of the aqueous solution remaining after the soap has been salted out by means of sodium chloride. It is used in large amounts for making alkyd resins, nitroglycerine (glyceryl trinitrate), and cosmetic preparations. Because it is hygroscopic and has a high boiling point (280° at 760 mm) glycerol finds application as a moistening agent for tobacco, printing inks, lipsticks, and textile processing. Since 1950 glycerol has been manufactured by a synthetic route, starting from propylene.

24.3 Detergents and Wetting Agents

Water and oil do not mix because water does not wet an oily surface. The forces of attraction between water and oil molecules are much smaller than the attraction of water for other water molecules, and not large enough to overcome the surface tension of the water. In the presence of small amounts of a third substance that contains a hydrophilic (water-attracting) *and* an oleophilic (oil-attracting) group in the same molecule, the surface tension is reduced so that droplets of water and oil can become commingled to form an emulsion. Detergents are cleaning agents that remove films of oil and dirt adhering to a surface by emulsification with water. Sodium and potassium soaps, the oldest detergents, are effective because of the presence of a long hydrocarbon group (lipophilic) and a carboxylate anion (hydrophilic). Synthetic detergents of similar type are produced commercially in a large scale. Examples are the sodium salts of long-chain alkyl sulfates ($C_{12}H_{25}$—O—SO_3Na) and sulfonates. An important advantage of the synthetic detergents is that they do not give precipitates with hard water.

The early synthetic detergents were *branched-chain* alkylbenzenesulfonates, but these compounds are not readily biodegradable. They caused difficulty in sewage treatment plants because they are not broken down by the microorganisms used for this purpose. Straight-chain alkyl sulfonates and sulfates, like ordinary soaps, are biodegradable. Methane gas from the biodegradation of sewage and garbage is becoming a practical energy source.

24.4 Preparations and Reactions

(A) *n*-Butyl Acetate: Esterification of Acetic Acid[1]

In a 100-mL round-bottomed flask provided with a water-cooled reflux condenser, mix *thoroughly* 15 mL (11 g, 0.15 mole) of *n*-butyl alcohol (1-butanol), 17 mL (18 g, 0.30 mole) of glacial acetic acid, and 2 mL (3.6 g) of concentrated sulfuric acid. Reflux the mixture on a wire gauze for 2 hr.

---

[1] Other esters such as *n*-propyl acetate or ethyl propionate may be prepared from the appropriate alcohol and acid combination by using proportionate molar quantities of reactants. A twofold excess of acetic acid is used to help drive the equilibrium reaction toward complete conversion of the alcohol into ester. The excess acid is easily removed by base extraction.

Cool the contents of the flask slightly, remove the condenser, and set it downward for distillation.

▶ *CAUTION*   *n*-Butyl acetate is a flammable liquid. Avoid fire hazards.

Add a boiling chip and distill the reaction mixture until the residue in the distilling flask amounts to only a few milliliters (avoid overheating the residue). The distillate consists of butyl acetate, together with *n*-butyl alcohol, acetic acid, sulfurous acid, and water. Place the distillate in a 500-mL Erlenmeyer flask or beaker, cool well by immersion in cold water or in an ice bath, and add carefully in small portions (**Caution**—*foaming!*) saturated aqueous sodium carbonate, until testing with blue litmus paper shows that the acid present is neutralized completely.

Transfer the mixture to a separatory funnel, remove the lower layer as completely as possible, and wash the *n*-butyl acetate layer with about 10 mL of water. Separate the water carefully and dry the *n*-butyl acetate with a small amount of anhydrous magnesium sulfate (or 8–10 g of Drierite).[2] Decant through a funnel into a dry distilling flask. Add a small boiling chip, distill on a wire gauze, and collect in a weighed bottle the portion boiling at 119–125°. The yield is 12–14 g.

The infrared spectrum of *n*-butyl acetate is shown in Figure 10.6.

Saponification of *n*-Butyl Acetate. In a 250-mL round-bottom flask provided with a reflux condenser, place 5 g of *n*-butyl acetate. To this add 25 mL of 10% aqueous sodium hydroxide and about 45 mL of water. Add two small boiling chips and boil until the odor of *n*-butyl acetate can no longer be detected (about 2 hr). Set the condenser downward for distillation, distill off about 30 mL of liquid, and examine the distillate. What does it contain? Add enough clean salt (sodium chloride) to saturate the liquid (about 10 g will be required), shake it thoroughly, and allow it to stand undisturbed for a short while.

(B) Methyl Benzoate[3]

In a 100-mL round-bottom flask place 12.2 g (0.1 mole) of benzoic acid and 40 mL (32 g, 1 mole) of methanol. Carefully pour 3 mL of concentrated sulfuric acid down the wall of the flask and swirl the flask to obtain good

---

[2] Anhydrous calcium chloride is not suitable because it forms addition complexes with esters.

[3] An interesting variation of this esterification experiment is to use a mixture of methanol and isopropyl alcohol in a 1:1 molar ratio and to determine by gas chromatography the composition of the esters formed. It should be noted that the boiling point of isopropyl benzoate is about 18° higher than methyl benzoate, so that the distillation is carried far enough to obtain the isopropyl ester.

An independent value for the composition of the mixed esters may be obtained by determination of the saponification equivalent.

mixing. Add two small boiling chips, attach an upright condenser, and reflux the mixture for 1 hr in a water bath. Cool the solution to room temperature and decant it from the boiling chips into a separatory funnel containing about 50 mL of water and 50 mL of methylene chloride (dichloromethane). Rinse the reaction flask with 10–15 mL of methylene chloride and pour this into the separatory funnel. Shake the mixture vigorously and separate the organic liquid from the aqueous layer, which contains sulfuric acid and methanol. Wash the organic liquid with 25 mL of water and then with 25 mL of 5% aqueous sodium carbonate (**Caution**—*foaming!*) to remove unesterified benzoic acid. Shake the mixture gently at first, then insert the stopper and shake more vigorously; invert the separatory funnel and release internal pressure by opening the stopcock. Separate the layers carefully and reserve the aqueous portion for recovery of benzoic acid.[4]

Shake the organic layer with a second 25-mL portion of sodium carbonate solution and separate the layers carefully (the aqueous layer may be discarded). Finally, wash the organic layer with water, separate the layers, and place the organic layer in a dry Erlenmeyer flask. Add a small amount

[4] The unconverted benzoic acid may be recovered by careful acidification of the sodium carbonate extract with hydrochloric acid (**Caution**—*foaming!*). The precipitated benzoic acid is collected with suction, washed with a little water, and dried. If an appreciable amount of the acid is recovered, this may be taken into account in calculating the yield of ester.

FIGURE 24.1 NMR Spectrum of Neat Methyl Benzoate

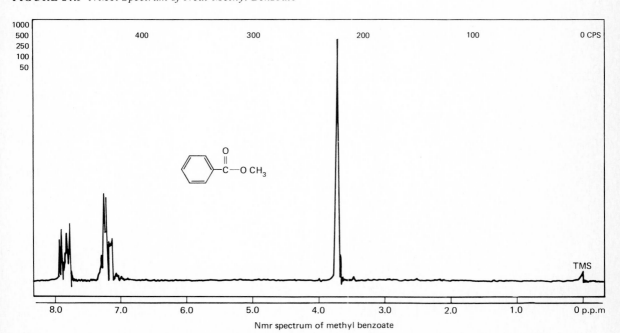

Nmr spectrum of methyl benzoate

of anhydrous magnesium sulfate,[2] cork the flask firmly, shake the mixture thoroughly, and allow it to stand for at least 20 min.

Filter the liquid into a dry flask, attach a condenser, add a boiling chip, and distill off the methylene chloride from a water bath or steam bath. Use a long condenser and collect the recovered solvent in a receiver cooled in an ice bath. When no more solvent distills over, decant the residual ester into a small dry distilling flask. Fit the side arm with a short air-cooled condenser tube (or an ordinary condenser without water in the jacket),[5] add a boiling chip, and distill the product by heating the flask directly with a small luminous flame kept in motion. Collect in a dry weighed bottle the fraction boiling above 190° (mainly in the range 192–196°, uncor). The yield is 9–10 g.

The saponification equivalent of the product may be determined by the procedure described in Section 9.8, using a 1-mL sample (1.09 g). Nitration of methyl benzoate furnishes a crystalline derivative, methyl *m*-nitrobenzoate, mp 78° (see Chapter 28).

Obtain IR and NMR spectra of your product. The NMR spectrum of neat methyl benzoate is shown in Figure 24.1.

Questions

1. What is the purpose of adding sulfuric acid in the preparation of *n*-butyl acetate? Would any of the ester be formed in the absence of sulfuric acid?
2. What procedures may be used to drive an esterification toward completion?
3. Assuming a K_E value of 4, calculate the percentage conversion of butyl alcohol to ester (at equilibrium) with the molar ratio of reactants used in this preparation. (Answer = ~85%.)
4. Assuming a K_E value of 3, calculate the percentage conversion of benzoic acid to methyl benzoate (at equilibrium) with the molar ratio of reactants used in this preparation. (Answer = ~96%.)
5. What is formed by the action of ammonia on methyl benzoate?
6. Suggest a method for preparing *t*-butyl benzoate from benzoic acid and *t*-butyl alcohol.
7. Contrast the behavior of primary, secondary, and tertiary alcohols in esterification with a carboxylic acid and in reaction with hydrobromic acid. Explain.
8. Why is the hydrolysis of an ester by an alkali called saponification? Why is this preferable to the use of an acid catalyst such as sulfuric acid?
9. In what way would the saponification equivalent of a sample of *n*-butyl acetate be affected by the presence of the following impurities?
    (a) water
    (b) butyl alcohol
    (c) acetic acid
10. What is the value of the saponification equivalent of the following compounds?
    (a) $CH_3CO-OCH_2CO-OCH_3$
    (b) $CH_3O-CH_2CH_2CO-OCH_3$
    (c) tripalmitin (glyceryl tripalmitate)

---

[5] For substances that boil above 160–170° an air-cooled condenser is used since the hot vapors may crack a water-cooled condenser.

25 Ionization of Carboxylic Acids

25.1 Introduction

Carboxylic acids contain the functional group $-\overset{\overset{O}{\|}}{C}-OH$, which ionizes to release a proton according to the equilibrium of equation (25.1).

$$R-\overset{\overset{O}{\|}}{C}-OH \rightleftharpoons R-\overset{\overset{O}{\|}}{C}-O^- + H^+ \qquad (25.1)$$

The acid dissociation constant, K_a, for this equilibrium is written in terms of the concentrations[1] of the species involved.

$$K_a = \frac{[RCO_2^-][H^+]}{[RCO_2H]} \qquad (25.2)$$

In water, typical carboxylic acids have K_a values of about 10^{-4} to 10^{-5}, which means that they are much weaker than acids such as sulfuric or hydrochloric but much stronger than very weak acids such as the alcohols ($K_a \approx 10^{-15}$) or the exceedingly weak acids such as the hydrocarbons ($K_a \approx 10^{-50}$).

---

[1] The value of K_a, defined in terms of concentrations, varies slightly with the concentration of the carboxylic acid or on addition of salts. The so-called thermodynamic constant substitutes *activities* for concentrations and is truly constant, but the estimation of activities is complex, and organic chemists normally employ the simpler concentration expression.

For convenience in expressing the wide range of observed acidities, it is common practice to substitute pK_a values, defined by analogy to $pH = -\log_{10}[H^+]$ as the negative base 10 logarithm of the K_a.

$$pK_a = -\log_{10} K_a$$

In these terms the pK_a values of typical carboxylic acids in water are 4–5, alcohols are about 15, and hydrocarbons are about 50. The important point is that larger pK_a values correspond to *weaker* acids.

It is convenient to take the negative logarithm of both sides of equation (25.2) to give equation (25.3), which in the biological literature is called the Henderson–Hasselbalch equation.

$$pK_a = -\log_{10}\left(\frac{[RCO_2^-]}{[RCO_2H]}\right) + pH$$

or
$$pH = pK_a + \log_{10}\left(\frac{[RCO_2^-]}{[RCO_2H]}\right) \quad (25.3)$$

Equation (25.3) shows that when the concentration of a carboxylic acid equals the concentration of the carboxylate anion, the logarithmic term is zero and the pH of the solution is the pK_a of the acid.

25.2 Inductive Effects

When electron-withdrawing substituents are attached to the R group of a carboxylic acid, the acid strength increases (larger K_a, smaller pK_a). For example, the pK_a of acetic acid in water is 4.74, whereas the pK_a of α-chloroacetic acid is 2.86.

$$\underset{pK_a = 4.74}{CH_3C\overset{O}{\underset{\|}{-}}OH} \qquad \underset{pK_a = 2.86}{\overset{Cl}{\underset{}{\diagdown}}CH_2-C\overset{O}{\underset{\|}{-}}OH}$$

The increase in acidity by 1.88 pK_a units is usually attributed to the electrostatic stabilization by the polar C—Cl bond of the negative charge formed on the anion during ionization.

$$\overset{\delta-}{Cl} \qquad \overset{\frac{1}{2}-}{O}$$
$$\diagdown \overset{\delta+}{C}-C \diagup$$
$$\diagup \qquad \diagdown$$
$$H \quad H \quad \overset{\frac{1}{2}-}{O}$$

Actually, the stabilizing interaction (inductive effect) is quite complex and the full explanation requires consideration of the solvent not shown in the equilibrium represented by equation (25.1). In the gas phase, where there is no solvent, inductive effects can be greater by a factor of 10. When the solvent is changed from water to one with more hydrocarbon character, which is poorer at solvating the carboxylate anion, the pK_as become much larger.

The acid-stabilizing inductive effect of a substituent falls off markedly with increasing distance from the anion centers. This phenomenon is exemplified by the chloropropionic acids.

$$CH_3CH_2CO_2H \qquad CH_3\underset{|}{\overset{Cl}{C}}HCO_2H \qquad ClCH_2CH_2CO_2H$$
$$pK_a = 4.87 \qquad pK_a = 2.83 \qquad pK_a = 3.83$$

The falloff can be understood as a consequence of the inverse distance dependence of the electrostatic interaction of charges.

When more than one substituent is included, the inductive effect is greater but not strictly additive. This is exemplified by the chloroacetic acids.

$$CH_3CO_2H \qquad ClCH_2CO_2H \qquad Cl_2CHCO_2H \qquad Cl_3CO_2H$$
$$pK_a = 4.74 \qquad pK_a = 2.86 \qquad pK_a = 1.26 \qquad pK_a = 0.64$$

25.3 Analysis of pH/Titer Data for pK

From a set of pH versus basic titer data for the titration of a carboxylic acid, there are several ways to analyze the data to determine the pK of the acid. The method chosen depends on the precision desired.

In all cases, one starts by plotting the pH as a function of the total volume of base added, as shown in Figure 25.1. The steep rise at the right of the pH curve is the region of neutralization of the acid; the volume of base required to produce the steep rise is the equivalence point of the acid. From this volume and the known normality of the base, one can calculate the equivalents of acid present in the original solution. This quantity can be compared with the amount of acid placed in the starting solution and is a useful check on the purity of the acid. The pK of the acid is (approximately) the pH of the solution at the base volume where half of the acid has been neutralized. This result follows directly from equation (25.3). At the half-neutralization point, the concentration of the remaining undissociated acid, [HA], is almost exactly equal to the concentration of the anion, [A$^-$], so that [A$^-$]/[HA] = 1 and K_a = [H$^+$], and thus pK_a = pH.

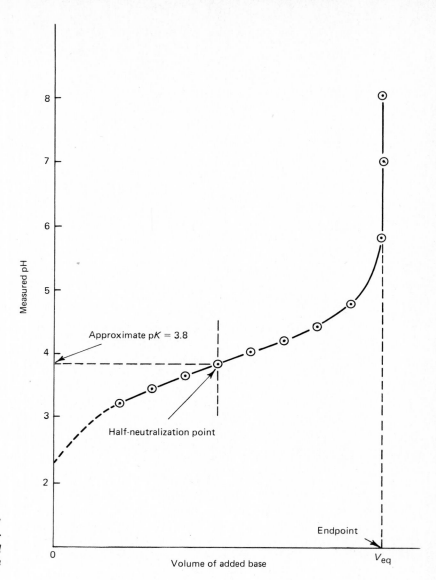

FIGURE 25.1
Titration Curve for Carboxylic Acid Neutralization

One problem with this simple approach is that it puts undue weight on a single pH measurement. If the measurement does not happen to fall exactly at the half-neutralization point, additional error is introduced. These problems can be overcome by applying equation (25.4) to all of the pH measurements.

$$\left(\frac{[A^-]}{[HA]}\right)_x \approx \frac{(\text{fraction of acid neutralized})_x}{1 - (\text{fraction of acid neutralized})_x} = \frac{V_x}{V_{eq} - V_x} \quad (25.4)$$

The required ratios of [A$^-$]/[HA] can be well approximated from the fraction of acid neutralized, which in turn is derived from the volume of added base, V_x, divided by the volume of base at the equivalence point, V_{eq}. On insertion of this expression into the Henderson–Hasselbalch equation, the pK evaluated at point x becomes

$$pK = pH_x - \log\left(\frac{V_x}{V_{eq} - V_x}\right) \qquad (25.5)$$

In applying equation (25.5), it is best to use only the measurements lying between 20 and 80% neutralization. For lower percentages of neutralization, equation (25.5) is slightly inaccurate because it does not include a correction for the ionization of water; at higher percentages of neutralization, the term $V_{eq} - V_x$ approaches zero and the resulting uncertainty can produce large errors in the calculated pK.

The pK of the acid is taken as the average of the several estimates obtained from applying equation (25.5) to the measurements. Even with error-free measurements, there will be some discrepancy among the different measurements because equation (25.5) does not include any corrections for the variation in concentration of ions that occurs during titration. However, if the measurements are carried out as described in the next section, the discrepancy from this source should be less than ± 0.003 pK unit.

25.4 Measurement of the pK of a Carboxylic Acid

In a 100-mL volumetric flask, weigh out a 1-g sample of a carboxylic acid. (**Caution**—*many carboxylic acids can be quite corrosive.*) Fill the flask about half full with distilled water and swirl the flask until the acid is dissolved; then fill it to the mark with distilled water. Transfer the solution as completely as possible to a 400-mL beaker.

Standardize your pH meter according to the directions provided with the meter, wash the electrodes with distilled water from a wash bottle, and gently dry them with a tissue. Clamp the electrodes so that they dip about 2–3 cm into the acid solution contained in the 400-mL beaker. Read and record the pH.

Fill a 50-mL buret about two-thirds full with 0.5 M NaOH solution. Read the volume and be sure to record the normality of the base. Add about 1.0 mL of base and stir the solution gently with a stirring rod. Read and record the buret volume and the pH. Repeat the addition, mixing, reading, and recording operations until the difference between successive increments becomes greater than 0.25 pH unit. When this point is reached, decrease the volume of added base to about 0.2 mL, and continue the titration until the pH rises above 9.

Clean the electrodes with distilled water from a wash bottle and leave them clamped in a beaker of fresh water with the tips dipping 2–3 cm below the surface. The pH electrodes should always be stored vertically and never allowed to dry out.

Analyze the volume/pH data as described in Section 25.3 and determine the pK of the acid. If a least-squares program is used in the analysis, determine the standard deviation of the pK as well.

Questions

1. Calculate the percent dissociation of monochloroacetic acid in the following solutions at 25°.
   (a) 0.1 M
   (b) 0.01 M
   (c) 0.001 M
   What is the pH of each of these solutions?
2. Which is the stronger base at 25°, $ClCH_2CO_2^-$ or $CH_3CO_2^-$?
3. The pK_as of very strong carboxylic acids such as trichloroacetic acid are difficult to determine on a 0.01 M solution. Why?

26 Side-Chain Oxidation of Aromatic Compounds

26.1 Oxidation of Side Chains

Oxidation of methyl groups and other side chains is an important method of preparing aromatic carboxylic acids. Side-chain oxidation is also useful in the identification of aromatic hydrocarbons and their derivatives because the acids are crystalline solids that can be purified and characterized readily.

Practically all side chains with a carbon atom directly attached to the benzene ring may be oxidized to carboxylic acid groups. Thus, the xylenes are oxidized to the corresponding phthalic acids, $C_6H_4(CO_2H)_2$. Common reagents for oxidizing side chains in aromatic compounds are chromic acid, nitric acid, potassium permanganate, and potassium ferricyanide. Since the oxidation of an alkyl side chain is initiated by attack of a C—H group adjacent to the ring, a t-butyl side chain is extremely resistant.

The presence of halogens, nitro, and sulfonic acid groups does not interfere with the oxidation of the alkyl group, but if hydroxyl or amino groups are present in the benzene ring, most oxidizing agents destroy the molecules completely. On the other hand, alkyl groups in the presence of alkoxy or acetamido groups can be satisfactorily oxidized to carboxylic acids (see Section 30.4).

Phthalic anhydride, a very important intermediate in the preparation of anthraquinone dyes and other chemicals, is prepared by the direct air oxidation of naphthalene in the presence of vanadium oxide as a catalyst. This procedure applied to benzene is a convenient commercial process for the preparation of maleic anhydride.

26.2 Preparations

(A) *p*-Nitrobenzoic Acid

In a 250-mL round-bottom flask, place 2.7 g (0.02 mole) of *p*-nitrotoluene, and a solution of 9 g (0.03 mole) of sodium dichromate crystals ($Na_2Cr_2O_7 \cdot 2H_2O$) in 20 mL of water. Add 13 mL (23 g) of concentrated sulfuric acid slowly, with constant swirling and *thorough mixing*, and attach a water-cooled reflux condenser. Thorough mixing is essential to avoid the danger of the reaction getting out of control during the heating. Heat the reaction mixture carefully until oxidation starts; then remove the flame until the vigorous ebullition subsides. When the mixture has ceased to boil from the heat of reaction, replace the flame under the flask and reflux the material vigorously for 2 hr (**Caution**—*bumping!*). Cool the reaction mixture and pour it into 30–40 mL of cold water. Collect the precipitate of crude *p*-nitrobenzoic acid with suction and wash it on the filter with two 10-mL portions of water.

Grind the precipitate thoroughly in a mortar to break up the lumps, and then transfer it to a 250-mL beaker. Add 15 mL of about 5% sulfuric acid, made by adding 1 mL of concentrated sulfuric acid to 35 mL of water. Warm on a steam bath, and stir thoroughly to extract the chromium salts as completely as possible from the *p*-nitrobenzoic acid. Cool, filter with suction, and wash the product with two 10-mL portions of water. Transfer the crude *p*-nitrobenzoic acid to a beaker, break up any lumps of the material, and treat it with 25–30 mL of 5% aqueous sodium hydroxide. The *p*-nitrobenzoic acid dissolves, and any unchanged *p*-nitrotoluene remains undissolved; chromium salts will be converted largely to chromium hydroxide. Add 0.5–1 g of decolorizing carbon, warm to 50° with stirring for about 5 min, and filter the alkaline solution with suction.[1] Precipitate the purified acid by pouring the alkaline solution with stirring, *into* 30 mL of about 15% sulfuric acid (prepared by adding 3 mL of concentrated sulfuric acid to 30 mL of water). Collect the purified acid with suction, wash it thoroughly with cold water, and dry. The yield is 2–3 g. The *p*-nitrobenzoic acid obtained in this way is sufficiently pure for most purposes. To obtain a product of high purity, a small sample may be crystallized from a large volume of hot water or from glacial acetic acid.

Because the melting point of *p*-nitrobenzoic acid is fairly high (about 240°), a bath of di-*n*-butyl phthalate or a MEL-TEMP apparatus is preferred. The purity of the acid may be checked also by determination of its neutralization equivalent, as described in Chapter 9, but with aqueous ethanol as the solvent.

---

[1] The addition of 3–4 g, about 10 mL in volume, of a filter aid (Celite) greatly facilitates the filtration.

The NMR spectrum of *p*-nitrobenzoic acid is interesting because of its simplicity.[2]

(B) *o*-Nitrobenzoic Acid[3]

In a 500-mL round-bottom flask prepare a solution of 9 g (0.057 mole) of potassium permanganate in 150 mL of warm water and add 3 mL (3.5 g, 0.025 mole) of *o*-nitrotoluene.[3] Attach an upright condenser and heat the flask over a wire gauze until the mixture boils vigorously (**Caution—bumping!**). Swirl the mixture frequently to minimize risk of breaking the flask.[4] Continue the refluxing for 2–3 hr; the time may be divided between two laboratory periods if necessary.

Filter the hot solution through a fluted filter. If the solution is colored purple by residual permanganate, add a pinch of solid sodium bisulfite and 3 mL of concentrated hydrochloric acid; stir the solution thoroughly. Add more bisulfite if necessary to decolorize the excess permanganate, but avoid an excess.

To complete the precipitation of the nitrobenzoic acid, pour the solution *into* a well-stirred mixture of 10 mL of concentrated hydrochloric acid and 25 mL of water. Dissolve the nitrobenzoic acid by heating, add about 0.5 g of decolorizing carbon, and filter the hot solution through a fluted filter. Collect the filtrate in a 250-mL Erlenmeyer flask. If crystals have separated during the filtration, redissolve them by heating, then cork the flask very *loosely*, and set it aside to cool. Collect the crystals with suction, wash them with a little water, and allow them to dry thoroughly. The yield is 2.5–3 g.

Questions

1. What procedure may be used to arrest the oxidation of a methyl side chain at the aldehyde stage? (See Chapter 20.)
2. What acid is formed by oxidation of **(a)** *p*-cymene (*p*-isopropyltoluene) and **(b)** naphthalene?
3. How may *o*-aminobenzoic acid (anthranilic acid) be prepared from *o*-xylene?
4. Compare the ease of oxidation of toluene, benzyl alcohol, and benzaldehyde.
5. If you were given a substance that might be *o*-, *m*-, or *p*-chlorotoluene (all three of which boil at about the same temperature), how would you identify the substance?

---

[2] The aromatic protons of *p*-nitrobenzoic acid appear as a singlet in the NMR spectrum, although the protons are coupled. The nitro group and carboxylic acid group cause similar downfield shifts of the ring protons although they are *chemically* nonequivalent. Such sets of protons are said to be *accidentally isosynchronous*.

[3] *o*-Chlorobenzoic acid may be prepared by this procedure, by substituting 3 mL (3.2 g) of *o*-chlorotoluene for the *o*-nitrotoluene.

[4] For this reaction, it is advantageous to use a magnetic stirrer if one is available. Another way of moderating the bumping is to provide the flask with a Claisen adapter containing a glass tube in the central arm, extending nearly to the bottom of the flask, through which a *very slow* stream of compressed air (or steam) can be introduced.

27 Friedel–Crafts Reactions

27.1 Alkylation of Benzene and Related Hydrocarbons

The Friedel–Crafts reaction, discovered in 1877, has become the most important method for introducing alkyl and acyl groups into benzene and other aromatic compounds. It depends upon the remarkable catalytic activity of anhydrous aluminum chloride. Other catalysts having similar activity (boron trifluoride and hydrogen fluoride) and useful modifications of the reaction have been developed in more recent studies.

Typical alkylation reactions make use of alkyl halides (including aralkyl types such as Ar—CH_2Cl, Ar—$CHCl_2$) and alkenes.

$$C_6H_6 + CH_3CH_2Cl \xrightarrow{AlCl_3} C_6H_5-CH_2CH_3 + HCl$$

$$C_6H_6 + CH_2=CH_2 \xrightarrow[HCl]{AlCl_3} C_6H_5-CH_2CH_3$$

Halogen atoms directly attached to an aromatic ring are inert and do not take part in the reaction. Aliphatic halides with two or more reactive halogen atoms furnish diarylated alkanes and higher arylated compounds.

The role of the catalyst is to generate a highly reactive intermediate (carbocation) that attacks the mobile electrons of the aromatic system, by an S_E2 electrophilic substitution process. Subsequent transformations lead to formation of the alkylbenzene and hydrogen chloride, and regeneration of the catalyst.

$$C_2H_5-Cl + AlCl_3 \rightleftharpoons C_2H_5{}^+[AlCl_4]^-$$
$$C_2H_5{}^+[AlCl_4]^- + C_6H_6 \rightleftharpoons [C_2H_5-C_6H_6]^+[AlCl_4]^-$$
$$[C_2H_5-C_6H_6]^+[AlCl_4]^- \rightleftharpoons C_2H_5-C_6H_5 + AlCl_3 + HCl$$

Further alkylation tends to occur because the ease of reaction (nucleophilic activity of the aromatic ring) increases with successive attachment of alkyl groups into the aromatic ring. Monoalkylation is favored by using a large excess of benzene.

As might be expected from the carbocation mechanism, primary alkyl halides and some secondary halides, give rise to products of rearrangement. *n*-Propyl halides give almost entirely isopropylbenzene; *n*-butyl halides, *sec*-butylbenzene; and isobutyl halides, *t*-butylbenzene. Similar products of rearrangement are formed from analogous pentyl and higher halides.

The orientation of attack in the alkylation of substituted benzenes is not very selective, presumably because the alkyl carbocations are so extremely reactive. Although the alkylation of toluene with isopropyl chloride under *very mild* conditions gives essentially *ortho/para* substitution (*ca.* 63% *ortho*, 12% *meta*, 25% *para*), under vigorous conditions all of the isomeric isopropyltoluenes rearrange to give entirely the *meta* isomer. This indicates that the initial products are the result of kinetic control, owing to the rapid attack by the alkyl carbocation, ion, but the reversibility of the reaction leads eventually to the most stable isomer (thermodynamic control).

In industry alkenes are used for alkylation of benzene, under conditions different from the typical Friedel–Crafts synthesis. The catalysts used may be a liquid complex of aluminum chloride–hydrogen chloride–hydrocarbon; phosphoric acid supported on a solid substrate; anhydrous hydrogen fluoride; or concentrated sulfuric acid. Propylene and benzene, over a phosphoric acid catalyst at 250° and under high pressure, give isopropylbenzene (cumene). Cumene is an intermediate for an efficient process for the manufacture of phenol (via cumene hydroperoxide). Cyclohexene and benzene, with sulfuric acid as catalyst, give cyclohexylbenzene.

In laboratory preparations using reactive tertiary halides, it is convenient to form anhydrous aluminum chloride directly in the reaction mixture, from amalgamated aluminum metal.

27.2 Friedel–Crafts Acylation

Acylation of aromatic compounds by means of the Friedel–Crafts reaction and its modifications is one of the chief synthetic methods for the preparation of aromatic ketones. Aliphatic and aromatic acid chlorides, in the presence of anhydrous aluminum chloride, react with aromatic compounds to furnish alkyl aryl ketones and diaryl ketones.

In acylations with acid chlorides a slight excess over 1 mole of aluminum chloride is used, since a 1:1 addition compound is formed by reaction of aluminum chloride with the ketone produced in the reaction. Acid anhydrides react in a similar way to produce ketones, usually in better yields than are obtained from acid chlorides, but it is necessary to use 2 moles of aluminum chloride because the organic acid formed in the reaction reacts with aluminum chloride.

$$C_6H_6 + CH_3CO-Cl \longrightarrow C_6H_5-\underset{CH_3}{\overset{+}{C}=\overset{-}{O}-\overline{A}lCl_3} + HCl$$

$$C_6H_6 + (CH_3CO)_2O + 2\,AlCl_3 \longrightarrow C_6H_5-\underset{CH_3}{\overset{+}{C}=\overset{-}{O}-\overline{A}lCl_3} + CH_3CO_2AlCl_2 + HCl$$

The function of the catalyst is to generate a reactive acyl cation (acylonium ion), which attacks the aromatic system in the same manner as other active electrophiles.

$$CH_3-CO-Cl + AlCl_3 \rightleftharpoons CH_3-\overset{+}{C}=O[AlCl_4]^-$$

Acylation does not occur in systems that have a deactivating substituent, such as NO_2, CO_2CH_3, $CO-CH_3$, or $C\equiv N$. For this reason, Friedel–Crafts acylations do not go beyond the introduction of more than one acyl group, since the carbonyl group of the ketone deactivates the molecule for further substitution.

Unlike the alkylation reactions, where rearrangements occur readily in the alkylcarbonium intermediates, the acyl groups do not undergo rearrangement in this reaction. The intermediate acylonium cation from an acid halide or an anhydride, $[R-C\equiv O^+ \leftrightarrow R-\overset{+}{C}=O]$, has enhanced stability resulting from distribution of its positive charge between the carbonyl carbon and the oxygen atom.

Friedel–Crafts acylations with anhydrides of dibasic acids, such as phthalic, maleic, and succinic anhydrides,[1] furnish ketonic acids that are important synthetic intermediates. Benzene and phthalic anhydride give *o*-benzoylbenzoic acid, which can be decarboxylated to obtain benzophenone or cyclized to produce anthraquinone (see Chapter 38). Derivatives of anthraquinone are used in the manufacture of important vat dyes.

The Fries reaction is a modification of the Friedel–Crafts reaction applicable to the preparation of *o*- and *p*-hydroxyaryl ketones. Esters of phenols with aliphatic or aromatic carboxylic acids, when heated with anhydrous aluminum chloride, undergo a rearrangement in which the *O*-acyl group enters the *ortho* or *para* position of the ring.[2] The distribution

---

[1] Berliner, *Organic Reactions*, **5**, 229 (1949).
[2] Blatt, "The Fries Reaction," *Organic Reactions*, **1**, 342 (1942).

of isomers is influenced by the conditions of reaction; at lower temperatures the *para* isomer is favored and at higher temperatures the *ortho* isomer predominates.

$$\text{Ph-O-CO-R} \xrightarrow{\text{AlCl}_3} \text{(2-HO-C}_6\text{H}_4\text{-CO-R)} + \text{R-CO-C}_6\text{H}_4\text{-OH}$$

Aromatic systems containing one or more alkoxyl groups (ethers of mono- and dihydric phenols, and naphthols) are acylated readily by heating with aliphatic acid anhydrides in the presence of iodine (1 mole percent) as a catalyst.[3] This method is simple and gives satisfactory yields. Alkoxy derivatives of benzophenone may be obtained by using aroyl chlorides, but the yields are lower.

For the acylation of aromatic compounds that cannot conveniently be used in excess as the reaction medium, or for solids such as naphthalene and biphenyl, it is desirable to employ an inert solvent. Tetrachloroethane and nitrobenzene have been used, but their boiling points are relatively high, making removal from the product troublesome. Dichloromethane (bp 40°) is well suited as an inert solvent. The following procedure, using acetylation of biphenyl as the example, is a general one that can be used for naphthalene, alkoxybenzenes, and other reactive hydrocarbons. With naphthalene this solvent of low polarity strongly favors acylation at the 1-position, whereas the highly polar nitrobenzene favors the 2-position.

27.3 Preparation of 4-Acetylbiphenyl

The preparation of the ketone 4-acetylbiphenyl is a specific example of a general procedure for acylation of aromatic compounds in the presence of an inert solvent.

$$\text{CH}_3\text{CO-C}_6\text{H}_4\text{-C}_6\text{H}_5$$

4-Acetylbiphenyl

In a hood, assemble an apparatus with a 250-mL round-bottom flask fitted with a Claisen adapter; place a separatory funnel in the central opening of the adapter and an upright condenser in the side arm. Attach a

---

[3] Chodroff and Klein, *J. Amer. Chem. Soc.*, **70**, 1647 (1948); Dominguez et al., *ibid.*, **76**, 5150 (1954).

drying tube filled with calcium chloride at the top of the condenser and connect the open end of the drying tube to a gas absorption trap (see Figure 8.2) to dispose of the hydrogen chloride evolved during the reaction.

▶ *CAUTION* Acetyl chloride is extremely corrosive and reacts vigorously with water to release toxic hydrogen chloride gas. Work only in a hood and be certain that all of the apparatus is kept scrupulously dry. If any of the acetyl chloride is spilled on your skin, wash the affected area immediately and thoroughly with water.

In the flask place 40 mL of dichloromethane and 16 g (0.12 mole) of pulverized *anhydrous* aluminum chloride. Replace the condenser and add dropwise, by means of the separatory funnel, 8 mL (8.7 g, 0.11 mole) of acetyl chloride, while cooling the flask in an ice bath. Over a period of about 15 min, add a solution of 15.5 g (0.10 mole) of biphenyl in 50 mL of dichloromethane. Allow the reaction mixture to stand for an hour, occasionally shaking it, and then reflux it gently in a water bath for about 20 min.

To decompose the ketone–aluminum chloride complex, pour the cooled reaction mixture cautiously into a well-stirred mixture of 50 g of chipped ice and 40 mL of concentrated hydrochloric acid. Transfer the mixture to a separatory funnel, shake well to extract the aluminum chloride from the organic layer, and separate the layers. Rinse the reaction flask with some of the aqueous layer and then extract it with 25 mL of dichloromethane. Wash the combined organic extracts with water, then with 5% aqueous sodium hydroxide, and finally with water. Dry the dichloromethane solution with a small amount of anhydrous magnesium sulfate and distill off the solvent (bp 40°). Recrystallize the crude ketone from ethanol, with addition of decolorizing carbon. The yield is 10–11 g. The reported melting point of 4-acetylbiphenyl is 120–121°.

Questions

1. What products would be formed by the reaction of the following alkenes with benzene, in the presence of aluminum chloride?
   (a) ethylene
   (b) isobutylene
   (c) cyclohexene
2. Explain why *n*-propyl bromide reacts with benzene, in the presence of aluminum chloride, to give mainly isopropylbenzene. Suggest a method for obtaining *n*-propylbenzene.
3. What products will be formed from the following reactants, in the presence of aluminum chloride?
   (a) benzene (in excess) + $ClCH_2CH_2Cl$
   (b) benzene (in excess) + CCl_4
   (c) chlorobenzene and propionic anhydride
   (d) toluene and phthalic anhydride
   (e) *p*-tolyl benzoate (Fries rearrangement)

4. Indicate how the following substances may be prepared by means of the Friedel–Crafts reaction.
   (a) diphenylmethane
   (b) deoxybenzoin (benzyl phenyl ketone)
   (c) *m*-bromobenzophenone
   (d) *p*-nitrobenzophenone.
5. Write equations for reactions suitable for conversion of acetophenone to the following compounds.
   (a) phenylacetic acid
   (b) 1,1-diphenylethanol
   (c) α-hydroxy-α-phenylpropionic acid
   (d) *m*-aminoacetophenone
6. How could 4-acetylbiphenyl be converted to each of the following?
   (a) 4-ethylbiphenyl
   (b) biphenyl-4-carboxylic acid
   (c) 4-amino-4-acetylbiphenyl

28 Nitration of Aromatic Compounds

28.1 Mechanism of Nitration

Nitration is one of the most important examples of S_E2 electrophilic aromatic substitution. Although aromatic nitro compounds have limited usefulness as such, mainly as high explosives or booster charges (TNT, Tetryl), they are exceedingly useful as intermediates for the preparation of the corresponding amines and, indirectly, many other functional groups (—OH, —CN, —I, —AsO$_3$H$_2$). Nitrobenzene (bp 209°) is a good solvent for organic substances and also dissolves many inorganic compounds (AlCl$_3$, ZnCl$_2$). It is used occasionally as a reaction medium and as a solvent for recrystallization. Most nitro compounds are *dangerously poisonous* and must be handled carefully.

Nitrations may be effected by means of pure nitric acid,[1] mixtures of concentrated nitric and sulfuric acids, and solutions of nitric acid in glacial acetic acid or acetic anhydride. Selection of the appropriate nitrating agent and the conditions of reaction is based upon factors such as the reactivity of the compound to be nitrated, its solubility in the nitrating medium,[2] and the ease of isolation and purification of the product.

---

[1] Anhydrous nitric acid (sp g 1.50, sometimes called *white* fuming nitric acid) is a colorless liquid boiling at 86°; ordinary concentrated nitric acid (sp g 1.42) is the water-nitric acid azeotrope, bp 120°, containing 70% by weight of nitric acid; yellow fuming nitric acid contains 85–90% nitric acid with small amounts of oxides of nitrogen; red fuming nitric acid contains relatively large amounts of dissolved oxides of nitrogen.

[2] For a compound that is sparingly soluble in the nitrating mixture (aromatic hydrocarbons, aryl halides, etc.) the rate of nitration may be governed by its rate of solution in the medium; hence good agitation hastens the reactions.

The mechanism of nitration has been studied extensively and it is known that the active electrophilic species is the nitronium ion, $[O=N=O]^+$. This is formed in the typical nitrating mixtures by a reversible interaction of nitric and sulfuric acids. Attack on the aromatic system by the nitronium ion, usually the slow rate-determining step, is followed by the rapid loss of a proton, leading to the nitro derivative.

$$HNO_3 + 2\, H_2SO_4 \rightleftharpoons [O=N=O]^+ + [H_3O]^+ + 2\,[HSO_4]^-$$

The ease of nitration depends on the nature of the substituents present; electron-releasing groups ($-OH$, $-NHCOCH_3$, $-CH_3$) facilitate the nitration and electron-withdrawing groups ($-NO_2$, $-CO_2H$) retard the reaction.

Since nitration is *not* a reversible reaction, the distribution of *ortho*, *meta*, and *para* isomers in the product is controlled by the *relative* rates of substitution at each position. In general, substituents may be divided into three broad categories.

1. Activating and *ortho/para* directing

   $-OH$ $-OCH_3$ $-NHCOCH_3$ $-CH_3$ $-CH_2CO_2CH_3$

2. Deactivating and *ortho/para* directing

   $-CH_2Cl$ $-Cl$ $-Br$ $-I$

3. Deactivating and *meta* directing

   $-CCl_3$ $-COCH_3$ $-CO_2CH_3$ $-NO_2$ $-(NR_3)^+$ $-CN$ $-SO_3H$

The observed kinetic order of the reaction (in organic solvents)[3] varies as the rate of nitration changes in relation to the rate of formation of the nitronium ion. For the more active compounds, in a large excess of nitric acid, the rate is independent of the concentration of the aromatic compound; the reaction is pseudo-zero order. Compounds of intermediate activity (C_6H_6, C_6H_5Cl) show kinetic orders intermediate between pseudo-

---

[3] For a detailed discussion of aromatic substitution see Ingold, *Structure and Mechanism in Organic Chemistry*, 2nd ed. (Ithaca, NY: Cornell University Press, 1969), and Lowry and Richardson, *Mechanism and Theory in Organic Chemistry* (New York: Harper and Row, 1976).

TABLE 28.1 *Partial Rate Factors for Nitration of Substituted Benzenes*

Substituent	Distribution of isomers			Activity vs. C_6H_6	Partial rate factors		
	ortho	*meta*	*para*		f_o	f_m	f_p
—CH_3	56	4	40	24	42	2.5	58
—$C(CH_3)_3$	12	8	80	16	5.5	4	75
—Cl	30	1	69	0.033	0.033	0.001	0.137
—CH_2Cl	32	15	53	0.300	0.29	0.14	0.95
—Br	36	1	63	0.030	0.03	0.001	0.112
—CO_2Et	28	68	4	0.0003	2.5×10^{-4}	6×10^{-4}	5×10^{-5}
—NO_2	6	93	1	1×10^{-7}	1.8×10^{-6}	2.8×10^{-5}	2×10^{-7}
—$N(CH_3)_3^+$	0	90	10				

zero and pseudo-first order. For compounds that undergo nitration with difficulty ($C_6H_5NO_2$, $C_6H_5CO_2CH_3$) the rate is proportional to the concentration of the aromatic compound; the reaction is pseudo-first order.

Studies of the nitration of mixtures of benzene and a second aromatic compound have led to evaluation of the reactivity of the *ortho*, *meta*, and *para* positions of the second compound relative to a position of the benzene ring. This permits calculation of partial rate factors for the positions of the compound under comparison, with the results shown in Table 28.1.

Although deactivating for electrophilic substitution, the electron-withdrawal effect of the nitro group brings about activation of the aromatic system for *nucleophilic substitution* at the *ortho* and *para* positions. In a similar way the halogen atom in 2,4-dinitrochlorobenzene can be displaced readily by the hydroxide ion or ammonia.

$$O_2N-C_6H_4-NO_2 + KOH \longrightarrow O_2N-C_6H_4-OH + KNO_2$$

When two nitro groups are present in the *ortho* or *para* positions, one of the nitro groups will undergo nucleophilic displacement (as nitrite).

28.2 Preparations

(A) *m*-Dinitrobenzene

In a 125-mL Erlenmeyer flask mix 3 mL (3.6 g, 0.03 mole) of nitrobenzene and 8 mL of concentrated sulfuric acid. After the nitrobenzene has dissolved add 5 mL (0.085 mole) of concentrated nitric acid (70%) and mix by gentle swirling.[4] Heat is evolved and the mixture warms up to 60–70° in about 5 min. After the initial heat evolution has subsided, place the flask in 100 mL of boiling water contained in a 400-mL beaker or heat it on a steam bath. Remove the flask occasionally and swirl the contents carefully. After 15 min of heating, pour the reaction mixture into 100 mL of cold water while stirring it. Collect the crude *m*-dinitrobenzene by suction filtration and wash it *thoroughly* on the filter with two or three 20-mL portions of water. It is important to wash out the admixed nitric acid before recrystallizing the product from ethanol. For effective washing, release the suction, mix the crystals thoroughly with the washing liquid, apply suction again, and press the crystals firmly.

▶ *CAUTION*   All nitro compounds are poisonous and must be handled carefully. If any nitro compound comes in contact with the skin it should be removed by washing with a little ethanol, followed by soap and water.

Crystallize the washed product from 25 mL of ethanol, reserving a very minute quantity of material to be used as seed crystals. Allow the filtrate to cool until a slight turbidity develops, then introduce a few tiny particles of the seed crystals; swirl the solution gently and set it aside to cool slowly. After crystallization is complete, mix the semisolid mass thoroughly, filter with suction, and press the crystals on the filter. Wash the crystals with two 5-mL portions of cold ethanol and allow them to dry in the air. The yield is about 3 g.

(B) *p*-Bromonitrobenzene

In a 125-mL Erlenmeyer flask, place 10 mL (14 g, 0.17 mole) of concentrated nitric acid and add carefully 10 mL (18 g) of concentrated sulfuric acid. Cool the mixture to room temperature and add, in two or three portions, 5.5 mL (8 g, 0.05 mole) of bromobenzene. Shake the reaction flask continuously and cool it in running water, if necessary, to keep the temperature between 50° and 60°. After all the bromobenzene has been added and the temperature no longer tends to rise from the heat of reaction, place the flask in a beaker of boiling water or a steam bath, and heat for 0.5 hr, with occasional swirling.

▶ *CAUTION*   All nitro compounds are poisonous and must be handled carefully. If any nitro compound comes in contact with the skin it should be removed by washing with a little ethanol, followed by soap and water.

---

[4] With small quantities it is permissible to add the nitric acid (or other reactant) all at once, but with large amounts of material one of the reactants should be added gradually in small portions so that the heat evolution can be controlled and the reaction temperature kept within the proper limits (note the procedures used in 28.2(B) and 28.2(C)).

Cool the flask to room temperature, and pour the reaction mixture into about 100 mL of cold water. Collect with suction the crude *p*-bromonitrobenzene (containing some *o*-bromonitrobenzene), wash it with water, and press on the filter with a clean cork or flat glass stopper. To purify the product, crystallize it from 50–60 mL of hot ethanol. The *ortho* isomer is more soluble and remains in solution in the ethanol whereas the *para* isomer crystallizes when the solution is cooled. The purification can be monitored by thin-layer chromatography or gas chromatography (see Chapter 7). The yield of pure *p*-bromonitrobenzene is 5–7 g.

(C) Methyl *m*-Nitrobenzoate

Place 14.5 mL (25 g) of concentrated sulfuric acid in a 125-mL Erlenmeyer flask, cool the acid to 0°, and add 6.2 mL (6.8 g, 0.05 mole) of methyl benzoate, while swirling the solution. While maintaining the internal temperature at 5–15°, by cooling as needed in an ice-water bath, add drop by drop a cold mixture of 5 mL (9 g) of concentrated sulfuric acid and 5 mL (7 g, 0.085 mole) of concentrated nitric acid. Swirl the solution during the addition and for 10 min after all of the acid has been added.

⮕ *CAUTION* All nitro compounds are poisonous and must be handled carefully. If any nitro compound comes in contact with the skin it should be removed by washing with a little ethanol, followed by soap and water.

Pour the reaction mixture, while stirring it, onto about 50 g of cracked ice to precipitate the crude methyl *m*-nitrobenzoate (which contains an appreciable amount of the *ortho* isomer and a trace of *para*). Collect the product with suction and wash it thoroughly on the filter with two or three 15-mL portions of water, to remove nitric and sulfuric acids. For effective washing, release the suction, mix the material thoroughly with the washing liquid, apply suction, and press the crystals firmly.

Wash the product finally with two 5-mL portions of *ice-cold* methanol, in the manner described earlier, and press the crystals thoroughly. Proper washing removes most of the more soluble *ortho* isomer. The crude product weighs about 6–7 g and melts at 74–76°. It may be purified by recrystallization from a small volume of hot methanol,[5] reserve a minute amount of material to be used as seed crystals. The recorded melting point of pure methyl *m*-nitrobenzoate is 78.5°.

Questions

1. Why is sulfuric acid used in nitration?
2. Mention several important reactions that are characteristic of aromatic compounds and that differentiate them from aliphatic compounds.

---

[5] To prevent contamination through ester exchange, it is good practice in recrystallization to avoid using an alcohol different from that corresponding to the alkoxyl group of the ester.

3. What is formed by the reduction of nitrobenzene in the presence of acids?
4. Write the structural formula of
   (a) picric acid
   (b) TNT
   (c) Tetryl
5. What position will be taken by the entering nitro group when the following substances are nitrated?
   (a) phenol
   (b) acetanilide
   (c) toluene
   (d) benzaldehyde
   (e) benzonitrile
   (f) benzoic acid
   (g) phenylacetic acid
6. In the nitration of methyl benzoate why doesn't the ester hydrolyze in the highly acidic medium?

29 Nitration of Anilines: Use of a Protecting Group

29.1 Protecting Groups

In synthetic procedures primary and secondary arylamines often are converted to their acetyl derivatives as a protective measure, to reduce their susceptibility to oxidative degradation and to moderate their high reactivity in electrophilic substitution reactions (especially in halogenations). In some instances the amino group is acetylated to prevent an undesired reaction with another functional group or a reagent, such as —COCl, —SO$_2$Cl, or HNO$_2$. At the end of a synthetic sequence, the amino group can be regenerated readily by hydrolysis with acids or bases.

Arylamines (and aliphatic amines) may be acetylated by means of acetic anhydride or acetyl chloride or by heating the amine with glacial acetic acid under conditions that permit removal of the water formed in the reaction. The last procedure is an economical one but requires a relatively long period of heating.

Acetic anhydride is the preferred acetylating agent. In some instances, the solution of the acetyl derivative in glacial acetic acid that results from the acetylation may be used for a subsequent reaction of the acetylated compound without the necessity of isolating it. If the temperature and reaction time are increased in acetylations with acetic anhydride, primary amines may form a bis-acetyl derivative, Ar—N(COCH$_3$)$_2$, but this can be hydrolyzed under mild conditions to the monoacetyl derivative.

The rate of hydrolysis of acetic anhydride is sufficiently slow to permit acetylation of amines to be carried out in buffered aqueous solutions (method of Lumière and Barbier). This is a general procedure that gives a product of high purity in good yield, but it is not suitable for acetylation of the nitroanilines and other extremely weak bases. By a similar procedure, acetic anhydride may be used to acetylate phenols in an aqueous alkaline solution.

29.2 Acetylation of Aniline

(A) Acetylation in Water—Lumière–Barbier Method[1]

Dissolve 5.5 mL (5.6 g, 0.06 mole) of aniline in 150 mL of water to which 5 mL (0.06 mole) of concentrated hydrochloric acid has been added. If the amine is discolored, add 1–2 g of decolorizing carbon, stir the solution for a few minutes, and filter it with suction. Meanwhile, prepare for use in the next step a solution of 9 g (0.065 mole) of sodium acetate crystals ($CH_3CO_2Na \cdot 3H_2O$) in 20 mL of water; if any insoluble particles are present, filter the solution.

Transfer the solution of aniline hydrochloride to a 250-mL flask. Add 8 mL (8.3 g, 0.15 mole) of acetic anhydride and swirl the contents to dissolve the anhydride. Add *at once* the previously prepared sodium acetate solution and mix the reactants thoroughly by swirling. Cool the reaction mixture in an ice bath and stir vigorously while the product crystallizes. Collect the crystals on a suction filter, wash with cold water, and allow them to dry. The yield is 6–7 g. The material obtained by this acetylation procedure is usually quite pure and of better quality than that prepared by the acetylation in acetic acid. If necessary, the product may be recrystallized from water, with addition of about 1 g of decolorizing carbon.

(B) Acetylation in Acetic Acid

In a 125-mL Erlenmeyer flask dissolve 7 mL (7.1 g, 0.075 mole) of aniline in 15 mL of glacial acetic acid. To the solution add 9 mL (9 g, 0.09 mole) of acetic anhydride and mix well by swirling. The solution becomes warm from the heat of reaction. Add two boiling chips, attach a reflux condenser, and boil the solution *gently* for 15 min to complete the acetylation. To hydrolyze the excess of acetic anhydride and any bis-acetyl derivative, add cautiously through the top of the condenser tube 5 mL of water, and boil gently for 5 min longer. Allow the reaction mixture to cool slightly and pour it *slowly*, stirring thoroughly, into 35–40 mL of cold water. After

---

[1] Other arylamines (toluidines, xylidines, anisidines, phenetidines) can be acetylated by this general procedure. It is advantageous to use the Lumière–Barbier method if the sample of amine is discolored because the procedure affords a preliminary treatment with decolorizing carbon. Very weak bases (nitroanilines, dihalogenated anilines) cannot be acetylated by this method. For preparation of phenacetin, see Chapter 30.

allowing the mixture to stand for about 15 min with occasional stirring, collect the crystals on a suction filter, and wash them with a little cold water. Recrystallize the crude product from hot water (about 20 mL/g) with addition of about 1 g of decolorizing carbon. The yield is 6–7 g.

(C) Direct Acetylation with Acetic Acid

This is the most economical method of effecting acetylation, but the operation requires a longer time than methods A and B, which are most suited for small-scale operations.

In a 100-mL round-bottom flask, place 9 mL (9.2 g, 0.1 mole) of aniline and 12 mL (12 g, 0.2 mole) of glacial acetic acid. Provide the flask with a short fractionating column fitted with a thermometer and connected with a condenser arranged for distillation. For the receiver use a small graduated cylinder. Add two boiling chips and heat the flask gently, so that the solution boils quietly and the vapor does not rise into the column.

After 15 min increase the heating slightly so that the water formed in the reaction, together with a little acetic acid, distills over very slowly at a *uniform* rate (vapor temperature 104–105°). After about an hour, when 5–6 mL of distillate has collected, increase the heating so that the temperature of the distilling vapor rises to about 120°. Continue the distillation slowly for about 10 min longer, to collect additional 1–2 mL of distillate (total volume, 6–7 mL), and then discontinue the heating. The distillate, consisting of 70–75% acetic acid, may be discarded.

As the reaction mixture will solidify upon cooling, pour it out at once into about 200 mL of ice and water in a large beaker. Stir the aqueous mixture vigorously to avoid formation of large lumps of the product. Collect the acetanilide with suction, wash with a little cold water, and press it firmly on the filter. Crystallize the moist product from hot water (about 20 mL/g) with addition of about 1 g of decolorizing carbon. For filtration use a large fluted filter and a large funnel with a short, wide stem. Cool the filtrate rapidly while stirring vigorously to obtain small crystals. Allow the material to stand for about 10 min in an ice-water bath and then collect the crystals with suction. Wash the product with a small amount of cold water and spread it on a clean paper to dry. If the material is dark colored it should be recrystallized. The yield is 8–9 g.

29.3 Nitration of Acetanilide and Deacetylation

When aniline nitrate or sulfate is nitrated at low temperature with concentrated nitric and sulfuric acids, the product consists of about 60% *m*-nitroaniline and 38% *p*-nitroaniline, with very little of the *ortho* isomer. The yield is not high because some of the aniline is lost through oxidation.

The formation of *m*-nitroaniline would be anticipated because the powerful inductive electron withdrawal of the positively charged —NH_3^+ group brings about strong deactivation, especially at the *ortho* and *para* positions. The formation of a substantial amount of *p*-nitroaniline has sometimes been attributed to the nitration of free aniline (*ortho*/*para* directive effect) that is present in a small amount in equilibrium with the anilinium cation and would undergo nitration at a much faster rate. Recent experimental evidence casts doubt on this explanation since the *meta*/*para* ratio changes less than expected with increase in acidity of the nitrating medium. The observed *meta*/*para* ratio evidently represents the actual directive effect of the —NH_3^+ group. The extent of *para* nitration diminishes with successive introduction of methyl groups on the nitrogen atom; C_6H_5—$N(CH_3)_3^+$ gives 89% *meta* and 11% *para* nitro derivative.

To diminish the susceptibility to oxidation and avoid the *meta*-directive effect of salt formation, it is customary to convert an aromatic amine to its acetyl derivative before carrying out nitration. The acyl group is removed subsequently by hydrolysis with aqueous acid or alkali. Thus, aniline is converted to acetanilide, which on nitration in the usual way furnishes almost exclusively *p*-nitroacetanilide, which is hydrolyzed to *p*-nitroaniline.

Since only a small amount of *o*-nitroacetanilide is obtained by nitration of acetanilide, an indirect method is used to secure *o*-nitroaniline as the principal product of a series of reactions. Aniline is converted to sulfanilic acid, in which the reactive *para* position is blocked by the SO_3H group; nitration of sulfanilic acid produces 4-amino-3-nitrobenzenesulfonic acid; hydrolysis of the latter by boiling with 60% sulfuric acid brings about elimination of the SO_3H group and yields *o*-nitroaniline in a state of high purity.

m-Nitroaniline is prepared most conveniently by the controlled partial reduction of *m*-dinitrobenzene, using ammonium sulfide or sodium hydrosulfide (NaSH) as the reducing agent.

The three nitroanilines are extremely weak bases but differ appreciably from one another in base strength: *meta* ≫ *para* > *ortho*. Mixtures of the nitroanilines can be separated by dissolving them in strong aqueous acid and precipitating successively the *ortho*, *para*, and *meta* isomers by progressive neutralization with dilute ammonia.

p-Nitroaniline is used to prepare *p*-phenylenediamine and also for the production of a special type of azo dye, such as para red (an ingrain color, Chapter 44). Cotton cloth may be soaked in a dilute alkaline solution of 2-naphthol (or similar coupling component), dried, and dipped into an ice-cold solution of diazotized *p*-nitroaniline; coupling takes place and the dye is formed within the pores of the cellulose fibers.

p-Nitroaniline is used as an intermediate for laboratory syntheses leading to compounds that cannot be obtained readily, or in a pure state, by direct substitution processes. Thus, diazotization of *p*-nitroaniline and subsequent treatment of the diazonium fluoborate with sodium nitrite (in the presence of copper powder) affords a route to *p*-dinitrobenzene.

$$[O_2N-C_6H_4-\overset{+}{N}\equiv N][\overset{-}{BF_4}] + NaNO_2 \xrightarrow{Cu} O_2N-C_6H_4-NO_2 + NaBF_4$$

Similarly, replacement of the diazonium group by $-C\equiv N$, using sodium cyanide and cuprous cyanide (the Sandmeyer reaction), furnishes *p*-nitrobenzonitrile. Neither of these compounds can be obtained by direct nitration because of the *meta*-directive effect of the $-NO_2$ and $-C\equiv N$ groups.

(A) *p*-Nitroacetanilide

In a 125-mL Erlenmeyer flask dissolve 6.8 g (0.05 mole) of pure acetanilide in 8 mL of glacial acetic acid by warming it gently. Cool the warm solution until crystals begin to form and then add slowly, while swirling the solution, 10 mL of ice-cold concentrated sulfuric acid. Prepare a nitrating mixture by adding 3.5 mL (5 g, 0.06 mole) of concentrated nitric acid to 5 mL of cold concentrated sulfuric acid; cool the solution to room temperature and transfer it to a small separatory funnel.

Cool the acetanilide solution to 5° in an ice bath, remove the flask from the bath, and add the nitrating mixture slowly, drop by drop. Swirl the reaction mixture to obtain good mixing in the viscous solution and do not permit the temperature to rise above 20–25°. After all of the nitrating mixture has been added, allow the solution to stand at room temperature for about 40 min (but not longer than 1 hr) to complete the reaction. Pour the solution slowly with stirring into a mixture of 100 mL of water and 20–25 g of chipped ice. Collect the product with suction, press it firmly on the filter, and transfer the filter cake to a beaker. Mix the crystals thoroughly with about 75 mL of water to form a thin paste, return them to the suction filter, and wash thoroughly with more water to remove the nitric and sulfuric acids. Press the material as dry as possible. The crude, moist *p*-nitroacetanilide is sufficiently pure to be used directly for hydrolysis to *p*-nitroaniline. The moist product is equivalent to about 6 g of dry material.

A small portion of the material may be purified by crystallization from 80% aqueous ethanol, with the addition of a little decolorizing carbon. The melting point of *p*-nitroacetanilide is about 215–216°; use a metal block or MEL-TEMP unit.

(B) *p*-Nitroaniline

In a 125-mL Erlenmeyer flask mix the moist, crude *p*-nitroacetanilide with 15 mL of water and 20 mL of concentrated hydrochloric acid. Reflux the mixture *gently* for 15–20 min. The material gradually dissolves and an orange-colored solution is formed.² When the hydrolysis is completed, add 30 mL of cold water and cool the mixture to room temperature. Crystals of the product may separate.

Pour the *p*-nitroaniline hydrochloride slowly, stirring thoroughly, into a mixture of 20 mL of concentrated aqueous ammonia, 75 mL of water, and 25–30 g of chipped ice. The mixture must be distinctly alkaline at the end of the mixing; test with litmus, and add a little more ammonia if necessary. Collect the orange-yellow precipitate of *p*-nitroaniline with suction and wash it with cold water. Recrystallize the product from a large volume of hot water; about 30 mL of water will be required per gram of material. The yield is 3–4 g; the recorded melting point is 147°.

Questions

1. Show a detailed mechanism for the reaction of aniline with acetic anhydride, including the possibility of a cyclic intermediate for elimination of acetic acid from the initial addition product.
2. What product would you expect to obtain by reaction of aniline with the mixed anhydride acetic-formic anhydride ($HCO-O-COCH_3$)? Explain.
3. How could the presence of a small amount of nitrobenzene in a sample of aniline be detected and how could it be removed?
4. (a) How may aniline be converted to benzoic acid? (b) How may benzoic acid be converted to aniline?
5. What may be the particular advantages of using ketene ($CH_2=C=O$, bp $-80°C$) as an acetylating agent?
6. When acetic acid is used for acetylation of an amine, why is it desirable to use an excess of the acid and to distill off the water formed in the reaction?
7. In the preparation of *p*-nitroaniline, why is aniline converted to acetanilide before nitration?
8. What product is obtained from aniline and each of the following reagents?
    (a) 1-naphthyl isocyanate
    (b) succinic anhydride, followed by heating
    (c) potassium cyanate + acetic acid (HNCO)
    (d) dimethylketene
9. When trimethylamine and isobutyryl chloride are mixed in dry ether, trimethylamine hydrochloride is precipitated but the amine is not acylated. What is the other product?
10. What products are formed by nitration of the following?
    (a) *p*-acetotoluidide
    (b) *m*-acetotoluidide
    (c) *m*-cresol
    (d) *p*-toluic acid

---

² If the reaction mixture has been boiled too vigorously, it may be necessary to add more hydrochloric acid to replace that lost by evaporation and to heat the solution longer to complete the hydrolysis.

11. Would benzoic acid undergo nitration satisfactorily under the mild conditions used in this experiment? Explain.
12. (a) Can *p*-nitroacetanilide be hydrolyzed by alkalis?
    (b) What accessory product may be formed by the action of hot aqueous alkalis on *p*-nitroaniline? (Consider the activating effect of the nitro group.)
13. Outline a series of reactions for each preparation.
    (a) *o*-nitroaniline from aniline
    (b) *m*-nitroaniline from benzene
14. Is *p*-nitroaniline a stronger or weaker base than aniline? Is *p*-nitrophenol a stronger or weaker acid than phenol? Explain.

30 Compounds of Medicinal and Biological Interest

Organic chemistry became a separate branch of chemistry about 150 years ago, probably because of an increasing interest in naturally occurring compounds. Studies in this field were facilitated by gradually improving methods of isolation and purification of organic compounds. In recent years great advances have been made through the development of chromatographic methods to achieve isolation of pure compounds and of a variety of spectroscopic methods to monitor purification and identify structural features of complex molecules. This has led to establishing the detailed structure of many important compounds from natural sources.

Much progress has been made also in the development of special reagents for the synthesis of such compounds as quinine, penicillins, and polypeptides. Syntheses of this type are far beyond the scope of this manual—only a few of the simplest examples are given here.[1] The compounds included in the following experiments are aspirin, phenacetin, ethoxyphenylurea (dulcin), p-aminobenzoic acid (PABA), and sulfanilamide.

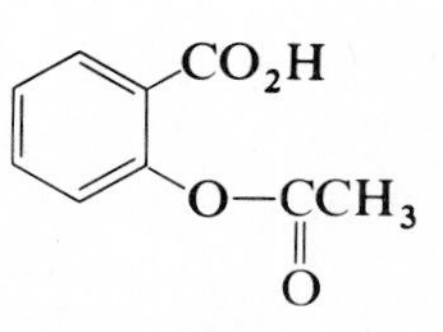

Aspirin

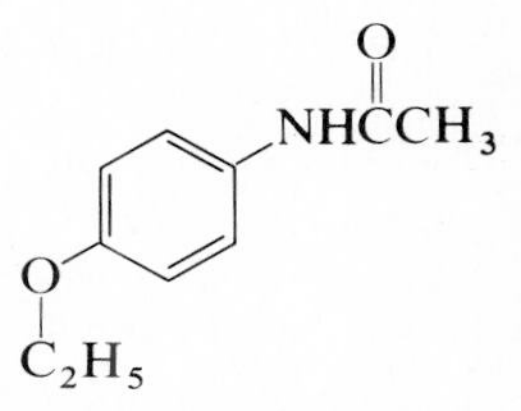

Phenacetin

---

[1] A multiple-step synthesis of dilantin (5,5-diphenylhydantoin) is given in Chapter 33.

Dulcin   PABA   Sulfanilamide

30.1 Acetylsalicylic Acid (Aspirin)

Derivatives of salicylic acid have been used in medicine for many years. Salicyclic acid occurs in nature in the form of esters in a variety of glycosides and essential oils. The methyl ester is present in oil of wintergreen and in many other fragrant oils from flowers, leaves, and bark.

The sodium salt of salicylic acid is prepared commercially by heating sodium phenoxide with carbon dioxide at 150° under slight pressure (the Kolbe synthesis). If potassium phenoxide is used in place of sodium phenoxide and the reaction carried out at 180–200°, the carboxyl function is introduced into the *para* instead of the *ortho* position; this affords a practical synthesis of *p*-hydroxybenzoic acid.

Salicylic acid is used for the manufacture of medicinal compounds, artificial oil of wintergreen, and certain dyes. Aspirin is used as an analgesic (to relieve pain) and antipyretic (to reduce fever). The phenyl ester of salicylic acid, known as Salol, is used as an intestinal antiseptic.

Other compounds related to salicylic acid are the correponding aldehyde and primary alcohol, salicylaldehyde and saligenin, which also occur in nature. Salicylaldehyde may be prepared, together with the *para* isomer, by the action of chloroform on phenol in the presence of excess alkali (the Reimer–Tiemann reaction). The reactive intermediate is *dichlorocarbene*, CCl_2, formed by stepwise abstraction of a proton and a chloride anion from chloroform.

Preparation of Acetylsalicylic Acid

In a 50-mL Erlenmeyer flask place 1.4 g (0.01 mole) of pure salicylic acid. Add 3 mL (3.1 g, 0.03 mole) of acetic anhydride in such a way as to wash down any material adhering to the walls of the flask, and then 5 drops of syrupy (85%) phosphoric acid. Heat the flask for 5 min on a steam bath or in a beaker of water heated to 85–90°. Remove the flask from the bath and, without allowing it to cool, add 2 mL of water in one portion. The excess acetic anhydride decomposes vigorously and the contents of the flask come to a boil (**Caution**—*hot acid vapors!*).

When the decomposition is complete, add 20 mL of water and allow the flask to stand at room temperature until crystallization begins. The crystallization can be hastened by occasionally scratching the walls of the flask at the surface of the solution with a glass stirring rod. When crystals begin to appear, place the flask in an ice bath, add 10–15 mL of cold water, and chill thoroughly until crystallization is complete. Collect the product on a Büchner funnel and press it firmly to remove the mother liquor. Wash the material with cold water and allow it to dry thoroughly. The yield is 1.5–1.6 g.

Purify the crude product in the following way: In an Erlenmeyer flask dissolve the *thoroughly dried* material in 20–25 mL of ether by warming it gently and stirring (*flammable solvent!*). To the solution (if not clear, filter into a clean flask) add 20 mL of petroleum ether (bp 30–60°), stopper the flask, place in an ice bath, and allow to cool undisturbed for about an hour. Collect the product on a Büchner funnel, wash with a little petroleum ether, and spread it on a clean paper to let traces of solvent evaporate.

30.2 *p*-Ethoxyacetanilide (Phenacetin)

Acetanilide is one of the oldest synthetic medicinals (1886) and was used for many years as an antipyretic (fever-reducing) and analgesic (pain-relieving) drug under the name Antifebrin. It was abandoned because of the toxicity of its metabolite, aniline. *p*-Ethoxyacetanilide, called phenacetin, has similar activity and is less toxic than acetanilide. Another drug of similar type, *p*-acetaminophenol (Tylenol, *p*-hydroxyacetanilide; generic drug name, acetaminophen), has a free phenolic group instead of an ethoxy group. In general these drugs have the effect of reducing the oxygen-carrying capacity of the bloodstream. The widely used antipyretic and analgesic drug, acetylsalicylic acid (aspirin), is one of the least toxic medicinals of this type.

Preparation of Phenacetin (see Chapter 29)

In a small beaker dissolve 3.3 mL (3.5 g, 0.025 mole) of *p*-phenetidine (*p*-ethoxyaniline) in 60 mL of water to which 2.5 mL of concentrated hydrochloric acid has been added. When the amine has dissolved, stir the solution with 1 g of decolorizing carbon for a few minutes, and filter the solution with suction. Meanwhile prepare a solution of 5 g of sodium acetate crystals ($CH_3CO_2Na \cdot 3H_2O$) in 15 mL of water; if necessary, filter the solution to remove any foreign particles.

Transfer the filtered solution of the amine hydrochloride to a 250-mL flask and warm it to 50°. Add 3 mL (3.25 g) of acetic anhydride and swirl the liquid to dissolve the anhydride. Add *at once* the previously prepared sodium acetate solution, and mix the reactants thoroughly by swirling them. After a few minutes cool the reaction mixture in an ice bath and stir vigorously during crystallization of the product. Collect the crystals with suction, wash them with *cold* water, and allow them to dry. The yield is 2.5–3 g; recorded mp is 134–135°. The phenacetin prepared in this way is usually quite pure. If desired, it may be recrystallized from hot water with the addition of a little decolorizing carbon.

Another derivative of p-phenetidine of some interest is the urea derivative, dulcin (p-ethoxyphenylurea), that has been used as an artificial sweetening agent.

30.3 p-Ethoxyphenylurea (Dulcin)

Artificial sweetening agents have been used for many years as sugar substitutes by persons suffering from diabetes or desiring low-calorie diets. There seems to be no systematic approach to the synthesis of sweetening agents. Two of the most important ones were found by chance observations of sweet taste noticed by research workers engaged in laboratory syntheses. Saccharin was discovered by a graduate student in the laboratory of Professor Ira Remsen in the Johns Hopkins University (1879), and the cyclamates by Michael Sveda in Professor Louis Audrieth's laboratory at the University of Illinois (1937). Dulcin was first prepared in Germany (1883), but information about the discovery of its sweet taste is lacking.

Saccharin sodium   Sucaryl sodium (Sodium cyclamate)   Dulcin

Soluble saccharin (saccharin sodium) is about 500–600 times as sweet as ordinary sugar (sucrose); sodium cyclohexylsulfamate (sucaryl sodium, sodium cyclamate), about 30–35 times; dulcin, about 150–200 times.

Prolonged use of Dulcin as a sugar substitute may lead to toxic effects resulting from p-aminophenol that is formed by chemical transformations in the body.

(A) Preparation by Cyanate Method[2]

$$NCO^- + H^+ \rightleftharpoons H-N=C=O$$
Cyanate

$$C_2H_5O-\underset{}{\underset{}{\bigcirc}}-\overset{..}{N}H_2 + H-N=C=O \longrightarrow C_2H_5O-\underset{}{\underset{}{\bigcirc}}-NH_2-\underset{N-H}{\overset{O}{C}} \longrightarrow dulcin$$

This reaction is analogous to the derivatization of amines with phenylisocyanate, $C_6H_5-N=C=O$ (see Chapter 9).

In a 50-mL Erlenmeyer flask, dissolve 1.4 mL (1.4 g, 0.01 mole) of *p*-phenetidine in 10 mL of water and 2 mL of glacial acetic acid. While swirling the solution vigorously, add drop by drop a solution of 1.6 g (0.02 mole) of potassium cyanate, KNCO (or 1.3 g of sodium cyanate) dissolved in 5 mL of water. As soon as a precipitate of the arylurea begins to appear, add the remainder of the cyanate solution at once and mix the contents thoroughly. Allow the mixture to stand with occasional shaking for an hour or longer, add 5 mL of water, and cool the mixture in an ice bath. Collect the product on a suction filter, wash it with a little cold water, and recrystallize it from hot water (~30 mL/g) with the addition of decolorizing carbon. Avoid prolonged boiling since this leads to the formation of a little *N,N'*-di(ethoxyphenyl)urea. The yield is about 1 g; the recorded melting point of dulcin is 173–174°.

(B) Preparation by Urea Method[3]

$$H_2N-CO-NH_2 + H^+ \rightleftharpoons NH_4^+NCO^- \rightleftharpoons H-N=C=O$$

$$C_2H_5O-\underset{}{\underset{}{\bigcirc}}-NH_2 + H-N=C=O \longrightarrow dulcin$$

Under the conditions of this experiment, urea is in equilibrium with ammonium cyanate, which reacts with *p*-ethoxyaniline as described in 30.3. The formation of ammonium cyanate from urea is the reverse of the famous 1828 experiment of Wöhler that led to the overthrow of the "vital force" theory of the origin of organic compounds.

In a 50-mL round-bottom flask place 1.4 mL (1.4 g, 0.01 mole) of *p*-phenetidine, 2.4 g (0.02 mole) of urea, and 5 mL of water. To this mixture add 1 mL of concentrated hydrochloric acid and 5–6 drops of glacial acetic acid. Adjust a reflux condenser, shake the material well, and boil the

---

[2] The cyan*ates* must not be confused with the extremely poisonous cyan*ides*.
[3] Kurzer, *Organic Syntheses*, Collective Volume **IV**, 51, 52 (1963).

mixture vigorously for 30 min, until the reaction is complete. At first the dark-colored solution remains clear, but in 15–20 min the product begins to separate rapidly. When the mixture sets to a semisolid crystalline mass, stop the heating *at once*.

After cooling the flask to room temperature, add 3–4 mL of cold water, cork the flask tightly, and shake it thoroughly to make a slurry of the crystals. Collect the product on a suction filter, wash it with cold water, and press firmly. Recrystallize the crude product from boiling water (~ 30 mL/g) with the addition of decolorizing carbon. Chill the material in an ice bath before collecting the purified crystals. The yield is about 1 g; recorded mp, 173–174°.

30.4 p-Aminobenzoic Acid (PABA) and Esters

p-Aminobenzoic acid is a member of the group of substances associated with the vitamin B complex. It is present as the central unit of folic acid, vitamin B_{10}, which is made up of a pteridine unit, a *p*-aminobenzoic unit, and a glutamic acid unit. *p*-Aminobenzoic acid is required for folic acid synthesis by some bacteria, and the sulfa drugs are thought to interfere with this synthesis.

The ethyl ester of *p*-aminobenzoic acid is a local anesthetic (Benzocaine) and a sunburn preventive. There is evidence that sodium *p*-hexadecylaminobenzoate ($C_{16}H_{33}NH-C_6H_4CO_2Na$) raises the level of a desirable cholesterol-controlling mechanism in the bloodstream of experimental animals.

p-Aminobenzoic acid is usually made from *p*-nitrotoluene by oxidation to *p*-nitrobenzoic acid (see Chapter 26) and subsequent reduction of the nitro group catalytically or by iron or zinc and hydrochloric acid. The present method is simpler for small-scale laboratory preparations and starts from *p*-toluidine ($CH_3C_6H_4NH_2$). This is acetylated, and the resulting *p*-acetotoluidide is oxidized by potassium permanganate, buffered with magnesium sulfate to avoid hydrolysis of the protective acetyl group. This furnishes *p*-acetamidobenzoic acid, which is hydrolyzed by heating with hydrochloric acid, to give the amino acid.

(A) Preparation of PABA

***p*-Acetotoluidide.** Acetylate 7.5 g (0.07 mole) of *p*-toluidine by the Lumière–Barbier method following carefully the directions given in Section 29.2(A). Dissolve the amine in 175 mL of water and 6.5 mL (0.07 mole) of concentrated hydrochloric acid, add 12 g of sodium acetate crystals, and then treat with 8 mL (8.7 g, 0.085 mole) of acetic anhydride. This should give somewhat more than the amount of material for the next step.

p-Acetamidobenzoic acid.[4] In a 1-L flask place 7.5 g (0.05 mole) of p-acetotoluidide, 20 g (0.08 mole) of magnesium sulfate crystals ($MgSO_4 \cdot 7H_2O$) and 500 mL of water. Heat the material to about 85° on a steam bath. Meanwhile, prepare a solution of 20.5 g (0.13 mole) of potassium permanganate in 70 mL of boiling water in a beaker or Erlenmeyer flask.

While swirling the acetotoluidide solution *vigorously*, add the hot permanganate solution in small portions over a period of about 30 min. Avoid a local excess of the oxidizing agent since this tends to destroy the product. After all of the permanganate has been added, swirl the mixture vigorously. Filter off the precipitated manganese dioxide from the hot solution, using a fluted filter and wash the manganese dioxide with a little water. If the filtrate is not colorless, add 5–10 mL of ethanol and boil the solution until the color has been discharged, and filter it again through a fresh paper.

Cool the colorless filtrate and acidify it to litmus with 20% aqueous sulfuric acid. Collect the p-acetamidebenzoic acid in a suction filter and press it as dry as possible. Dry a small portion for melting-point determination; the recorded melting point is 250–252°. The yield is 60–75%. It is not necessary to dry the main portion of the product thoroughly for the next step. Use a MEL-TEMP apparatus to determine the melting point.

p-Aminobenzoic acid ($H_2N-C_6H_4-CO_2H$). Weigh the acetamidobenzoic acid from the preceding step and for each gram use 5 mL of 18% hydrochloric acid (1 volume of concentrated acid to 1 volume of water) for the hydrolysis. Place the materials in a 200-mL round-bottom flask, add a reflux condenser, and boil the mixture *gently* for 25–30 min. Cool the reaction mixture, add half its volume of cold water, and make the solution just *barely alkaline* to litmus with 10% aqueous ammonia; do not go beyond the end point! For each 30 mL of solution add 1 mL of glacial acetic acid, stir vigorously, and cool the solution in an ice bath. If necessary, induce crystallization by scratching with a glass rod or adding a small seed crystal. Collect the product with suction, allow it to dry, and take the melting point; recorded mp, 186–187°. The yield is about 40–50% of the weight of the p-acetamidobenzoic acid used.

(B) Esterification of PABA

Butesin. In a 100-mL round-bottom flask place 2.7 g (0.02 mole) of p-aminobenzoic acid, 20 mL of n-butyl alcohol and 2 mL (3.7 g, 0.04 mole) of sulfuric acid (*add cautiously!*). Attach a reflux condenser and heat under reflux for 1 hr. Cool the solution to room temperature, neutralize with sodium carbonate (*foaming!*), and extract with two 25-mL portions of

---

[4] This method is a modification of the procedure described by Kremer, *J. Chem. Educ.*, **33**, 71 (1956).

methylene chloride. Extract the combined organic layers twice with 50-mL portions of water and dry them over anhydrous magnesium sulfate. Remove the methylene chloride by distillation using a steam bath as a heat source and recrystallize the residue from methanol–water. The yield of white crystals is about 1.6–1.8 g, mp 57–58°.

▶ *CAUTION*   Butesin may cause dermatitis in sensitive individuals.

The methyl ester (mp 114–115°), the ethyl ester (*benzocaine*, mp 91–92°), and the *n*-propyl ester (*propaesin*, mp 73–74°) can be prepared in a similar fashion with about the same yield by substituting 20 mL of the appropriate alcohol. The lower members of the series are easier to work with because of their higher melting points and the greater solubility of the unreacted alcohol in water.

30.5 Sulfanilamide

A convenient synthesis of sulfanilamide (*p*-aminobenzenesulfonamide) and related aminoarylsulfonamides makes use of *p*-acetamidobenzenesulfonyl chloride. It is not feasible to convert sulfanilic acid to the sulfonyl chloride having a free amino group, since *p*-aminobenzenesulfonyl chloride contains two functional groups that interact and lead to polymers of the type $+NH-C_6H_4-SO_2NH-C_6H_4-SO_2+$.

Protecting the amino group of aniline by conversion to acetanilide permits direct chlorosulfonation with chlorosulfonic acid to obtain *p*-acetamidobenzenesulfonyl chloride. A slight excess over 2 moles of chlorosulfonic acid is used, since chlorosulfonation is a two-step process. The sulfonic acid is formed in the first stage and converted to the sulfonyl chloride by reaction with a second molecule of chlorosulfonic acid.

Reaction of *p*-acetamidobenzenesulfonyl chloride (I) with an excess of aqueous ammonia produces the corresponding sulfonamide (II). For reaction with aminothiazole, aminodiazole, and similar compounds, a tertiary base such as pyridine is added to combine with the hydrogen chloride formed. Since a carboxylic amide is hydrolyzed more easily than a sulfonamide, it is a simple matter to deacetylate the acetamido compound by selective hydrolysis under controlled conditions and secure the aminobenzenesulfonamide (III).

Hundreds of compounds related to sulfanilamide have been prepared. Of these, two have proved of great therapeutic value against pneumococci, gonococci, and a variety of streptococcal infections. The two substances are sulfathiazole and sulfadiazine, and they are prepared in a manner similar to sulfanilamide, by substituting for ammonia and heterocyclic bases aminothiazole and aminodiazine.

$$H_2N-\underset{Sulfathiazole}{\underline{\bigcirc}}-SO_2NH-C\underset{N}{\overset{S-CH}{\underset{\parallel}{\diagdown}}}CH \qquad H_2N-\underset{Sulfadiazine}{\underline{\bigcirc}}-SO_2NH-\underset{N}{\overset{N}{\diagdown}}$$

It has been suggested that the effectiveness of sulfa drugs results from their interference with folic acid synthesis in certain microorganisms. Sites on the enzyme involved with folic acid synthesis that are usually occupied by *p*-aminobenzoic acid become occupied by the structurally related *p*-aminobenzenesulfonic derivative and prevent formation of the essential folic acid (pteroylglutamic acid, vitamin B_{10}). Compounds that act in this way are called antimetabolites.

$$\underset{H_2N}{\overset{OH}{\underset{N}{\bigcirc}}}-CH_2-NH-\underline{\bigcirc}-CO-NH-CH-CH_2CH_2CO_2H$$
$$|$$
$$CO_2H$$

Folic acid

Preparation of Sulfanilamide

***p*-Acetamidobenzenesulfonyl chloride.**[5] In a *dry* 50-mL Erlenmeyer flask that is clamped securely, place 10 mL (17 g, 0.15 mole) of chlorosulfonic acid,[6] $Cl-SO_3H$ (**Caution**—*handle carefully! the acid causes severe burns if dropped on the skin*). Cool the acid to 10–15°, but *not below 10°*, in a water bath containing a few pieces of ice. Then add 3.4 g (0.025 mole) of finely powdered, *dry* acetanilide, in small portions and with thorough mixing, so that the temperature does not rise above 20°. After the acetanilide has dissolved (a few small particles may remain undissolved), place the flask in a

---

[5] Since hydrogen chloride is evolved, it is desirable to conduct the reaction in a hood or to use an inverted large funnel connected to an aspirator pump.

[6] If a poor grade of chlorosulfonic acid is used, the yield will be lowered. Discolored and impure specimens of the acid may be purified by distillation from an all-glass aparatus (with ground-glass joints), collecting the material at bp 148–150° (760 mm).

400-mL beaker containing just enough water to reach about to the level of the reaction mixture in the flask. Heat the water to 60–70° and maintain this temperature for an hour. Take care that the water level is maintained during the heating so that the bath does not go dry.

Pour the reaction mixture *slowly and carefully* in a thin stream onto a well-stirred mixture of 150 g of finely cracked ice and enough water to make stirring easy. *It is important to carry out this step cautiously to avoid spattering of the chlorosulfonic acid.* The reaction product separates as a white or pinkish white gummy mass that soon becomes hard and can be broken up with a stirring rod. Break up any lumps that have formed, collect the product on a Büchner funnel, and wash it with several portions of cold water. The crude product, sucked as dry as possible, is used directly in the next step.[7]

p-Acetamidobenzenesulfonamide. Place the crude, damp product from the first step in a 100-mL Erlenmeyer flask and add 15 mL of concentrated aqueous ammonia (28%) diluted with 10 mL of water. An immediate exothermic reaction ensues. Rub the mixture with a glass stirring rod until a smooth, thin paste is obtained, and then heat it at 70° for 30 min. Remove the flask to an ice bath and, after cooling it, add dilute sulfuric acid until the mixture is acid to Congo Red test paper.[8] After thorough chilling in the ice bath, collect the product on a Büchner funnel, wash with cold water, and dry. The yield of crude product is 3–3.5 g. This material is sufficiently pure for the next step.

Recrystallize a small sample of the sulfonamide from hot water, using a little decolorizing carbon, and take the melting point of the purified material. Test the solubility of the product in dilute acid and dilute alkali. Explain the result. The recorded melting point for pure p-acetamidobenzenesulfonamide is 219°.

p-Aminobenzenesulfonamide. Place the crude, dry p-acetamidobenzenesulfonamide obtained in the previous preparation in a 100-mL round-bottom flask and add an amount of dilute hydrochloric acid (1 volume of concentrated acid to 2 volumes of water) equal to *2 mL* of dilute acid *per gram* of substance. Boil the mixture under reflux for 30–40 min, when all of the solid will have dissolved. Care should be taken at the start of the heating period to avoid charring of the initially pasty mixture.

---

[7] Moist p-acetamidobenzenesulfonyl chloride can be purified by dissolving it in a mixture of acetone and toluene (1:1), separating any water, and allowing the solvent to evaporate until crystallization takes place. In this form the product keeps well; the crude product tends to decompose on standing.

[8] Congo Red test paper may be prepared by dipping strips of filter paper into a solution of Congo Red and allowing them to dry. This indicator changes to *blue* with mineral acids, but is not affected by weak acids or acidic salts. Why is it necessary that the solution should be *strongly* acid at this point in the experiment?

Pour the solution into a 100-mL Erlenmeyer flask, and dilute it with an equal volume of water. Add a little decolorizing carbon, heat the solution to boiling, and filter through a fluted filter into a clean 400-mL beaker. Add solid sodium bicarbonate (**Caution**—*frothing!*) in small portions, stirring continuously, until the solution is just alkaline to litmus. During the neutralization, the free amine separates as a white, crystalline precipitate. After cooling it thoroughly in an ice bath, collect the product by suction filtration, wash with cold water, and dry. The yield is about 0.70–0.72 g per gram of *p*-acetamidobenzenesulfonamide.

Recrystallize the product from water, using about 12 mL of water per gram of sulfanilamide. Add a little decolorizing carbon (about 0.2 g) to the hot solution, boil for a few moments, and filter it through a fluted filter paper in a short-stemmed funnel. The filter paper and funnel should be preheated by pouring boiling water through them just before filtering the solution, to prevent the material from crystallizing in the stem of the funnel. The sulfanilamide[9] separates on cooling as long, silky white needles. Test the solubility of sulfanilamide in dilute acid and dilute alkali. Explain the result.

The recorded melting point of pure *p*-aminobenzenesulfonamide is 163°.

Questions

1. Compare the ease of hydrolysis of **(a)** benzenesulfonamide and benzamide and **(b)** benzenesulfonyl chloride and benzoyl chloride.
   - **(c)** Compare the behavior of sodium benzenesulfonate and sodium benzoate on fusion with sodium hydroxide.
2. What are **(a)** saccharin and **(b)** chloramine-T?
   - **(c)** How are they made commercially?
3. Why is it impractical to synthesize sulfanilamide by any process that would require *p*-aminobenzenesulfonyl chloride as an intermediate? What reaction would occur if *p*-aminobenzenesulfonyl chloride were formed momentarily in a synthetic process?

---

[9] Sulfanilamide is a powerful therapeutic agent and must be taken only with the advice and supervision of a physician.

31 Heterocyclic Aromatics: 3-Phenylsydnone

31.1 Mesoionic Compounds

The resonance structures of many compounds (vinyl ethers and halides, amides, and the like) involve one or more contributors that carry formally separated electrical charges. 3-Phenylsydnone, a derivative of 1,2,3-oxadiazole, is an example of the special class of compounds described as mesoionic, in which *all of the important resonance structures* have a separation of charges. The name *sydnones* was given to them by their discoverers at the University of Sydney (Australia).

A general method of synthesis consists in eliminating the elements of water from the *N*-nitroso derivatives of *N*-arylaminoacetic acids (I) by means of acetic anhydride or thionyl chloride.[1] The sydnones undergo numerous electrophilic substitutions and also 1,3-cycloaddition reactions (involving positions 2 and 4) with quinones, alkenes, and alkynes. The resulting adducts on loss of carbon dioxide furnish derivatives of pyrazoline or pyrazole ($C_3H_4N_2$; 1,2-diazole).

$$Ar-NH-CH_2-CO_2H + HO-N=O \longrightarrow Ar-N-CH_2-CO_2H + Ac_2O \longrightarrow$$

I

[structures of resonance contributors of the sydnone ring shown]

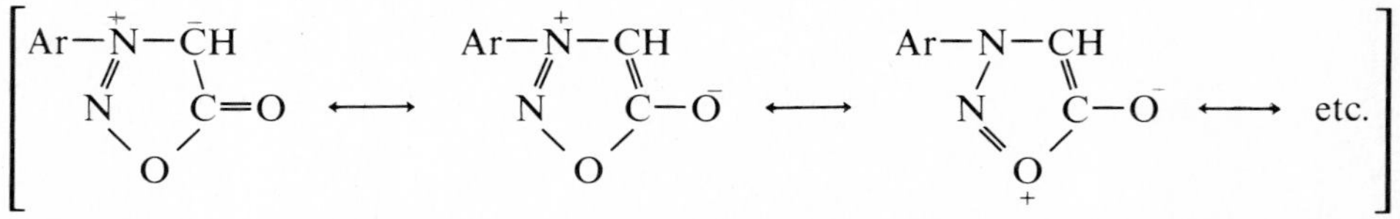

[1] The procedure given is that of Thoman and Voaden, *Organic Syntheses*, Collective Volume **V**, 962 (1973); references to other methods of preparation and to reactions of sydnones are given there.

31.2 Preparation of 3-Phenylsydnone

N-Phenylglycine. In a 250-mL round-bottom flask place 9.5 g (0.10 mole) of chloroacetic acid and 20 mL of water. *Neutralize* the acid by careful, slow addition of 10% aqueous sodium hydroxide solution (~40 mL, 0.1 mole), while shaking and cooling the solution. To this sodium chloroacetate solution add 10.0 mL (10.2 g, 0.11 mole) of aniline, adjust a reflux condenser, and boil the mixture gently for 20 min. Cool the solution to room temperature in an ice bath, add 5 g of sodium hydroxide pellets, and swirl the flask in the ice bath until all of the solid has dissolved. Extract the basic solution with two 50-mL portions of methylene chloride to remove the unreacted aniline. Acidify the aqueous solution by adding, drop by drop, concentrated hydrochloric acid until the pH is about 4 (Hydrion or Congo Red paper). If the product separates as an oil, induce crystallization by scratching with a glass rod.

Collect the crystals on a suction filter, wash them sparingly with ice-cold water, and spread them to dry in the air. The yield of crude product is 9–10 g. This may be used directly for the next step; if desired, it may be crystallized from 40% aqueous ethanol. Pure N-phenylglycine forms white crystals (mp 125–127°).

N-Nitroso-N-phenylglycine. In a 250-mL Erlenmeyer flask prepare a slurry of 9.1 g (0.06 mole) of finely pulverised N-phenylglycine in 100 mL of water, and cool the mixture in an ice–salt bath, swirling it constantly, until the inner temperature falls below 0°. Without allowing the temperature to rise above 0° add, drop by drop, over a period of 15–20 min, with thorough mixing, an ice-cold solution of 4.6 g (0.066 mole) of sodium nitrite in 30 mL of water. Filter the turbid, colored solution with suction on a large Büchner funnel, preferably using a filter aid, such as Hyflo Supercel or Celite. Transfer the filtrate to an Erlenmeyer flask, add a small amount of decolorizing carbon, shake well, and filter again. Swirl the solution vigorously while adding 9 mL (0.1 mole) of concentrated hydrochloric acid, and continue to swirl the slurry of crystals for several minutes after all of the acid has been added. Collect the crystals on a suction filter, wash them with two 15-mL portions of cold water, and allow the product to dry *thoroughly* in the air. The yield is 7–9 g; reported mp is 103–104°.

3-Phenylsydnone. Place the dried N-nitroso-N-phenylglycine in a 250-mL round-bottom flask and add 50 mL (0.5 mole; a large excess) of acetic anhydride (**Caution**—*acetic anhydride can cause painful burns, especially when hot*). Attach a reflux condenser fitted with a drying tube at the top, and heat the reaction mixture in a bath of boiling water or on a steam bath for about 40 min, with occasional swirling. Cool the solution to 20–30° and pour it slowly, with vigorous stirring, into about 250 mL of water in a large beaker. After the excess acetic anhydride has been hydrolyzed, collect the

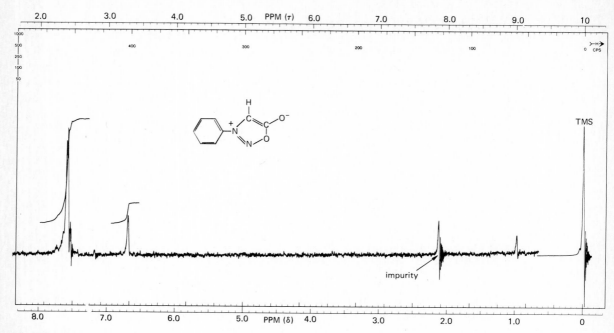

FIGURE 31.1 NMR Spectrum of 3-Pheynylsydnone in Deuterochloroform

crystals on a suction filter and wash them thoroughly with several portions of cold water. After drying, the product weighs 5–6 g; reported mp is 136–137°. On crystallization from boiling water the sydnone forms light tan needles.

The NMR of 3-phenylsydnone is shown in Figure 31.1. The lone proton on the sydnone ring comes much further downfield than would have been expected for an alkene proton or an enolate anion proton and suggests that a ring current exists in the sydnone ring (see Section 10.3).

Questions

1. Write equations for another method of converting aniline to N-phenylglycine.
2. Show the tautomeric forms and the resonance structures of indoxyl, C_8H_7NO, which is made commercially by cyclization of N-phenylglycine with sodium amide.
3. Devise structural formulas for two or more analogs of 3-phenylsydnone in which the heterocyclic system has been altered. Show their resonance structures.
4. Write a structure for the cycloaddition of methyl propiolate ($H-C\equiv C-CO_2CH_3$) to 3-phenylsydnone and show the formula of the substituted pyrazole that could be formed by elimination of CO_2 from the adduct. (The final product should be 3-carbomethoxy-1-phenylpyrazole.)

32 Aldol Condensation

32.1 Introduction

In its simplest form the aldol condensation is a self-addition involving two molecules of the same aldehyde or ketone and results in the formation of a new carbon–carbon bond joining the carbonyl carbon of one molecule to the α-position of the second.[1]

$$CH_3CH_2-CH=O + H-\underset{\underset{CH_3}{|}}{CH}-CH=O \xrightarrow[\text{or } H^+]{OH^-} CH_3CH_2-\underset{\underset{OH}{|}}{CH}-\underset{\underset{CH_3}{|}}{CH}-CH=O$$

To react in this way it is necessary that the molecule possess a reactive carbonyl function *and* one or more reactive hydrogens in the α-position.

An aromatic aldehyde has no hydrogen atoms in the α-position but is capable of participating in a mixed (crossed) aldol condensation with another aldehyde or ketone that can furnish an active methylene group.

$$C_6H_5-\overset{O}{\overset{\|}{C}H} + CH_3-\overset{O}{\overset{\|}{C}H} \xrightarrow{OH^-} C_6H_5-\underset{\underset{}{|}}{\overset{OH}{C}H}-CH_2-\overset{O}{\overset{\|}{C}H} \longrightarrow$$

$$[C_6H_5-\overset{+}{C}H-CH=\overset{O^-}{\underset{|}{C}H} \longleftrightarrow C_6H_5-CH=CH-\overset{O}{\overset{\|}{C}H} + H_2O]$$

[1] Nielsen and Houlihan, *Organic Reactions*, **16**, 1 (1968).

Usually the mixed aryl aldol undergoes dehydration spontaneously to form the resonance stabilized α,β-unsaturated aldehyde or ketone. The reaction is known as the Claisen-Schmidt condensation.

Aldol-type condensations of aromatic aldehydes extend to other reactants containing an active methylene group—such as malonic ester, acid anhydrides, nitriles, nitroalkanes, and similar compounds.

Either acids or bases will catalyze the aldol condensation but basic catalysts are generally preferred. Dilute aqueous or ethanolic sodium hydroxide, sodium ethoxide, and secondary amines (diethylamine or piperidine) are effective catalysts. The first step in the process is the formation of the enolate anion of the active methylene component by action of the base.

$$CH_3-CH=O \underset{H_2O}{\overset{OH^-}{\rightleftarrows}} [\bar{C}H_2-CH=O \longleftrightarrow CH_2=CH-\bar{O}]$$

$$C_6H_5-CH=O + :\bar{C}H_2-CH=O \underset{OH^-}{\overset{H_2O}{\rightleftarrows}} C_6H_5-\underset{OH}{\underset{|}{C}H}-CH_2-CH=O$$

The resulting carbanion combines with the carbonyl reactant and proton interchange with the solvent leads to the mixed aldol, which then undergoes dehydration to the α,β-unsaturated compound.

The kinetics of aldol-type condensations vary with the character of the reactants and the experimental conditions. In the self-condensation of acetaldehyde under typical conditions, with aqueous sodium hydroxide as catalyst, the enolization step is slow (and irreversible) and the second step is very fast. In the condensation of benzaldehyde and acetophenone, with sodium ethoxide as catalyst, the enolization of acetophenone is fast (and reversible) and combination of the carbanion with benzaldehyde is the slow step. The rate of this reaction is proportional to the concentrations of the carbanion and the benzaldehyde. With less acidic ketones the rate will be slower because of the lower equilibrium carbanion concentration; with carbonions substituted on the methyl group, the rate may be slower because of steric hindrance.

$$C_6H_5-CH=O + H_3C-\overset{O}{\underset{\|}{C}}-C_6H_5 \xrightarrow{OH^-} C_6H_5-\underset{H}{\overset{H}{C}}=\underset{C_6H_5}{\overset{O}{C-C}}$$

Benzaldehyde   Acetophenone

Mixed aldol condensations furnish intermediates for synthetic procedures used to obtain aromatic compounds having a variety of functional groups in the side chain and also structures having several aromatic rings attached to an aliphatic system. With ketones having two methylene groups, the aldol condensation is complicated by the possibility of mono or di substitution. For example, with benzaldehyde and excess acetone the mono-condensation product, benzalacetone, is obtained. However, with benzaldehyde and acetone in a 2:1 mole ratio, dibenzalacetone is obtained instead.

$$C_6H_5-\overset{O}{\overset{\|}{C}}-H + CH_3-\overset{O}{\overset{\|}{C}}-CH_3 \longrightarrow \underset{\text{Benzalacetone}}{C_6H_5-\overset{H}{\underset{H}{C}}\diagdown_{C-C-CH_3}^{O\|}} \xrightarrow{C_6H_5CHO} \underset{\text{Dibenzalacetone}}{C_6H_5-\overset{H}{\underset{H}{C}}\diagdown_{C-C-C}^{O\|}\diagup\overset{H}{\underset{H}{C}}-C_6H_5}$$

32.2 Preparation of Dibenzalacetone

In a 125-mL Erlenmeyer flask prepare a solution of 2 g (0.005 mole) of sodium hydroxide in 10 mL of water and 10 mL of 95% ethanol. After the solution has cooled, add 1.5 mL (1.2 g, 0.02 mole) of acetone and then 4 mL (4.1 g, 0.04 mole) of benzaldehyde. A yellow turbidity will appear almost immediately, which quickly turns into a flocculent precipitate. Swirl the flask from time to time over a 15-min period. Collect the mushy reaction product on a Büchner funnel and wash it first with water and then a little chilled 95% ethanol. Continue to draw air through the funnel until the product is dry and then recrystallize it from ethyl acetate (**Caution—** *flammable solvent!*) using about 2.5 g of solvent per gram of product. The yield of purified product is about 3 g, mp 110–111°.

Questions

1. What product would be formed by mixed aldol condensation of benzaldehyde with propionaldehyde? with acetone (in excess)?
2. Write equations, stepwise, showing the mechanism of addition of hydrogen chloride to benzalacetophenone.
3. What products are formed by reaction of benzalacetophenone with the following reagents?
    (a) phenylmagnesium bromide, followed by water and dilute acid
    (b) diethylamine
    (c) phenylhydrazine, followed by cyclization with sulfuric acid
4. Write projection formulas for the stereoisomeric forms of benzalacetophenone and its dibromide.
5. Benzalacetophenone can be nitrated to give a mononitro derivative.
    (a) What structure would you expect for this compound?
    (b) If a second nitro group were introduced, where would it enter?

33 The Benzoin Condensation

33.1 Introduction

Two molecules of an aromatic aldehyde, when heated with a catalytic amount of sodium or potassium cyanide in aqueous ethanol, react to form a new carbon–carbon bond between the carbonyl carbons. The product is an α-hydroxy ketone (a benzoin).

$$2\ C_6H_5\overset{O}{\underset{}{\overset{\|}{C}}}-H \xrightarrow{NaCN} C_6H_5-\underset{H}{\overset{OH}{\underset{|}{C}}}-\overset{O}{\overset{\|}{C}}-C_6H_5$$
$$\text{Benzoin}$$

This remarkably facile condensation was discovered accidentally by Wöhler and Liebig in 1832 when they attempted to extract the cyanohydrin of benzaldehyde with base to remove acid impurities.

$$C_6H_5-\underset{H}{\overset{OH}{\underset{|}{\overset{|}{C}}}}-CN \xrightarrow{\text{base}} \text{benzoin}$$

Benzaldehyde cyanohydrin
(Mandelonitrile)

The mechanism for benzoin formation involves a rather long sequence of steps. It starts with reversible cyanide ion addition to the carbonyl group of one benzaldehyde to form the anion of the cyanohydrin (step 1), which in aqueous ethanol rapidly equilibrates with the neutral cyanohydrin (step 2). The acidity of the C—H bond adjacent to the cyano group is enhanced by resonance stabilization of the anion, and under the basic conditions of the reaction (NaCN is basic) the isomeric carbanion is formed (step 3). This adds to a second molecule of benzaldehyde (step 4); proton interchange and loss of cyanide ion (steps 5 and 6) lead to benzoin. The rate-determining step appears to be step 4.

$$C_6H_5-C(=O)H \underset{(1)}{\rightleftharpoons} C_6H_5-\underset{H}{\overset{O^-}{C}}-C\equiv N \underset{(2)}{\rightleftharpoons} C_6H_5-\underset{H}{\overset{OH}{C}}-C\equiv N \underset{(3)}{\rightleftharpoons}$$

$$\left[\begin{array}{c} \text{(resonance structures)} \\ C_6H_5-\overset{OH}{\underset{..}{C}}-C\equiv N \longleftrightarrow C_6H_5-\overset{OH}{C}=C=\bar{N} \end{array} \right] \underset{(4)}{\overset{C_6H_5C(=O)H}{\rightleftharpoons}}$$

$$C_6H_5-\underset{C\equiv N}{\overset{OH\;O^-}{C-CH}}-C_6H_5 \underset{(5)}{\rightleftharpoons} C_6H_5-\underset{C\equiv N}{\overset{O^-\;OH}{C-CH}}-C_6H_5 \underset{(6)}{\rightleftharpoons} C_6H_5-\overset{O\;\;OH}{C-CH}-C_6H_5 + {}^-C\equiv N$$

The benzoin condensation is fairly general. Substituted benzaldehydes, such as the tolualdehydes and methoxybenzaldehydes, form the corresponding *symmetrical* benzoins. Two different aldehydes can react to form *unsymmetrical* or mixed benzoins such as anisbenzoin or benzfuroin.

$$R-\overset{O}{CH} + R'-\overset{O}{CH} \xrightarrow{CN^-} R-\overset{O}{\underset{H}{C}}-\overset{OH}{\underset{}{C}}-R' \quad or \quad R-\overset{HO}{\underset{H}{C}}-\overset{O}{C}-R'$$

Although two isomeric mixed benzoins and two symmetrical benzoins could be formed in these reactions, it is not uncommon to obtain a single mixed benzoin as the principal product. Generally, the carbonyl group is bound to the more electron-rich ring.

Some substituted benzaldehydes, such as *p*-chlorobenzaldehyde and *p*-dimethylaminobenzaldehyde, react poorly or not at all to form symmetrical benzoins but will form mixed benzoins in conjunction with another aldehyde. *p*-Chlorobenzaldehyde and *p*-nitrobenzaldehyde have good carbonyl reactivity, but the nucleophilicity of the carbanion needed for step 4 is reduced (or almost completely lost for the *p*-nitro compound) through the electron attraction of the substituent. *p*-Chlorobenzaldehyde does form a mixed benzoin with an appropriate partner.

p-Dimethylaminobenzaldehyde represents the opposite circumstance. It fails to form a simple benzoin because the electrophilic activity of its carbonyl group (needed for step 3) is greatly reduced through electron release by the *p*-amino group. With benzaldehyde it forms a mixed benzoin: $(CH_3)_2N-C_6H_4-CO-CH(OH)-C_6H_5$.

One must be careful in assigning nucleophilic and electrophilic roles to the partners of a mixed benzoin condensation. Benzoin and similar α-hydroxy aldehydes and ketones (including the simple sugars) are capable of existing in tautomeric enediol structures, which destroys the structural information about which aldehyde behaved as the nucleophile. Under basic conditions the two possible benzoins interconvert readily.

$$R-\underset{\underset{H}{|}}{\overset{\overset{O}{\|}}{C}}-\underset{}{\overset{OH}{\underset{|}{C}}}-R' \underset{H_2O}{\overset{OH^-}{\rightleftarrows}} \underset{HO}{\overset{R'}{\diagdown}}C=C\underset{R'}{\overset{OH}{\diagup}} \underset{OH^-}{\overset{H_2O}{\rightleftarrows}} R-\underset{\underset{H}{|}}{\overset{\overset{OH}{|}}{C}}-\underset{}{\overset{O}{\underset{\|}{C}}}-R'$$

The benzoin condensation also occurs with some aliphatic ketones (to give acyloins), but it is not a common reaction because the basic conditions favor the more rapid aldol condensation if that is possible.

33.2 Vitamin B₁ Catalysis

There are two requirements for an effective catalyst of the benzoin condensation. First, the catalyst must give significant amounts of carbonyl adduct (steps 1 and 2) but not form such a strong bond that the catalyst is not easily lost in the last step. Second, the catalyst must stabilize the anion sufficiently to allow the C—H bond to be broken readily but not so much that the anion becomes unreactive. For years, the only species found that

Thiazolium salt   Conjugate base

Resonance-stabilized α-anion

satisfied these requirements was the cyanide ion. However, Breslow (1958) discovered that the conjugate base of a thiazolium salt was also an effective catalyst; it added reversibly to aldehydes and stabilized the α-anion by resonance.

What gave Breslow's study broader significance was his recognition that thiamine (Vitamin B_1) contains a thiazole unit and that a number of important biochemical reactions requiring it as a coenzyme could be understood as analogs of the benzoin condensation.

Thiamine pyrophosphate (Cocarboxylase)

The preparation of benzoin using thiamine hydrochloride as the catalyst is described next. The conversion of benzoin to benzil and the use of these compounds as intermediates for syntheses are discussed later (Section 33.5).

33.3 Preparation and Reactions of Benzoin

In a 125-mL Erlenmeyer flask prepare a solution of 5.2 g (0.015 mole) of thiamine hydrochloride in 15 mL of water. When all of the thiamine hydrochloride has dissolved, add 40 mL of 95% ethanol, 12 mL of 10% sodium hydroxide (0.030 mole), and 15 mL (16 g, 0.15 mole) of benzaldehyde, thoroughly mixing the solution between each addition. Stopper the flask and allow it to stand at room temperature at least overnight (longer periods do no harm).

At the end of the reaction period, the benzoin should have separated as fine crystals. Cool the flask in an ice-water bath to complete the crystallization, collect the product with suction, and wash the crystals thoroughly with two 15-mL portions of *cold* 50% ethanol and several portions of water. Press the crystals as dry as possible and spread them on a fresh filter paper to dry in the air. The yield is 9–10 g (dry weight).

The product may be used, without careful drying or recrystallization, for the preparation of derivatives or for conversion to benzil or benzilic acid. Benzoin may be purified, with loss of 10–15%, by recrystallization from methanol (12 mL/g of benzoin) or from ethanol (8 mL/g). Determine the infrared spectrum of the purified product and demonstrate its purity by TLC.

Reactions. Benzoin and other acyloins (α-hydroxy ketones) may be characterized through reactions of the secondary alcohol function (acetylation, oxidation) or of the carbonyl group (formation of oximes, hydrazones, semicarbazones). Benzoin, like the keto sugars, is converted to an osazone by warming with excess phenylhydrazine. By reduction with sodium amalgam or sodium borohydride (NaBH$_4$), benzoin is converted to *meso*-hydrobenzoin (I); with zinc and hydrochloric acid, the reduction product is desoxybenzoin (II).

$$C_6H_5-CH(OH)-CH(OH)-C_6H_5 \qquad C_6H_5-CH_2-CO-C_6H_5$$
$$\text{I} \qquad\qquad\qquad\qquad\qquad \text{II}$$

Benzoin acetate. In a test tube place 1 g of benzoin, 1 mL of glacial acetic acid, and 1 mL of acetic anhydride. Introduce *one drop* of concentrated sulfuric acid and shake well to ensure good mixing; the mixture becomes warm and the benzoin dissolves. Heat the tube in a beaker of boiling water for 5–10 min (not longer), allow the solution to cool slightly, and pour it carefully into 25 mL of water. Collect the crystals with suction, wash them thoroughly with water, and press dry. Recrystallize the product directly from methanol or ethanol. The recorded melting point of benzoin acetate is 82–83°.

Benzoin α-oxime. In a small Erlenmeyer flask place 1 g of benzoin, 5 mL of ethanol, 1 mL of a 35% aqueous solution of hydroxylamine hydrochloride, and 2 mL of 30% aqueous sodium hydroxide. Add a small boiling chip, attach a short condenser, and reflux the mixture gently for 1.5–2 hr. Pour the warm reaction mixture, with stirring, into 50 mL of cold water to which 2 mL of concentrated sulfuric acid has been added. Collect the product on a suction filter, wash it well with water, and press dry. The oxime prepared in this way contains about 10% of the β-isomer (oximino hydroxyl in *syn*-configuration in relation to the —CHOH—C$_6$H$_5$ group). Recrystallize the crude product from ethanol, or, after thoroughly drying the material, from benzene. The recorded melting points are α-oxime, 152–155°; β-oxime, 99°.

$$\underset{\beta\text{-Benzoinoxime}}{C_6H_5-\underset{\underset{\text{HO}-\text{N}}{\|}}{\overset{\overset{\text{OH}}{|}}{\text{CH}}-\text{C}}-C_6H_5} \qquad \underset{\alpha\text{-Benzoinoxime}}{C_6H_5-\underset{\underset{\text{N}-\text{OH}}{\|}}{\overset{\overset{\text{OH}}{|}}{\text{CH}}-\text{C}}-C_6H_5}$$

The acetates of these oximes exhibit stereospecific behavior toward cold 5% aqueous sodium hydroxide. The α-oxime acetate is cleaved readily to benzaldehyde, benzonitrile, and sodium acetate; the β-oxime acetate is not cleaved but merely hydrolyzed to regenerate the β-oxime and sodium acetate. The α-oxime forms chelate derivatives with metal ions such as Cu^{2+} and is useful as an analytical reagent (known as *Cupron*).

33.4 Preparation and Reactions of Benzil

Benzoin can be oxidized to the diketone, benzil, in a number of ways, of which the most interesting is by a "coupled oxidation" using Cu^{2+} as the catalytic transfer oxidant. In a coupled oxidation the overall oxidation proceeds in two distinct stages. In the present procedure, cupric acetate is used in catalytic amount (less than 1% of the stoichiometric requirement) and is continuously reoxidized from the reduced (cuprous) state by ammonium nitrate, which is present in excess. The latter is reduced to ammonium nitrite, which decomposes in the reaction mixture into nitrogen and water. It is convenient to represent this two-stage oxidation in the manner used by biochemists, who commonly deal with multiple coupled reactions.

$$NH_4^+NO_3^- \longrightarrow Cu^{2+} \longrightarrow C_6H_5-\underset{\underset{Benzil}{}}{\overset{O}{\underset{\|}{C}}}-\overset{O}{\underset{\|}{C}}-C_6H_5$$

$$NH_4^+NO_2^- \longleftarrow Cu^+ \longleftarrow C_6H_5-\underset{\underset{Benzoin}{}}{\overset{O}{\underset{\|}{C}}}-\underset{\underset{H}{|}}{\overset{OH}{\underset{|}{C}}}-C_6H_5$$

$$\downarrow$$

$$N_2 + 2\,H_2O$$

Cupric salts are mild oxidizing agents that do not attack the diketone product. In the absence of Cu^{2+}, ammonium nitrate will not oxidize benzoin (or benzil) at a significant rate. The reaction is general for α-hydroxy ketones (acyloins) and is the basis for Fehling's test for reducing sugars.

Benzoin may also be oxidized by means of nitric acid.

(A) Oxidation of Benzoin by Cupric Salts

In a 100-mL round-bottom flask place 8.5 g (0.04 mole) of benzoin, 25 mL of glacial acetic acid, 4 g (0.05 mole) of pulverized ammonium nitrate, and

5 mL of a 2% solution of cupric acetate.[1] Add one or two small boiling chips, attach a reflux condenser equipped with a gas trap (see Figure 8.2), and heat the flask gently on a wire gauze, occasionally swirling it. As the reactants dissolve, evolution of nitrogen begins. Boil the green solution for 1.5 hr to complete the reaction. Cool the solution to 50–60° and pour it into 40 mL of ice water, while stirring it. After crystallization of the benzil is complete, collect the crystals on a suction filter and wash them thoroughly with water. Press the product as dry as possible on the filter. The yield is 7.5–8 g (dry weight). Benzil obtained in this way is usually sufficiently pure for conversion to derivatives or to benzilic acid. It may be purified by recrystallization from methanol or 75% aqueous ethanol.

Test for the Presence of Benzoin. Dissolve a few crystals of the product in 1 mL of ethanol and add a drop of sodium hydroxide solution. A purple coloration develops if any unoxidized benzoin is present. The color fades when the solution is shaken with air, but reappears when it stands undisturbed.

(B) Oxidation of Benzoin by Nitric Acid (*Alternative Procedure*)

In a 250-mL Erlenmeyer flask place 8.5 g (0.04 mole) of benzoin and 40 mL of glacial acetic acid. While swirling the flask to obtain good mixing, add 20 mL (28 g, 0.34 mole) of concentrated nitric acid. Heat the reaction mixture on a steam bath, in a hood, for 2 hr. Cool the flask in an ice-water bath, add 150 mL of water, mix thoroughly, and allow the yellow precipitate of benzil to settle. Collect the product with suction and wash it thoroughly with water to remove nitric acid (test with moist litmus paper). Press the crystals as dry as possible with a clean cork or flat glass stopper. The yield is 7–8 g (dry weight). Benzil obtained in this way is usually sufficiently pure for conversion to derivatives or to benzilic acid. It may be purified by recrystallization from methanol or 75% aqueous ethanol.

Test the product for the presence of unoxidized benzoin as described with the preceding preparation.

Reactions. 1,2-Diketones may be characterized by conversion to a variety of crystalline derivativesk such as the mono- and dioximes, hydrazones, and semicarbazones. The monosemicarbazone undergoes cyclization readily upon heating, or treatment with cold aqueous alkali, to give a triazine derivative, 5,6-diphenyl-1,2,4-triazin-3-one (I).

---

[1] The catalyst solution may be prepared by dissolving 2.5 g of cupric acetate monohydrate in 100 mL of 10% aqueous acetic acid, stirring well, and filtering to remove any basic copper salts that have precipitated.

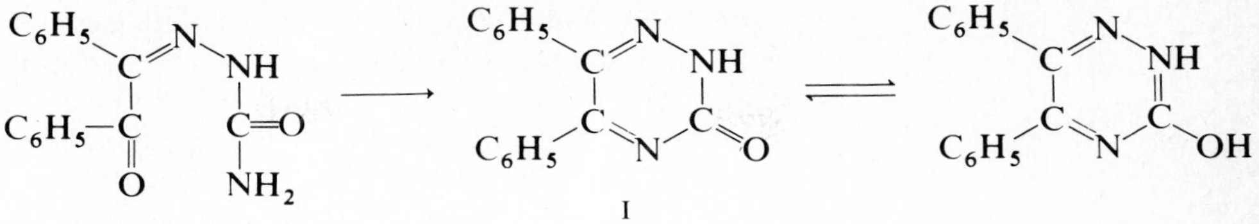

Sec. 33.4] PREPARATION AND REACTIONS OF BENZIL

[Structures showing conversion of benzil semicarbazone-type intermediate to triazine I]

Pyrazine derivatives are formed by reaction of 1,2-diketones with aliphatic and aromatic 1,2-diamines; the latter give quinoxalines (benzopyrazines, II) that may be used for purposes of identification. Benzil undergoes other cyclization reactions, such as the formation of triphenylimidazole (lophine, III) on treatment with ammonia and benzaldehyde, and conversion to tetraphenylcyclopentadienone (tetracyclone, IV) by reaction with dibenzyl ketone.[2]

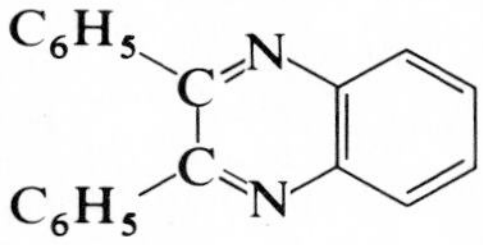

[Structures II (quinoxaline), III (lophine), IV (tetracyclone)]

II 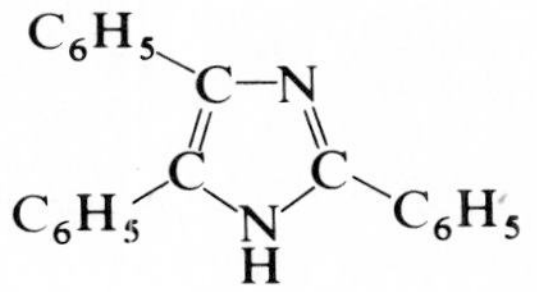   III 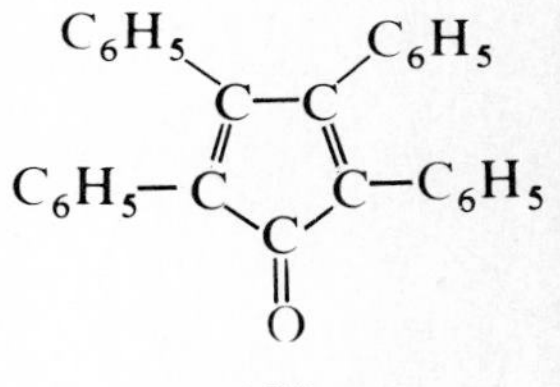   IV

The most interesting reaction of benzil, discovered by Justus Liebig in 1838, is its transformation to benzilic acid when heated with strong alkalis (Chapter 34).

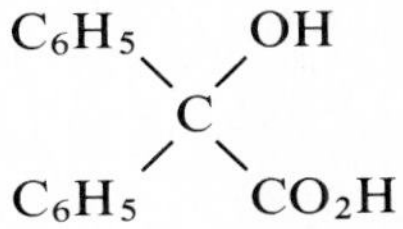

[Structure of benzilic acid: C_6H_5, OH, C_6H_5, CO_2H on central C]

Benzilic acid

The mechanism of this rearrangement involves addition of hydroxyl ion to one of the carbonyl groups, followed by a cycle of intramolecular shifts leading to the stable benzilate anion. An analogous rearrangement occurs when benzil is warmed with urea and alkali; the principal product is 5,5-diphenylhydantoin. The sodium salt of this hydantoin is an anticonvulsant medicinal, *dilantin sodium*, used in the treatment of epilepsy.

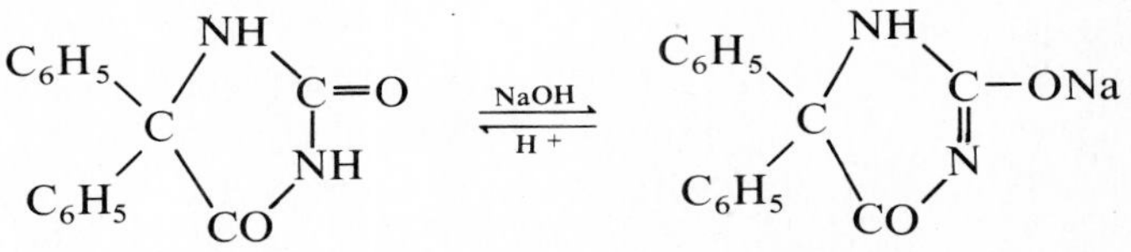

[Structures of 5,5-Diphenylhydantoin and Dilantin sodium, interconverted by NaOH / H^+]

5,5-Diphenylhydantoin   Dilantin sodium

---

[2] *Organic Syntheses*, Collective Volume **III**, 806 (1955); see also Fieser and Williamson *Organic Experiments*, 4th ed. (Lexington, MA: Heath, 1979).

Benzil monohydrazone. Dissolve 1 g of benzil in 5 mL of warm ethanol and add dropwise, with shaking, 1 mL of a 25% aqueous solution of hydrazine hydrate (5 M solution of NH_2-NH_2). Boil the solution for 5–10 min in a water bath, add 10 mL of 50% aqueous ethanol, and allow the reaction mixture to cool. Collect the crystals, wash them with cold 50% ethanol, and dry them in the air. The product may be recrystallized from hot ethanol. The recorded melting point of the monohydrazone is 149–150° (dec).

▶ *CAUTION* Hydrazine hydrate is a violent poison! It causes delayed eye irritation. It is also a cancer suspect agent.

When the monohydrazone is heated at the melting point, nitrogen is evolved and desoxybenzoin ($C_6H_5-CO-CH_2-C_6H_5$) is formed. Oxidation with mercuric oxide converts the monohydrazone to phenylbenzoyldiazomethane, which loses nitrogen at 100° and undergoes rearrangement to give diphenylketene,[3] $(C_6H_5)_2C=C=O$.

Benzil α-monoxime. In a test tube place 1 g of finely pulverized benzil, 5 mL of ethanol, and 1 mL of a 35% aqueous solution of hydroxylamine hydrochloride. Cool the mixture in an ice bath and add dropwise, with shaking, 2 mL of 30% aqueous sodium hydroxide solution. Allow the tube to stand in the cooling bath, occasionally shaking it, until a drop of the reaction mixture when placed in water is *almost* completely soluble (2–3 hr is required). Pour the reaction mixture into 20 mL of water, filter the solution, and acidify the filtrate with 20% sulfuric acid. Collect the product on a suction filter, wash it with water, and recrystallize from 30% ethanol or from methanol. The recorded melting point of the α-monoxime is 137–138°; it corresponds in configuration to benzoin α-oxime (oximino hydroxyl in *anti* configuration relative to the $-CO-C_6H_5$ group). The stereoisomeric β-monoxime melts at 105–108°.

The α-monoxime acetate is cleaved by cold 5% sodium hydroxide to sodium benzoate, benzonitrile, and sodium acetate; the β-monoxime acetate is hydrolyzed to regenerate the oxime. In alkaline solution the α-monoxime forms an insoluble blue chelate complex on treatment with ferrous sulfate; the β-monoxime does not form such complexes. The α-monoxime dissolves in 10% aqueous potassium hydroxide to form a deep orange solution; the β-monoxime forms a canary yellow solution.

Benzil monosemicarbazone. In a 50-mL Erlenmeyer flask, dissolve 1 g of benzil in 20 mL of warm ethanol and cool the solution rapidly, while swirling it, to obtain a fine suspension of small crystals. To this add a solution of 0.5 g of semicarbazide hydrochloride and 1.5 g of sodium acetate crystals ($CH_3CO_2Na \cdot 3H_2O$) in 5 mL of water. Cork the flask firmly

---

[3] *Organic Syntheses*, Collective Volume **III**, 356 (1955).

and shake the mixture vigorously. Allow the flask to stand for several days at room temperature, then add 15–20 mL of water, and cool it in an ice bath. Collect the product with suction, wash it with water, and press it dry on the filter. The yield is 0.5 g. The crystals melt with decomposition in the range 170–175°, when heated rapidly.

To effect cyclization of the product, dissolve a small portion in 5% aqueous sodium hydroxide, shaking it, and then acidify the solution with glacial acetic acid. Collect the yellow crystals of 5,6-diphenyl-1,2,4-triazin-3-one, and wash them with water. This compound sinters at 190° and melts at 224–226°.

5,5-Diphenylhydantoin: Dilantin. In a small round-bottom flask place 1 g of benzil, 0.5 g of urea, 15 mL of ethanol, and 3 mL of 30% aqueous sodium hydroxide. Attach an upright condenser, add a boiling chip, and boil the mixture gently for 2 hr. Cool the reaction mixture, add 25 mL of water, and filter the solution to remove a sparingly soluble side product. Acidify the filtrate with hydrochloric acid, collect the product on a suction filter, and wash it thoroughly with water. This hydantoin may be recrystallized from ethanol. The yield is 0.7–0.8 g. Do not attempt to determine the melting point (recorded mp 286–295°). **Warning:** *Dilantin is a powerful therapeutic agent and must be taken only with the advice and supervision of a physician!*

Questions

1. Compare the mechanism of the benzoin condensation with those of the aldol condensation and the Cannizzaro reaction, including the role of the reagents employed.
2. Write the formula for the product formed by the action of excess phenylmagnesium bromide, followed by treatment with the water and mineral acid, on each compound.
   (a) benzoin
   (b) benzil
   (c) mandelonitrile (benzaldehyde cyanohydrin)
3. Write configurational formulas for the stereoisomers of benzil dioxime.
4. Application of the benzoin condensation to an equimolar mixture of benzaldehyde and 4-methoxybenzaldehyde (anisaldehyde) furnishes the compound $CH_3O-C_6H_4-CO-CH(OH)-C_6H_5$ (mp 106°).
   (a) Suggest a method for establishing definitely that the carbonyl group is in the assigned position rather than adjacent to the phenyl group.
   (b) Devise a synthesis that would furnish the isomeric methoxybenzoin (mp 89–90°).
5. When benzil is heated with urea and alkali to prepare 5,5-diphenylhydantoin, a sparingly soluble compound of the molecular formula $C_{16}H_{14}N_4O_2$ is produced in small amount by a side reaction. Devise a structural formula for this compound. (The side reaction does not involve rearrrangement of benzil.)

34 The Benzilic Acid Rearrangement

34.1 Introduction

When benzil is warmed with strong alkalis, it is converted into a salt of α-hydroxydiphenylacetic acid (benzilic acid). The 1,2-molecular rearrangement is initiated by addition of hydroxyl ion to the diketone (step 1), followed by transfer of the aryl group with its bonding electrons (carbanion rearrangement) to the adjacent carbon atom (step 2). By concurrent proton interchange the stable benzilate anion is formed.

$$\begin{array}{c} C_6H_5-C=O \\ | \\ C_6H_5-C=O \end{array} \underset{(1)}{\overset{OH^-}{\rightleftharpoons}} \begin{array}{c} O^- \\ | \\ C_6H_5-C-OH \\ | \\ C_6H_5-C=O \end{array} \xrightarrow{(2)} \begin{array}{c} O \\ \parallel \\ C_6H_5-C-OH \\ | \\ C \\ / \setminus \\ C_6H_5 \quad O^- \end{array} \xrightarrow{(3)} \begin{array}{c} O \\ \parallel \\ C_6H_5-C-O^- \\ | \\ C \\ / \setminus \\ C_6H_5 \quad OH \end{array}$$

It has been established by means of oxygen exchange with ^{18}O-labeled water that step 1 is reversible and faster than step 2. With methoxide ion in methanol, a similar rearrangement occurs, and the product is the methyl ester of benzilic acid.[1] An analogous reaction is the conversion of benzil to dilantin sodium with urea and base (see Section 33.3).

The benzilic acid rearrangement extends to many substituted diaryl 1,2-diketones but not to simple aliphatic analogs (such as

---

[1] For a discussion of the benzilic acid and related rearrangements see Lowry and Richardson, *Mechanism and Theory in Organic Chemistry* (New York: Harper and Row, 1976), and Selham and Eastman, *Quarterly Reviews* (London), **14**, 221 (1960).

CH_3—CO—CO—CH_3). The latter undergo complex aldol condensations under the influence of alkaline reagents.

Cyanide ion in ethanolic solution, in catalytic amounts, causes a rapid and complete cleavage of benzil at the central carbon–carbon bond; the products are ethyl benzoate and benzaldehyde.

$$\begin{array}{c} C_6H_5-C=O \\ | \\ C_6H_5-C=O \end{array} + C_2H_5OH \xrightarrow{CN^-} C_6H_5-\underset{OC_2H_5}{\overset{C=O}{|}} + C_6H_5-\underset{H}{\overset{C=O}{|}}$$

The mechanism of the cleavage is uncertain, but it is likely that the process involves addition of cyanide ion to a hemiketal formed by interaction of the diketone with ethanol.

34.2 Preparation of Benzilic Acid

(A) From Benzil

In a 100-mL flask dissolve 5 g (0.075 mole) of 85% pure solid potassium hydroxide pellets in 10 mL of water, add 15 mL of ethanol, and mix well by swirling. To the solution add 5 g (0.024 mole) of pure benzil (a bluish coloration is developed), attach an upright condenser, and reflux the solution on a steam bath for 10–15 min. Transfer the contents of the flask to a small beaker or a porcelain dish and cover it with a watch glass. Allow the reaction mixture to stand for several hours, preferably overnight, until crystallization of the potassium salt of benzilic acid is complete. Collect the crystals on a suction filter and wash them sparingly with ice-cold ethanol. The ethanolic mother liquor will furnish a small additional quantity of potassium benzilate if allowed to stand overnight.

Dissolve the potassium salt in about 150 mL of water and add to the solution, with stirring, *two drops* of concentrated hydrochloric acid. A reddish brown, slightly sticky precipitate is formed. Add a small amount of decolorizing carbon and filter off the solid material. If the procedure has been performed successfully, the filtrate will be colorless or only faintly yellow. Pour the clear filtrate slowly, with stirring, into a solution of 8 mL of concentrated hydrochloric acid in 50 mL of water. Collect the precipitated benzilic acid with suction, wash it thoroughly with water to remove chlorides, and press it dry. The crude product is usually light pink or yellow and weighs 4–4.5 g. Crystallize the material from hot water, with addition of a little decolorizing carbon. The yield of purified benzilic acid is 3.5–4 g.

(B) From Benzoin (*Alternative Procedure*)

In a small porcelain dish prepare a solution of 7 g (0.175 mole) of solid sodium hydroxide and 1.5 g (0.01 mole) of sodium bromate (or 1.7 g of potassium bromate) in 15 mL of water. To the warm solution, add in portions 6–6.5 g of the slightly moist benzoin obtained in Chapter 33 (this amount corresponds to about 0.025 mole of dry benzoin). During and after the addition of the benzoin, heat the reaction mixture on a steam bath, and stir it constantly. The mixture should not be heated above 90–95°, since higher temperatures favor decomposition to form benzohydrol (diphenylmethanol). From time to time add small portions of water (in total about 12–15 mL) to keep the mixture from becoming too thick. Continue the heating and stirring until a small test portion is completely or almost completely soluble in water. This usually requires 1.5–2.0 hr.

Dilute the reaction mixture with 60 mL of water, and allow it to stand, preferably overnight. Filter the solution to remove the oily or solid side product (diphenylmethanol). To the filtrate add slowly, with stirring, sufficient 40% sulfuric acid (prepared by adding 5 mL of concentrated sulfuric acid to 15 mL of water) to reach a point just short of the liberation of bromine.[2] Usually about 17 mL of the sulfuric acid is required. Collect the benzilic acid with suction, wash it well with water, and press dry. The yield is 4.5–5 g, and the product is usually quite pure. Benzilic acid may be crystallized from hot water.

34.3 Reactions of Benzilic Acid

Benzilic acid may be characterized by acylation of its alcohol group or through reactions of the carboxyl group (esterification, amide formation). Oxidation with dichromate mixture converts it to benzophenone and carbon dioxide; reduction with hydriodic acid leads to diphenylacetic acid, $(C_6H_5)_2CH-CO_2H$.

The tertiary alcohol group also enters readily into condensation reactions with aromatic hydrocarbons and phenols; for example, benzilic acid reacts with benzene in the presence of stannic chloride as catalyst to give triphenylacetic acid.

Benzilic acid is converted by warming with phosphorus pentachloride (2 moles) into diphenylchloroacetyl chloride, but a different reaction occurs upon treatment with thionyl chloride, $SOCl_2$: benzophenone and carbon monoxide are produced, together with sulfur dioxide and hydrogen chloride (see Question 4). Thionyl chloride (3 moles) in carbon tetrachloride solution, converts benzilic acid to diphenylchloroacetic acid.

---

[2] To minimize the danger of passing the end point, it is advisable to set aside in a test tube, 8–10 mL of the filtrate and add sulfuric acid to the remainder until a *trace* of bromine is liberated. This is removed by adding the small portion of the solution from the test tube.

(A) Benzophenone from Benzilic Acid

In a 50-mL round-bottom flask, place 2.3 g (0.01 mole) of benzilic acid and attach a water-cooled condenser connected to a gas trap to absorb the gaseous by-products (SO_2 and HCl) (see Figure 8.2). Through the condenser carefully add 6 mL (9.9 g, 0.08 mole) of thionyl chloride, which causes an immediate vigorous reaction. Warm the reaction mixture gently on a steam bath until the benzilic acid has dissolved, and then reflux the solution for 30 min longer.

After the heating period, add *cautiously* 15 mL of water in small portions, through the condenser, to decompose the excess thionyl chloride. Add 5 mL of methylene chloride, transfer the mixture to a separatory funnel, and separate the organic layer. Wash the organic layer with water and then with 5% sodium bicarbonate solution (**Caution**—*frothing!*). Dry the organic layer with a small amount of anhydrous sodium sulfate and filter it into a small beaker. Remove the solvent on a steam bath, and set the beaker aside to cool. Benzophenone exists in two crystalline forms: one melting at 26° and the other at 48°. The low-melting form may require standing or a seed crystal for conversion to the higher-melting form. Once they are formed, wash the crystals with a little petroleum ether (bp 30–40°) (**Caution**—*flammable!*). The yield of purified benzophenone is about 0.5 g; mp 48°.

(B) Acetylbenzilic Acid (α-Acetoxydiphenylacetic Acid)

In a large test tube place 0.5 g of benzilic acid, 1 mL of glacial acetic acid, and 1 mL of acetic anhydride. Add *one drop* of concentrated sulfuric acid, mix well, and heat the tube in a bath of boiling water for 2 hr. Cool the solution to 25° and add 5 mL of water, drop by drop, while shaking it. Allow the reaction mixture to stand overnight or longer to permit crystallization of the acetyl derivative. Collect the product with suction, wash it well with water, and press it dry on the filter. The air-dried material is a monohydrate (mp 96–98°). Prolonged drying in a vacuum desiccator over sulfuric acid is necessary to obtain anhydrous acetylbenzilic acid (mp 104–105°).

(C) Methyl Benzilate

In a large-diameter test tube place 0.5 g of benzilic acid, 5 mL of methanol, and 0.5 mL of concentrated sulfuric acid. Addition of the acid causes the development of a red color that disappears when the tube is shaken. Attach a short condenser, add a boiling chip, and reflux the solution for 30 min. Cool the reaction mixture and pour it in small portions into 20 mL of 5% aqueous sodium carbonate solution (**Caution**—*foaming!*). Chill the mixture in an ice bath, collect the crystals with suction, and wash them well with water. The recorded melting point of methyl benzilate is 74–75°.

Questions

1. Account for the difference in type of reagent used to bring about the benzilic acid and the pinacol–pinacolone rearrangements.
2. What products are formed in the following reactions?
   (a) benzilic acid + toluene (+ $SnCl_4$ catalyst)
   (b) benzilic amide + NaOBr + NaOH (Hofmann reaction)
   (c) methyl benzilate + excess C_6H_5—MgBr (followed by H_2O + acid)
   (d) methyl benzilate + ammonia (in methanol)
   (e) methyl benzilate + phenyl isocyanate (C_6H_5—N=C=O)
3. In the presence of a trace of sulfuric acid, benzilic acid reacts with acetone to form a crystalline product (mp 48°) with molecular formula $C_{17}H_{16}O_3$. Suggest a structural formula for this compound.
4. The formation of benzophenone upon treatment of benzilic acid with thionyl chloride, $SOCl_2$, presumably involves two steps: conversion of benzilic acid to diphenylhydroxyacetyl chloride, followed by decomposition of the acid chloride through a cyclic transition state. Write a detailed set of mechanisms for this transformation.

35 Triphenylmethanol

35.1 Triarylmethanols

Triphenylmethane and its derivatives may be synthesized conveniently by means of the Friedel–Crafts reaction or the Grignard reaction. Hydroxyl and amino derivatives, used in the manufacture of triphenylmethane dyes, are obtained by condensation of aromatic aldehydes and diaryl ketones with phenols and arylamines in the presence of acid catalysts.

In the presence of anhydrous aluminum chloride, chloroform reacts stepwise with benzene (in excess) to form benzylidene chloride, benzohydryl chloride, and finally triphenylmethane. With carbon tetrachloride the end product is triphenylchloromethane (also called trityl chloride).

$$CHCl_3 \xrightarrow[AlCl_3]{C_6H_6} C_6H_5-CHCl_2 \xrightarrow{C_6H_6} (C_6H_5)_2CHCl \xrightarrow{C_6H_6} (C_6H_5)_3CH$$

$$CCl_4 \xrightarrow[AlCl_3]{C_6H_6} C_6H_5-CCl_3 \xrightarrow[20°]{C_6H_6} (C_6H_5)_2CCl_2 \xrightarrow[70°]{C_6H_6} (C_6H_5)_3CCl$$

The successive steps of arylation require increasingly vigorous conditions, and it is not possible to introduce a fourth aryl group by the Friedel–Crafts reaction. Tetraphenylmethane has been obtained, in low yield, by heating trityl chloride with phenylmagnesium bromide. Trityl chloride is extremely reactive and is hydrolyzed rapidly by cold water to form triphenylmethanol.

Triphenylmethanol can be synthesized readily by the Grignard reaction. (See Chapter 17 for a discussion of the Grignard reaction.) Phenylmagnesium bromide is prepared by direct reaction of bromobenzene with metallic magnesium in the presence of anhydrous diethyl ether (or tetrahydrofuran).

$$C_6H_5-Br + Mg + (C_2H_5)_2O \longrightarrow C_6H_5-Mg-Br \cdot 2(C_2H_5)_2O$$

Usually a crystal of iodine is added to aid in starting the reaction, which must be carried out with carefully purified reagents and under anhydrous conditions. A small amount of biphenyl, $C_6H_5-C_6H_5$, is formed through coupling of the aryl groups (Wurtz–Fittig reaction).

The magnesium bromide salt (I) of the tertiary alcohol triphenylmethanol may be obtained by reaction of phenylmagnesium bromide with any one of several reagents, for example, benzophenone (equation 35.1), an ester of benzoic acid (equation 35.2), and dimethyl or diethyl carbonate (equation 35.3).

$$C_6H_5-\underset{II}{\overset{O}{\overset{\|}{C}}}-C_6H_5 + C_6H_5-MgBr \longrightarrow \underset{I}{(C_6H_5)_3C-O-MgBr} \qquad (35.1)$$

$$C_6H_5-\underset{III}{\overset{O}{\overset{\|}{C}}}-OCH_3 + C_6H_5-MgBr \longrightarrow II \longrightarrow I \qquad (35.2)$$

$$CH_3O-\overset{O}{\overset{\|}{C}}-OCH_3 + C_6H_5-MgBr \longrightarrow III \longrightarrow II \longrightarrow I \qquad (35.3)$$

Subsequent hydrolysis of the salt yields the neutral triphenylmethanol. These reactants require, respectively, 1, 2, and 3 equivalent(s) of the Grignard reagent. Benzophenone gives the salt of triphenylmethanol directly; esters of benzoic acid add one equivalent of Grignard reagent to produce benzophenone, which then adds a second equivalent of Grignard reagent; dimethyl carbonate first produces a benzoate ester, which then reacts sequentially with two more equivalents of Grignard reagent. It is not feasible to arrest the reaction at an intermediate stage because the reactivities toward phenylmagnesium bromide decrease in the sequence: benzophenone > methyl benzoate > dimethyl carbonate, corresponding to the diminishing electrophilic activity of the carbonyl group of these reactants. This sequence is the reverse of the reactivities observed in the Friedel–Crafts arylation of carbon tetrachloride: $CCl_4 > C_6H_5-CCl_3 > (C_6H_5)_2CCl_2$. The relative reactivities in this series correspond to the diminishing electrophilic activity of the corresponding carbocations: $[CCl_3]^+ > [C_6H_5-CCl_2]^+ > [(C_6H_5)_2CCl]^+$.

Triphenylmethanol, like other tertiary alcohols, is not acetylated by reaction with acetyl chloride but is converted to triphenylchloromethane. The colorless alcohol dissolves in cold concentrated sulfuric acid to form a yellow solution of the halochromic salt, containing the relatively stable triphenylmethyl carbocation, $(C_6H_5)_3C^+$. On dilution with water the alcohol is regenerated.

Triphenylmethanol is the parent structure of the color bases of triphenylmethane dyes, such as malachite green and crystal violet. It is related in a similar way to the rhodamine dyes, prepared from *m*-aminophenols and phthalic anhydride, and to the phthalein and sulfonphthalein acid–base indicators (see Chapter 44).

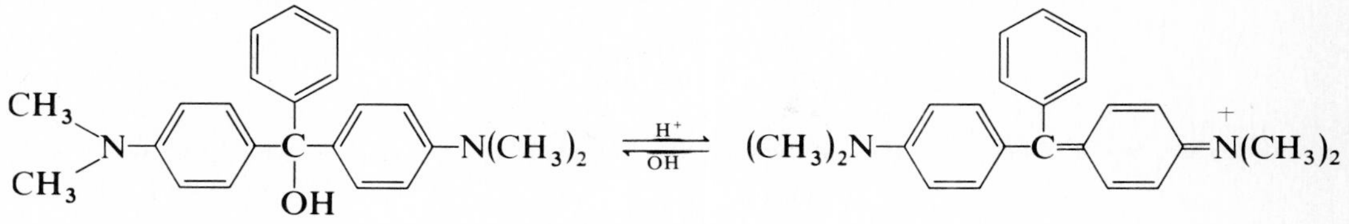

Malachite green

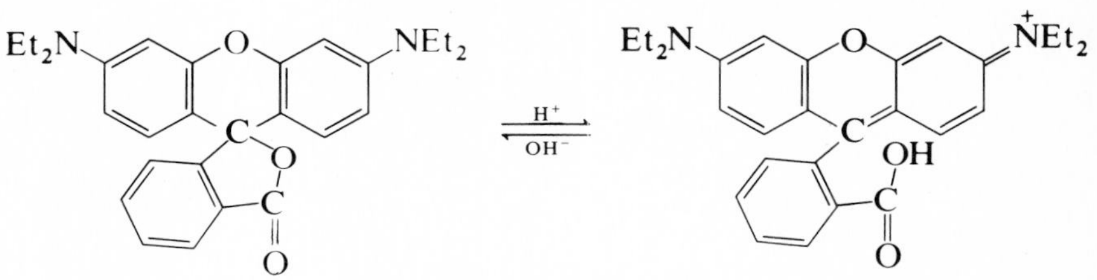

Rhodamine

The colored compounds are salts of amino or hydroxyl derivatives that have quinonoid structures of the types shown. Reduction of these compounds gives the colorless leuco bases, which are derivatives of triphenylmethane.

35.2 Preparation and Reactions

(A) Grignard Synthesis of Triphenylmethanol

In Grignard reactions it is essential that the reagents be free from ethanol and water, and the apparatus perfectly clean and dry. *Do not use* rubber stoppers because they contain extractable, deleterious sulfur compounds.

▶ *CAUTION*   It is advisable to have a bath of ice and water at hand during this preparation as the reaction may start suddenly with vigorous ebullition of the ether. Take care that no flame is nearby.

Phenylmagnesium bromide solution. In a 250-mL round-bottom flask provided with a Claisen adapter bearing an addition funnel and a vertical condenser, place 2.4 g (0.1 mole) of magnesium turnings (*carefully weighed*). Introduce directly into the flask a mixture of 2 mL (3 g) of bromobenzene, 7 mL of anhydrous diethyl ether[1] and a small crystal of iodine. If a reaction

---

[1] It is essential that the ether be of the anhydrous grade.

does not start at once, warm the flask gently in a bath of warm water. *After the reaction has started*, as evidenced by disappearance of the iodine color, appearance of turbidity, and spontaneous boiling, add 70 mL of anhydrous diethyl ether. For the success of the experiment it is essential that the reaction begin before the main portions of the ether and bromobenzene are added.

▬▶ *CAUTION* Ether that has been exposed to air and light for some time while being stored may contain an unstable peroxide that can explode violently when heated, especially toward the end of a distillation. This peroxide can be detected by liberation of iodine when a test portion of the ether is shaken with a little 2% potassium iodide solution that has been acidified with hydrochloric acid.

The peroxide can be removed by washing the ether with an equal volume of dilute, weakly acidified ferrous sulfate solution.

Place in the addition funnel 9 mL (13.5 g) of bromobenzene (a total of 0.105 mole) and allow it to flow drop by drop into the previously activated reaction mixture at such a rate that the ether refluxes without external heating.

After all of the halide has been added, reflux the mixture gently for 30 min on a steam bath. Do not heat the material so vigorously that ether vapors traverse the condenser. The reaction is complete when the magnesium has dissolved; some dark particles of impurities will remain undissolved. Remove the heating bath and proceed without delay to the next step.

Reaction of phenylmagnesium bromide with methyl benzoate.[2] Cool the reaction flask containing the Grignard reagent to 15–20° and place in the addition funnel a solution of 6.5 mL (7 g, 0.05 mole) of pure methyl benzoate in about 25 mL of anhydrous ether. Allow the methyl benzoate solution to flow slowly into the Grignard reagent, with continuous swirling, and cool the flask from time to time to control the reaction. The bromomagnesium derivative of the alcohol separates as a white precipitate. After all of the methyl benzoate has been added, allow the mixture to stand at room temperature for 30 min or longer.

Pour the contents of the flask as completely as possible into a mixture of about 50 g of ice, 100 mL of water, and 5–6 mL of concentrated sulfuric acid, contained in a 500-mL flask. Add 4–5 mL of strong sodium bisulfite solution to remove any free iodine. Shake the mixture thoroughly to complete the decomposition of the magnesium derivative and rinse the reaction flask with the acid mixture to remove material that adheres to the wall of the flask. Add about 75 mL of *ordinary* diethyl ether to aid in extracting the product completely. Separate the ether layer, wash it with two 20-mL por-

---

[2] In place of methyl benzoate, 8 mL (8 g) of ethyl benzoate may be used.

tions of 5–10% sulfuric acid and once with saturated salt (NaCl) solution. Finally, wash the ether layer with aqueous sodium bicarbonate solution and once more with saturated salt solution.[3]

Dry the ether solution over anhydrous magnesium sulfate and transfer it to a 500-mL round-bottom flask arranged for distillation. Distill off the ether as completely as possible, using a steam bath (*Avoid fire hazards!*). The residual crude product contains the impurities biphenyl and unreacted bromobenzene and methyl benzoate, along with triphenylmethanol.

Stir the residue with petroleum ether or ligroin (about 5 mL/g of residue) and collect the solid on a suction filter; this process removes most of the impurities. Complete the purification by recrystallization from 2-propanol (about 7 mL/g), and collect the product on a suction filter. The yield is 7–8 g; mp 160–172°.

The ligroin extract may be evaporated to dryness and examined by chromatographic and spectroscopic methods to study its components.

In a small test tube place 1–2 mL of concentrated sulfuric acid. Add to this a very small pinch of triphenylmethanol and observe. What is formed?

(B) β,β,β-Triphenylpropionic Acid from Triphenylmethanol

In a small Erlenmeyer flask place 1.3 g (0.005 mole) of triphenylmethanol and 2.6 g (0.025 mole) of finely pulverized malonic acid, $HO_2C-CH_2-CO_2H$. Mix the solids thoroughly by shaking and heat the flask, under an inverted funnel connected to an aspirator pump, on a hot plate to about 150–160° for about 10 min. After allowing the flask to cool, dissolve the product in 3 mL of toluene and add 10 mL of petroleum ether (bp 60–90°) (**Caution**—*flammable solvents!*).

$$C_6H_5-\underset{\underset{C_6H_5}{|}}{\overset{\overset{C_6H_5}{|}}{C}}-OH + \underset{\underset{CO_2H}{|}}{\overset{\overset{CO_2H}{|}}{CH_2}} \xrightarrow{\Delta} C_6H_5-\underset{\underset{C_6H_5}{|}}{\overset{\overset{C_6H_5}{|}}{C}}-CH_2CO_2H + CO_2 + H_2O$$

β,β,β-Triphenylpropionic acid

Allow the solution to stand until crystals of the acid have separated. Collect them on a small suction filter and wash them with a little petroleum ether. The yield is 0.8–1.0 g; mp 174–176°.

Questions 1. Write equations for the action of phenylmagnesium bromide on the following compounds, including hydrolysis of the reaction mixture with dilute acid.
    (a) carbon dioxide
    (b) ethanol

---

[3] For a discussion of the uses of saturated salt solution as a preliminary drying agent see Ellern, *J. Org. Chem.*, **47**, 3569 (1982).

(c) oxygen
(d) *p*-tolunitrile
(e) ethyl formate

2. How may the following compounds be prepared from phenylmagnesium bromide?
   (a) 1,2-diphenylethanol
   (b) benzaldehyde
   (c) benzyl alcohol
   (d) benzopinacol

3. Indicate a series of reactions for the conversion of triphenylmethanol to hexaphenylethane. Cite evidence showing that hexaphenylethane undergoes dissociation (homolysis) to form triphenylmethyl radicals. Why is this radical more stable than a free methyl radical?

4. Write equations for the synthesis of a typical triphenylmethane dye, starting from simple aromatic compounds. Explain the terms leuco base and color base.

36 Pheromones and Insect Repellents

36.1 Chemical Communication

Pheromones are compounds excreted by animals for communication among members of the same species. Such chemical messages include attraction of the opposite sex, marking trails to food, and warning of danger. In the case of sex pheromones, the amount of material required can be phenomenally small. A particularly striking example is the female gypsy moth's ability to attract a mate by releasing as little as 1×10^{-9} g of pheromone. Structurally, sex pheromones differ widely from species to species and can be complex or simple, as can be seen from the following examples.

Sex Pheromones

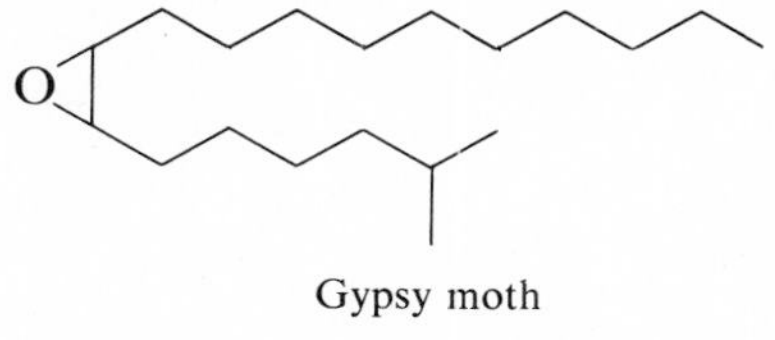

Gypsy moth

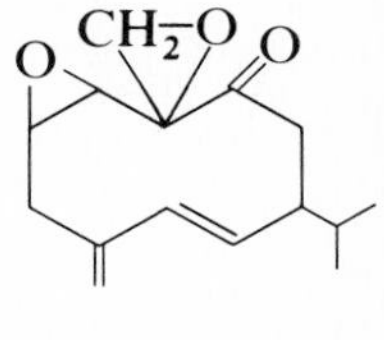

American cockroach

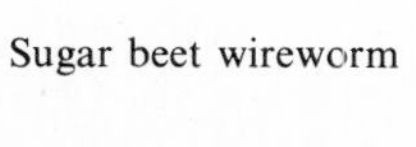

Sugar beet wireworm

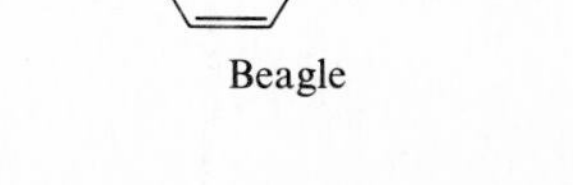

Beagle

There is evidence that some animals may require more than one chemical for transmission of a chemical message and that the relative amount of each component is important as well. The use of a multicomponent message would reduce the "cross-talk" between different species. It is known that although the chemical signal of one species can be detected by members of a different species, the response is usually different and much less intense than an intraspecies response.

36.2 Insect Repellents

Strictly speaking, current commercial insect repellents should not be classified as pheromones because they are artificial substances that are not used for chemical communication in nature. It is not even clear that they interact with the pheromone receptors. However, because their effects are transmitted by chemical vapors, repellents and pheromones are commonly considered together.

A widely used repellent is N,N-diethyl-m-toluamide, which is effective against mosquitoes, fleas, gnats, and many other insects. This substance is present in Off! as a 14% solution and in Cutter as a pure liquid. This author can testify to the effectiveness of both brands against very hungry mosquitoes.

The antennae of mosquitoes are covered with receptors that detect the currents of carbon dioxide and water vapor rising from warm-blooded hosts. They can track these currents, much like a heat-seeking missile, to their source and proceed to feed. However, when the mosquito encounters an atmosphere filled with the repellent, the signals from its receptors are distorted in some fashion and it has difficulty recognizing or finding the host.

In this preparation[1] you will make the active ingredient of the commercial repellent Off!. The starting material is m-toluic acid, which is converted to the acid chloride with thionyl chloride.

$$\underset{m\text{-Toluic acid}}{\underset{CH_3}{\bigodot}-CO_2H} \xrightarrow{SOCl_2} \underset{CH_3}{\bigodot}-\underset{Cl}{\overset{O}{\underset{\|}{C}}} + SO_2 + HCl$$

[1] The preparation is patterned after the one described by Wang, *J. Chem. Educ.*, **51**, 631 (1974).

The acid chloride is converted to the amide by addition of diethylamine.

$$\underset{CH_3}{\underset{|}{C_6H_4}}-\overset{O}{\underset{Cl}{C}} + H-N(C_2H_5)_2 \longrightarrow \underset{CH_3}{\underset{|}{C_6H_4}}-\overset{O}{C}-N(C_2H_5)_2 + (C_2H_5)_2\overset{+}{N}H_2Cl^-$$

N,N-Diethyl-m-toluamide

36.3 Preparation of N,N-Diethyl-m-toluamide

m-Toluyl chloride. Assemble the apparatus shown in Figure 8.1(b) using thoroughly dried glassware and a 250-mL round-bottom flask. The top of the reflux condenser should be connected to a water aspirator by means of a T-tube, as shown in Figure 8.2(b). In the flask place 2.7 g (0.020 mole) of m-toluic acid and 3 mL (5.0 g, 0.042 mole) of thionyl chloride ($SOCl_2$). Add a boiling chip, and reflux the mixture gently on a steam bath in the hood for 15 min. During the heating, the toluic acid goes into solution and sulfur dioxide and hydrogen chloride are evolved. In some procedures for preparing acid chlorides the excess thionyl chloride would be removed at this point by evaporation under reduced pressure, but that is not necessary in this case because the excess will be destroyed in the next step.

▶ *CAUTION* Handle thionyl chloride with great care. The liquid burns the skin, and the vapor is an irritant and harmful to breathe. If any is spilled on the skin, wash the affected area thoroughly with soap and water.

N,N-Diethyl-m-toluamide (Off!). Remove the steam bath from the reflux setup and replace it with a pan of ice water. While the reaction cools, prepare a solution of 7 mL of diethylamine (5.0 g, 0.068 mole) (**Caution—Stench**, use hood!) in 25 mL of diethyl ether and place the solution in the dropping funnel.

▶ *CAUTION* Ether is extremely volatile and highly flammable. Extreme care must be taken to avoid fire hazards.

Add the amine solution drop by drop to the acid chloride with constant agitation. At first, a vigorous reaction occurs with each drop accompanied by evolution of heat and formation of a white cloud of the amine hydrochloride that fills the apparatus. After a few milliliters of the amine solution have been added, however, the reaction will become more moderate. After all of the amine solution has been added, remove the ice bath and allow the reaction mixture to stand for about 10 min to ensure complete reaction. Transfer the mixture to a separatory funnel with the aid of 5–10 mL of dichloromethane, and wash it successively with 25-mL portions of water, 5% aqueous sodium hydroxide, 3 M hydrochloric acid, and finally with water. Dry the light brown dichloromethane/ether solution with a little anhydrous magnesium sulfate and, after filtration, remove most of the solvent on a steam bath in the hood.

The crude amide can be purified by column chromatography using about 15 g of alumina or silica gel and dichloromethane as the elution solvent. The first compound to come off the column is N,N-diethyl-m-toluamide. Evaporation of the solvent gives about 2.5–3.0 g of the purified amide as a colorless to light tan oil (bp 160–165 at 20 mm pressure).

If an IR or NMR instrument is available, obtain the spectrum of the amide and verify that the assigned structure is correct.

Questions

1. Propose syntheses for the gypsy moth and beagle sex pheromones.
2. What is the mechanism for conversion of carboxylic acids to acid chlorides using thionyl chloride?
3. What is the mechanism for conversion of the m-toluyl chloride to the amide using diethylamine?
4. In the reaction of diethylamine with m-toluyl chloride, a small amount of m-methylbenzaldehyde is formed. How can you account for this? What might the other products be?

37

The Pinacol-Pinacolone Rearrangement

37.1 Introduction

Ketones may be reduced by means of amalgamated magnesium, or a mixture of magnesium and iodine (magnesious iodide), to form bimolecular reduction products called pinacols, of the type $R_2C(OH)-C(OH)R_2$ (equation 37.1). Aryl ketones, such as benzophenone, may be reduced photochemically by exposing a solution of the ketones in ethanol or 2-propanol to ultraviolet illumination (equation 37.2). The reduction is a one-electron process that involves free radical intermediates.

$$2 \begin{array}{c} CH_3 \\ \diagdown \\ C=O + 2H \\ \diagup \\ CH_3 \end{array} \xrightarrow{Mg+Hg} CH_3-\underset{\underset{OH}{|}}{\overset{\overset{CH_3}{|}}{C}}-\underset{\underset{OH}{|}}{\overset{\overset{CH_3}{|}}{C}}-CH_3 \quad (37.1)$$

$$\begin{array}{c} C_6H_5 \\ \diagdown \\ C=O \\ \diagup \\ C_6H_5 \end{array} + \begin{array}{c} CH_3 \\ \diagdown \\ CHOH \\ \diagup \\ CH_3 \end{array} \xrightarrow{h\nu} \begin{array}{c} C_6H_5 OH \\ \diagdown \diagup \\ C\cdot \\ \diagup \\ C_6H_5 \end{array} + \begin{array}{c} CH_3 \\ \diagdown \\ \cdot C-OH \\ \diagup \\ CH_3 \end{array}$$

$$\begin{array}{c} C_6H_5 \\ \diagdown \\ C=O \\ \diagup \\ C_6H_5 \end{array} + \begin{array}{c} CH_3 \\ \diagdown \\ \cdot C-OH \\ \diagup \\ CH_3 \end{array} \longrightarrow \begin{array}{c} C_6H_5 OH \\ \diagdown \diagup \\ C\cdot \\ \diagup \\ C_6H_5 \end{array} + \begin{array}{c} CH_3 \\ \diagdown \\ C=O \\ \diagup \\ CH_3 \end{array}$$

$$2 \begin{array}{c} C_6H_5 OH \\ \diagdown \diagup \\ C\cdot \\ \diagup \\ C_6H_5 \end{array} \longrightarrow C_6H_5-\underset{\underset{OH}{|}}{\overset{\overset{C_6H_5}{|}}{C}}-\underset{\underset{OH}{|}}{\overset{\overset{C_6H_5}{|}}{C}}-C_6H_5 \quad (37.2)$$

When heated with strong acids, or a catalyst such as iodine, substituted 1,2-diols of the pinacol type undergo dehydration and rearrangement to ketones (the pinacol–pinacolone rearrangement). The transformation involves protonation of one of the hydroxyl groups (I), followed by elimination of water to give a carbocation (II). Migration of an alkyl or aryl group from the adjacent carbon atom, with its bonding electrons, leads to the protonated form of the pinacolone (III).

$$R_2C(^+OH_2)-CR_2(OH) \longrightarrow R_2C^+-CR_2(OH) \longrightarrow R_3C-C(^+OH)=R$$

$$\quad\quad\quad\text{I} \quad\quad\quad\quad\quad\quad \text{II} \quad\quad\quad\quad\quad \text{III}$$

Carbocation rearrangements of the pinacol type are encountered with 1,2-halohydrins, 1,2-amino alcohols, α-hydroxyaldehydes, and similar structures. The mechanism of the rearrangement has been studied extensively.[1]

Iodine is also capable of catalyzing the pinacol rearrangement. The first step is believed to be formation of the covalent hypoiodite followed by loss of hypoiodite ion to give the same carbocation formed with acid.

$$R\text{—}\ddot{\text{O}}\text{H} + \text{I—I} \longrightarrow \longrightarrow R\text{—O—I} + \text{HI}$$
$$\text{Hypoidoite}$$
$$\downarrow$$
$$R^+ + OI^-$$

Studies of the rearrangement of a series of symmetrical pinacols (type IV) have disclosed that substituents exert a marked effect on the selectivity of migration of the aryl group.

$$R\text{—}C_6H_4\text{—}\underset{C_6H_5}{\overset{OH}{C}}\text{—}\underset{C_6H_5}{\overset{OH}{C}}\text{—}C_6H_4\text{—}R \quad\quad (R\text{—}C_6H_4)_2\overset{OH}{C}\text{—}\overset{OH}{C}(C_6H_5)_2$$

$$\quad\quad\quad\quad\quad\text{IV} \quad\quad\quad\quad\quad\quad\quad\quad\quad\quad\quad\quad \text{V}$$

The *migration aptitudes* of a few substituted groups, in relation to C_6H_5— are *p*-methoxy-, 500–1000; *p*-methyl-, 15–18; *p*-chloro-, 0.7; *m*-methoxy-, 1.6. The rearrangement of unsymmetrical pinacols (type V) does not involve selectivity of migration but is governed by selectivity in the formation of

---

[1] Collins, "The Pinacol Rearrangement," *Quart. Revs.* (*London*), **14**, 357 (1960); Lowry and Richardson, *Mechanism and Theory in Organic Chemistry* (New York: Harper & Row, 1976); Hine, *Physical Organic Chemistry*, 2nd ed. (New York; McGraw-Hill, 1962).

one of the two possible carbocation intermediates (VI and VII). In this situation the ease of formation of the carbocation is the dominant factor, and the effect of substituents is different from that observed with the symmetrical pinacols.

$$(R-C_6H_4)_2\overset{+}{C}-\underset{\underset{VI}{|}}{\overset{\overset{OH}{|}}{C}}(C_6H_5)_2 \qquad (R-C_6H_4)_2\underset{\underset{VII}{|}}{\overset{\overset{OH}{|}}{C}}-\overset{+}{C}(C_6H_5)_2$$

Electron-releasing groups favor carbocation formation at the carbon atom to which they are directly attached, and this leads to preferential migration of a group from the *adjacent* carbon atom. In the unsymmetrical pinacol (type VI) when R is *p*-methoxyphenyl, this group migrates only to the extent of 30% and the unsubstituted phenyl group to 70%. With the *p*-chlorophenyl analog its migration is about 40% versus 60% for the phenyl group.

Benzopinacolone formed from the rearrangement of benzopincol can be cleaved by base to yield benzoic acid and triphenylmethane by the following mechanism. With substituted benzopinacolones the cleavage reaction can be used to identify the phenyl ring that migrates.

$$\underset{C_6H_5}{\overset{C_6H_5}{C_6H_5}}\!\!\!\!\!>\!\!C-\overset{O}{\overset{\|}{C}}-C_6H_5 + O\bar{H} \longrightarrow \underset{C_6H_5}{\overset{C_6H_5}{C_6H_5}}\!\!\!\!\!>\!\!C-\underset{OH}{\overset{O^-}{\overset{|}{C}}}-C_6H_5 \longrightarrow$$

$$\underset{C_6H_5}{\overset{C_6H_5}{C_6H_5}}\!\!\!\!\!>\!\!C^- + \underset{OH}{\overset{O}{\overset{\|}{C}}}-C_6H_5 \longrightarrow \underset{C_6H_5}{\overset{C_6H_5}{C_6H_5}}\!\!\!\!\!>\!\!C-H + \underset{^-O}{\overset{O}{\overset{\|}{C}}}-C_6H_5$$

37.2 Preparations

(A) Benzopinacol by Photochemical Reduction

Place 2.75 g (0.015 mole) of benzophenone in a 50-mL flask (or large test tube), add 20 mL of 2-propanol (isopropyl alcohol), and dissolve the solid by warming it on a steam bath. To the solution add exactly 1 drop of glacial acetic acid[2] and sufficient 2-propanol to fill the flask almost completely. Stopper the flask firmly with a good cork that fits tightly and

---

[2] The acid is added to neutralize traces of alkali from the glass vessel. Alkali is deleterious because it catalyzes cleavage of the pinacol to benzophenone and diphenylmethanol ($C_6H_5-CHOH-C_6H_5$).

projects about half its length into the neck of the flask. Wire the cork firmly to the neck of the flask by means of soft copper wire.

Invert the flask, support it firmly by means of a condenser or buret clamp, and expose it to direct sunlight or place it in close proximity to an ultraviolet lamp. Benzophenone is activated by absorption of light in the near-ultraviolet region, which is partially transmitted through ordinary glass. As the reduction progresses, benzopinacol separates in colorless, dense crystals. By occasional tapping and swirling, the crystals may be made to settle in the neck of the flask. Five or six days' exposure to moderate sunlight will furnish an abundant crop of crystals; usually a much shorter time is required with an ultraviolet lamp. Collect these with suction and, if necessary, return the filtrate for further exposure to complete the reaction. Under favorable conditions, the yield will attain at least 90%. The product is quite pure (mp 186–188°) and may be used directly for conversion to benzopinacolone.

(B) Benzopinacolone

In a 100-mL round-bottom flask place 1.5 g (0.07 mole) of benzopinacol, 8 mL of glacial acetic acid, and a *small* crystal of iodine. Attach a short reflux condenser and reflux the solution for 10 min. Allow the solution to cool slightly, add 8 mL of ethanol, swirl the mixture thoroughly, and allow it to cool. Collect the crystals with suction and wash them with cold ethanol to remove iodine. Benzopinacolone forms colorless crystals (mp 179–180°). The yield is 1.2–1.4 g.

Alkaline Cleavage of Benzopinacolone. In a small flask place 1 g (0.003 mole) of benzopinacolone, 6 mL of anhydrous ethylene glycol, and 1 g (0.02 mole) of solid 85% potassium hydroxide. Attach a reflux condenser, add a boiling chip, and reflux the two-phase mixture fairly vigorously for 1 hr. Cool the mixture, add 50 mL of water, and extract the triphenylmethane with 50 mL of methylene chloride. Some triphenylmethane may collect on the inside of the condenser and can be removed with small portions of methylene chloride. Save the aqueous layer for isolation of the benzoic acid.

Wash the methylene chloride layer with two 50-mL portions of water, dry it with anhydrous magnesium sulfate, and distill off the solvent. Recrystallize the triphenylmethane from 60% aqueous ethanol. The yield is about 0.5 g; mp 92–93°.

Acidify the aqueous alkaline layer cautiously with concentrated hydrochloric acid, cool if necessary, and extract the benzoic acid with two 25-mL portions of ether (**Caution**—*flammable!*). Dry the combined ether extracts over anhydrous magnesium sulfate, and distill off the ether from a steam bath. Recrystallize the benzoic acid from a small amount of water (about 25 mL/g) and take its melting point.

Questions

1. Indicate a stepwise mechanism for the alkali-catalyzed cleavage of benzopinacol.
2. Write equations for the reaction of *p*-tolylmagnesium bromide (in excess) on the following compounds, including hydrolysis of the reaction mixture with dilute acid.
   (a) $C_6H_5-CO-CO-C_6H_5$ (benzil)
   (b) methyl ester of benzilic acid
   (c) dimethyl oxalate
3. Write the structures of the isomeric pinacolones that could be obtained by pinacol rearrangement of symmetrical 4,4′-dimethylbenzopinacol. This reaction actually furnishes one of the pinacolones to the extent of more than 90%. How may its structure be determined?

38 Polycyclic Quinones

38.1 Quinones

Anthracene and phenanthrene can be oxidized directly at the 9,10-positions by means of chromic acid to give the corresponding quinones. These condensed polycyclic quinones are more stable to oxidation than the simple benzoquinones and naphthoquinones. In the oxidation reaction, some complex products are formed by coupling and disproportionation of the reactive intermediates and, with phenanthrene, some further oxidation of the quinone (I) leads to diphenic acid (II, 2,2′-biphenyldicarboxylic acid).

The crude phenanthrenequinone obtained by oxidation of technical phenanthrene is purified by conversion to the water-soluble sodium bisulfite addition product, from which the pure quinone is regenerated easily by treatment with base. Anthraquinone, derived from anthracene present as an impurity in technical phenanthrene, is a less reactive carbonyl compound and does not form a sodium bisulfite adduct.

Phenanthrenequinone bears a strong resemblance to the 1,2-diketone benzil and undergoes a benzilic acid rearrangement (see Chapter 34) when warmed with concentrated aqueous alkalies; the product is 9-hydroxyfluorene-9-carboxylic acid.

A general method for the preparation of anthraquinone and its derivatives consists of cyclization of *o*-benzoylbenzoic acids, which are obtained readily by the reaction of phthalic anhydride with benzene, toluene, chlorobenzene, and similar aromatic compounds, in the presence of anhydrous aluminum chloride. The cyclization can be effected by means of concentrated sulfuric acid or polyphosphoric acid.

Derivatives of anthraquinone are used in the manufacture of important vat dyes (*Indanthrene Brown*, *Caledon Jade Green*, etc.) that are unusually stable in light and with washing.

Anthraquinone is reduced easily to anthrahydroquinone (III), which dissolves in aqueous alkalies to form a deep red solution; air or mild oxidizing agents regenerate the quinone.

Reduction of anthraquinone with sodium dithionite, $Na_2S_2O_4$ (also called sodium hydrosulfite), in alkaline medium, or with stannous chloride in glacial acetic acid, gives anthrone (IV). On heating with alkalis, anthrone is deprotonated to a deep yellow anion (V) that has a number of resonance structures, a few of which are shown.

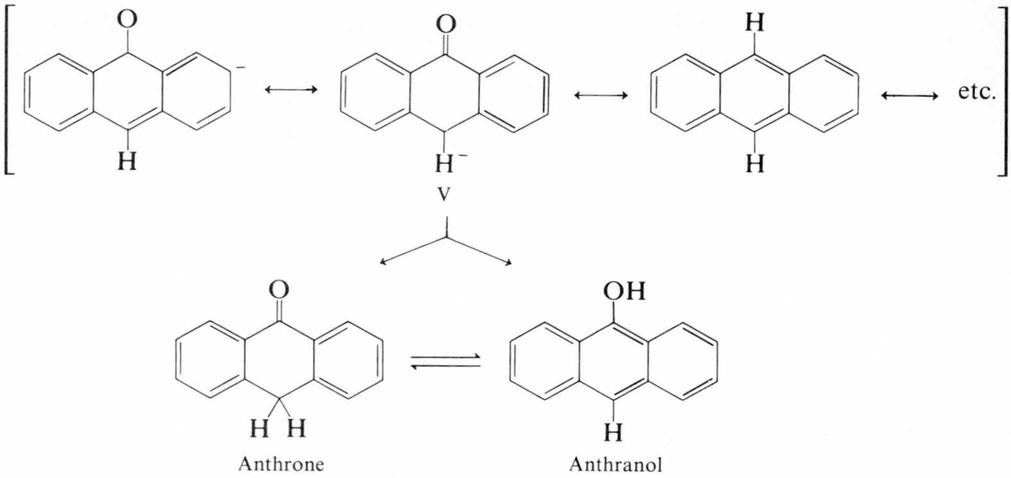

Anthrone   Anthranol

On acidification at low temperatures, protonation occurs most rapidly on oxygen to give anthranol, even though anthrone is thermodynamically more stable. Solutions of anthrone or anthranol slowly tautomerize into an equilibrium mixture of the two. The methylene group in anthrone enters into condensation reactions with active carbonyl compounds. Anthrone is a test reagent for carbohydrates: in sulfuric acid solution it gives a blue-green color with mono- and polysaccharides and glycosides.

Nitration of anthraquinone gives the 1-nitro derivative, which can be reduced to 1-aminoanthraquinone. Sulfonation at 140–150° with fuming sulfuric acid gives anthraquinone-2-sulfonic acid. Ammonolysis of the sodium salt of the sulfonic acid furnishes 2-aminoanthraquinone.

Many derivatives of anthraquinone occur in nature as pigments and active principles of plants, fungi, lichens, and insects. An example is emodin, the cathartic principle of cascara sagrada and of rhubarb, which is 1,3,8-trihydroxy-6-methylanthraquinone.

38.2 Preparations

(A) Anthraquinone

In a 250-mL round-bottom flask place 5 g (0.022 mole) of anhydrous *o*-benzoylbenzoic acid[1] and 25 mL of concentrated sulfuric acid. Heat the flask on a steam bath and swirl the mixture until the solid dissolves. By means of a clamp, support the flask firmly in a steam bath; place a towel or

---

[1] The anhydrous acid melts at 127–128°; it forms a monohydrate (mp 94–95°). The pure anhydrous acid can be purchased from chemical supply firms.

cloth around the bath and flask (to reduce heat loss), and heat the material at 100° for 30 min. With a medicine dropper or a pipet, add 5 mL of water, drop by drop, while swirling the solution. The product will begin to separate. Remove the flask from the steam bath and allow it to cool. Dilute the reaction mixture with about 150 mL of water and transfer it to a 600-mL beaker. Add enough chipped ice to bring the total volume to 300 mL and stir the mixture thoroughly. Collect the product by suction filtration, preferably on a hardened filter paper, and wash it well with water. To remove any unchanged starting material wash the product carefully with 10 mL of concentrated aqueous ammonia diluted with 50 mL of water, followed by a washing with water. To facilitate drying, wash the quinone finally with a little ice-cold acetone, and then spread it on a clean paper. Do not determine the melting point (284–286°). The yield is 4–4.5 g. A small sample may be purified by sublimation, at 230–250°, using the apparatus shown in Figure 5.7 or 5.8.

Reduction to Anthrone. In a 500-mL flask place 4.1 g (0.02 mole) of anthraquinone, 100 mL of water, 12.2 g (0.06 mole; based on 85% pure material) of sodium dithionate, $Na_2S_2O_4$, and a solution of 5 g (0.12 mole) of sodium hydroxide in 10 mL of water. Adjust a reflux condenser and attach a tube leading to an aqueous alkali trap (impurities in commercial sodium dithionite have a powerful stench) and boil the mixture gently for 40–50 min. The quinone is reduced first to anthrahydroquinone, which dissolves to give a deep red solution of the sodium salt. Further reduction gives the pale yellow, sparingly soluble anthrone.

Cool the reaction mixture, collect the product with suction, and wash it thoroughly with water. Dissolve the moist product in hot ethanol and filter the solution through a fluted filter. Chill the filtrate and collect the purified material. Concentrate the mother liquor and obtain a second crop of crystals. The yield is 3–3.5 g. Pure anthrone forms pale yellow crystals, mp 150–155° (dec.).

(B) Phenanthrenequinone

In a 500-mL Erlenmeyer flask place 30 mL of glacial acetic acid and 100 mL of water, and add carefully 60 mL of concentrated sulfuric acid. Swirl the solution, add 6 g (0.03 mole) of technical 90% phenanthrene, and heat the mixture to 95° (internal temperature) in a bath of boiling water. Prepare a solution of 36 g (0.12 mole) of sodium dichromate dihydrate in 25 mL of warm water and add this in portions, with swirling, to the hot suspension of phenanthrene. Watch the temperature of the reaction mixture carefully to observe the onset of a strongly exothermic reaction. When this occurs, stop the addition of the oxidizing solution, remove the flask from the heating bath, and swirl the mixture vigorously. The temperature will rise to 110–120°. As soon as the temperature begins to fall, resume addition of the dichromate solution and do not allow the temperature to drop below 85–90°. Dip the flask in the boiling water bath, when required, to maintain the desired temperature. After all of the oxidizing agent has been added, heat the mixture for 30 min in the boiling water bath, swirling it frequently.

Cool the reaction mixture and pour it into a well-stirred mixture of about 300 mL of water and 100 g of chipped ice. Break up any lumps of the product and collect the crude quinone on a large suction filter. Wash the crystals thoroughly with water, until the green chromous sulfate has been removed. Transfer the moist product to a 500-mL Erlenmeyer flask, add 65 mL of ethanol, swirl the mixture vigorously, and add a solution of 30 g (0.16 mole) of sodium bisulfite in 60 mL of water. The yellow color of the quinone is discharged as the bisulfite adduct is formed. Allow the mixture to stand for about 20 min, occasionally swirling it, to complete the reaction. Add water until the flask is nearly filled, cork it firmly, and shake the mixture thoroughly to dissolve the sodium bisulfite addition compound. Filter the mixture through a fluted filter to remove anthraquinone and other impurities. After the insoluble residue is washed with 20 mL of 50% aqueous ethanol, it may be discarded.

Place the filtrate and washings in a large beaker and add saturated aqueous sodium carbonate in small portions (**Caution**—*foaming!*), stirring thoroughly, until the solution is distinctly alkaline. Collect the phenanthrenequinone on a suction filter, wash it thoroughly with cold water, and press it as dry as possible on the filter. Place the damp product in a 250-mL round-bottom flask, add 55–60 mL of glacial acetic acid, and heat the mixture under a short reflux condenser to dissolve the quinone. Cool the solution to room temperature and allow the material to stand for 20 min or longer, occasionally swirling it before collecting the orange crystals of the purified product. Wash the crystals thoroughly with cold water and allow them to dry in the air. The yield is about 3 g; the recorded mp 206°.

Questions

1. What is the source of technical anthracene and phenanthrene? What methods are used to obtain these hydrocarbons in a state of high purity?
2. Write equations for the following syntheses of 2-methylanthraquinone.
   (a) starting from naphthalene and toluene
   (b) starting from naphthalene and aliphatic compounds, and making use of the Diels–Alder reaction (Chapter 41)
3. Write equations for the reaction of phenanthrenequinone with the following reagents.
   (a) hot aqueous potassium hydroxide solution
   (b) *o*-phenylenediamine
   (c) hydrogen peroxide, in glacial acetic acid

39
Enamine Synthesis of a Diketone: 2-Acetylcyclohexanone

39.1 The Enamine Reaction

Michael additions and many other acylation and alkylation reactions of less reactive carbonyl compounds involve the formation and reaction of an intermediate enolate anion. The scope of these reactions is limited by three factors.

To form the enolate anion it is necessary to use a strong base like sodium amide, triphenylmethide ion, or t-alkoxides, which may react competitively with the acylating or alkylating reagent. A second complication is the unwanted base-catalyzed aldol condensation of the aldehydes and the self-condensation of ketones. Finally, owing to the rapid equilibration of the intermediate anion with the product as it is formed, further acylation or alkylation may occur. An extreme case is 6-methoxy-2-tetralone, which on attempted monomethylation gives largely a one-to-one mixture of dimethylated ketone and unreacted starting material.

In 1954 Stork[1] introduced a new method for the acylation and alkylation of aldehydes and ketones via enamines (I), derived by condensation with a secondary amine. The anamine undergoes preferential attack at carbon as would the enolate anion, but being neutral it does not condense with itself nor exchange a proton with the product to form a reactive intermediate. The enamines are formed readily by refluxing a mixture of the amine and the carbonyl compound in a solvent such as dioxane, acetonitrile, toluene, or ethyl alcohol. The water formed in the reaction is removed by azeotropic distillation using an apparatus such as is shown in Figure 39.1. After acylation or alkylation the product is hydrolyzed by heating with water.

[1] Stork, Terrell, and Szmuszkovicz, *J. Amer. Chem. Soc.*, **76**, 2029 (1954); Stork, Brizzolara, Landesman, Szmuszkovicz, and Terrell, *ibid.*, **85**, 207 (1963).

FIGURE 39.1
Water Separator

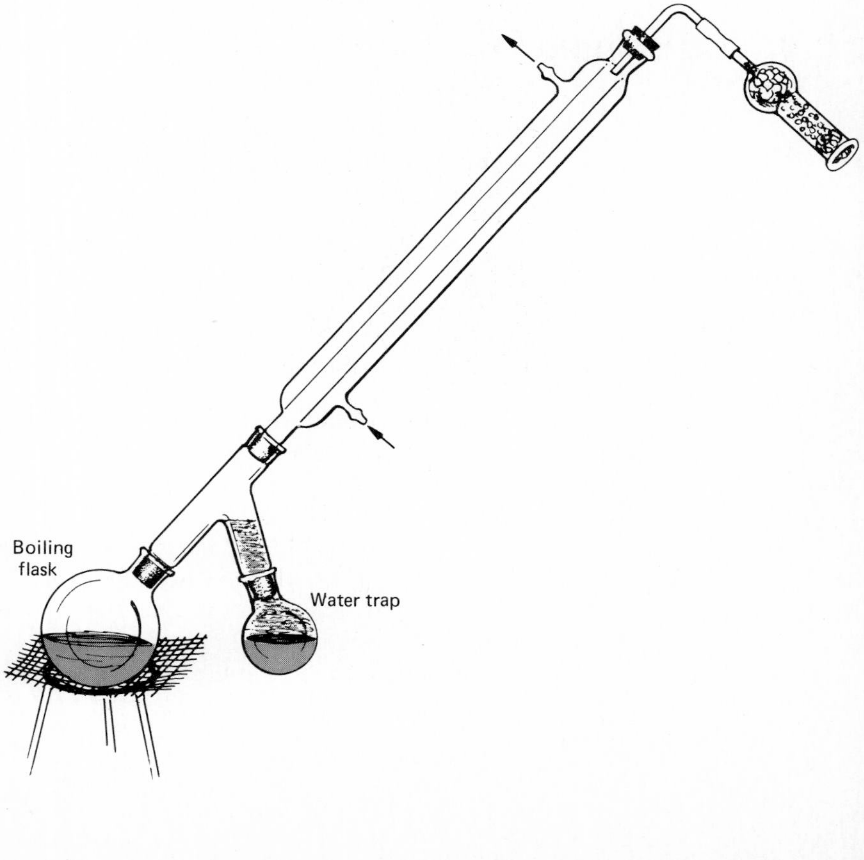

Sec. 39.1] THE ENAMINE REACTION

$$R_2NH + R-CO-CH_2-R' \rightleftharpoons H_2O + \left[R_2\overset{+}{N}-\underset{\underset{R}{|}}{C}=\underset{\underset{R'}{|}}{CH} \longleftrightarrow R_2\overset{+}{N}=\underset{\underset{R}{|}}{C}-\underset{\underset{R'}{|}}{\overset{-}{C}H} \right]$$

$$\xrightarrow{CH_3-X} R_2\overset{+}{N}=\underset{\underset{R}{|}}{C}-\underset{\underset{R'}{|}}{CH}-CH_3 \xrightarrow{H_2O} O=\underset{\underset{R}{|}}{C}-\underset{\underset{R'}{|}}{CH}-CH_3$$

With the enamine synthesis it is possible to prepare compounds that would otherwise require circuitous routes.

The enamine also can undergo Michael-type addition to α,β-unsaturated carbonyl or nitrile derivatives. This reaction differs from the earlier acylation and alkylation reactions in that a proton shift occurs in the adduct to regenerate a new enamine, which in ethanol (but not in benzene) can react with an excess of the α,β-unsaturated ketone.

39.2 Preparation of 2-Acetylcyclohexanone

N-1-Cyclohexenylpyrrolidine (II). Arrange the assembly shown in Figure 39.1 using a 250-mL round-bottom flask to hold the reagents and a 25-mL flask to serve as a water trap. In the 250-mL flask place 9.8 mL (9.8 g, 0.10 mole) of cyclohexanone,[2] 9.2 mL (7.8 g, 0.11 mole) of pyrrolidine, and 0.1 g of p-toluenesulfonic acid. Attach the flask to the apparatus and add slowly about 60 mL of toluene through the condenser, so that the 25-mL flask and side arm will be filled with toluene and the reaction flask slightly less than half filled. Attach a drying tube filled with calcium chloride to the condenser, and heat the reagents under reflux for 1.5 hr. During this time the water produced in the reaction will steam distill and collect as a lower layer in the trap.

Allow the reaction mixture to cool slightly, then replace the adapter, attached water trap and condenser, by an assembly for distillation. Remove the excess pyrrolidine by distilling the mixture until the thermometer reaches 105–108°. The residual enamine can be used for the next step without further purification.[3]

2-Acetylcyclohexanone. Add a solution of 10.4 mL (11.2 g, 0.11 mole) of acetic anhydride in 20 mL of toluene to the enamine at room temperature. Mix the reagents well, attach a drying tube to the flask, and set it aside for 24 hr or longer.

Add 10 mL of water to the reaction flask, arrange a condenser, and reflux the mixture for 0.5 hr. Wash the cooled toluene solution with water, then with 5% hydrochloric acid, and finally with water. Dry the solution with 2–3 g of anhydrous calcium chloride, and remove the toluene by distillation. Distill the residual liquid under reduced pressure; bp 97–104° at 12–14 mm. The yield of 2-acetylcyclohexanone is 5–8 g.

Questions

1. It is found that enamines prepared from pyrrolidine are more reactive than those from piperidine. Explain.
2. With unsymmetrical ketones the enamine synthesis introduces the alkyl or acyl group into the least substituted position. Explain.
3. Suggest a possible reason why use of ethanol as a solvent leads to dialkylated products in the reaction of enamines with α,β-unsaturated carbonyl compounds.
4. Devise syntheses for the following molecules that make use of the enamine reaction.
   (a) 2-carbomethoxycyclopentanone
   (b) pimelic acid
   (c) 1,15-pentdecanedioic acid

---

[2] The material prepared in Chapter 20 is suitable.
[3] Somewhat better yields are obtained if the cyclohexenylpyrrolidine (bp 106° at 13 mm) is purified by distillation before acylation.

40 Wagner–Meerwein Rearrangements: Camphor from Camphene

40.1 Introduction

An implicit guiding principle used in predicting products of organic reactions is that minimum structural change occurs in the reaction. Important exceptions to the principle are those carbocation reactions in which an alkyl or aryl group adjacent to the developing positive charge migrates to the positive carbon atom and gives rise to products with rearranged carbon skeletons (Wagner–Meerwein rearrangements).

$$\underset{\underset{R}{|}}{\overset{\overset{R}{|}}{R-C}}-\underset{\underset{H}{|}}{\overset{\overset{H}{|}}{C}}-X \longrightarrow \underset{\underset{R}{|}}{\overset{\overset{R}{|}}{R-C}}-\overset{\overset{H}{|}}{C^+} \longrightarrow \underset{+}{R-C}-\overset{H}{\underset{R}{C}}-H \longrightarrow \text{products}$$

The confusion that can arise from an unsuspected Wagner–Meerwein rearrangement is amplified in many of the reactions of bicyclic molecules because the rearranged products may retain the same bicyclic structure and differ from the nonrearranged products only in the position of substituents. The commercial conversion of camphene to isobornyl acetate is a classic example of a reaction of a bicyclic molecule proceeding with a Wagner–Meerwein rearrangement.[1]

---

[1] In studying this reaction, you are urged to prepare ball and stick models of each stage. It is difficult to see from two-dimensional drawings how the transfer of one bond causes the movement of so many groups.

[Reaction scheme: camphene → (protonated intermediate) → I, showing Wagner–Meerwein rearrangement]

[Reaction scheme: carbocation (identical to I) + CH₃CO₂H → Isobornyl acetate]

The product formed is a secondary acetate rather than the tertiary acetate that might have been expected from the structure of camphene. The explanation is that these acid-catalyzed reactions are reversible and the isobornyl acetate is isolated because it is more stable as a result of lower steric congestion.

The remaining steps of the synthesis of camphor do not involve Wagner–Meerwein rearrangements since neither the saponification of isobornyl acetate to isoborneol nor the chromic acid oxidation of isoborneol to camphor produces a carbocation in the bicyclic ring.

[Reaction scheme: Isobornyl acetate → (base) → Isoborneol → (CrO_3) → Camphor]

40.2 Preparation of Camphor

Isobornyl acetate. In a 125-mL flask dissolve 3 g (0.025 mole) of camphene in 7 mL (7.6 g, 0.13 mole) of glacial acetic acid and add 1 mL of 30% sulfuric acid. Warm the flask on a steam bath at 90–95° for 15 min with frequent swirling. Add 5 mL of water, mix well, and allow to cool.

Transfer the material to a separatory funnel and rinse the flask with a little water. Separate the ester layer, wash it well first with water and then with 10% aqueous sodium carbonate, and dry it over calcium chloride. The crude product is suitable for conversion to isoborneol without further purification. The yield is about 3–3.5 g.

PREPARATION OF CAMPHOR

Isoborneol. In a 100-mL round-bottom flask add 2.5 g (0.013 mole) of isobornyl acetate to a solution of 1 g (0.016 mole) of 85% pure solid potassium hydroxide in 5 mL of ethanol and 1.6 mL of water, and heat the mixture under reflux on a steam bath for 1 hr. Pour the solution slowly onto about 10 g of ice contained in a 100-mL beaker. Stir the mixture for several minutes until the isoborneol solidifies. Collect the solid on a Büchner funnel, wash it well with cold water, and press it dry. The yield of crude isoborneol is about 2 g. The crude product is sufficiently pure for oxidation to camphor.

Camphor. In a 100-mL round-bottom flask prepare a solution of 1.5 g (0.01 mole) of isoborneol in 3 mL of dry acetone and cool the solution in a beaker of ice-water to 15–25°. Add to the solution drop by drop 2.4 mL (0.007 mole) of Jones' reagent.[2] Allow the reaction mixture to stand for 30 min after the addition has been completed.

Add 20 mL of water to the reaction flask, attach a water-cooled condenser, and distill about 5 mL of distillate. This first portion of distillate contains most of the organic impurities and should be discarded. Continue the distillation into a chilled receiver until no more product is collected. If the camphor collects in the condenser to the point where the condenser might become plugged,[3] the distillation should be interrupted and the accumulation of camphor pushed out with a long glass rod.[4] The solid product is collected on a Büchner funnel and pressed dry. The yield is about 1 g. Camphor may be purified by sublimation (see Figure 5.7 or 5.8). Pure camphor has a melting point of 175°.[5]

Questions

1. How many asymmetric carbon atoms (chiral centers) are present in camphor? How many optically active forms of camphor exist?
2. Explain why dehydration of borneol with aqueous acid gives camphene.
3. Explain why optically active *exo*- and *endo*-bicyclo[2.2.1]heptanol-2 (norborneol) undergo racemization when treated with acids.
4. Optically active camphene also undergoes racemization on treatment with acid but by a different process than for norborneol. Write a mechanism that accounts for the racemization.

---

[2] Jones' reagent is available on the side shelf. It is prepared by dissolving 27 g of chromium trioxide in 23 mL of concentrated sulfuric acid, followed by cautious dilution with water to 100 mL.

[3] A warning that this dangerous state is being approached comes when the camphor is not completely condensed during passage through the condenser.

[4] An alternative isolation procedure that avoids this hazard is to stop the distillation after the organic impurities have been removed and collect the solidified camphor by filtration. The product should be washed well with cold water.

[5] Because camphor has an unusually high change of melting temperature with pressure the observed melting point in an open capillary is several degrees below the melting point taken in a sealed tube. For the same reason, the melting point is particularly sensitive to traces of impurities and the melting point of mixtures of camphor with known amounts of foreign materials is useful in determining molecular weights (Rast method).

41 The Diels–Alder Reaction

41.1 Introduction

One of the most interesting synthetic reactions of unsaturated compounds is the 1,4-addition of a conjugated diene to a molecule containing an active ethylenic or acetylenic bond (the dienophile), to form an adduct having a six-membered unsaturated ring, a 4 + 2 cycloaddition. This cyclization process, known as the Diels–Alder reaction or diene synthesis,[1] is of exceedingly broad scope and has been applied to syntheses of important medicinal products, insecticides, terpene derivatives and intermediates for the manufacture of industrial chemicals.

Diene reactants include alkyl, halogen, and alkoxy derivatives of 1,3-butadiene, and also cyclic 1,3-dienes such as cyclopentadiene, 1,3-cyclohexadiene, and some terpenes (α-terpinene, α-phellandrene). Furan and a number of furan derivatives, and the inner ring of anthracene, also participate in the reaction as dienes. In general, the presence in the diene of substituents that facilitate electron-release favors the reaction.

---

[1] For reviews of the Diels–Alder reaction, see Kloetzel, *Organic Reactions*, **4**, 1 (1948); Holmes, **4**, 60 (1948); Butz and Rytina, **5**, 136 (1949).

The most typical dienophiles are α,β-unsaturated carbonyl compounds and nitriles. Examples are acrolein, p-benzoquinone[2] and 1,4-naphthoquinones, maleic anhydride, esters of acetylenedicarboxylic acid, acrylic esters, and acrylonitrile. Under vigorous conditions vinyl ethers and halides, and even ethylene and acetylene, can be made to act as dienophiles toward the more reactive dienes.

The Diels–Alder reaction is a stereospecific *cis-cis* type of addition: the diene is obliged to assume a *cis* conformation to permit ring closure and it is observed that the configuration of substituents (*cis* or *trans*) in the dienophile is retained in the adduct. Thus, 1,3-butadiene reacts with maleic anhydride (a *cis*-dienophile) to give 4-cyclohexene-*cis*-1,2-dicarboxylic anhydride (II).

Since butadiene (bp −4.4°) is a gas at room temperature, it may be prepared conveniently for small-scale reactions by thermal decomposition of 3-sulfolene (butadiene sulfone, I), which gives butadiene and sulfur dioxide. This source of butadiene is used in the present experiment.

Active dienophiles are useful reagents for detecting the presence of a conjugated diene system and for analytical purposes. Diels–Alder adducts have been of value in establishing the structure of 1,3-dienes and in characterizing known dienes. *N*-Phenylmaleimide (maleanil, III) is a convenient reagent for identification purposes since it forms crystalline adducts that can be isolated and purified readily (IV).

[2] For benzoquinone reactions see Chapter 42.

N-Phenylmaleimide is synthesized readily from maleic anhydride in the present preparation and allowed to react with 1,3-butadiene and with cyclopentadiene to form adducts. These dienes react also with *p*-benzoquinone and with this dienophile give mono- and di-adducts.

With cyclic dienes such as cyclopentadiene and also furan derivatives, the 4 + 2 cycloaddition product might have either of two possible configurations. The ring system of the dienophile could have a *trans* (*endo*) disposition (V) or a *cis* (*exo*) relationship (VI) to the newly formed methylene (or oxygen) bridge of the adduct. The case of cyclopentadiene and maleic anhydride is illustrated here. In practice, it is observed that the *endo* configuration is favored in these reactions.

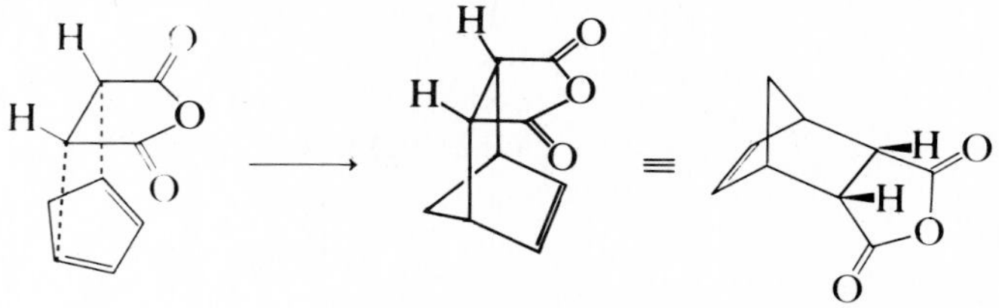

V
Endo adduct

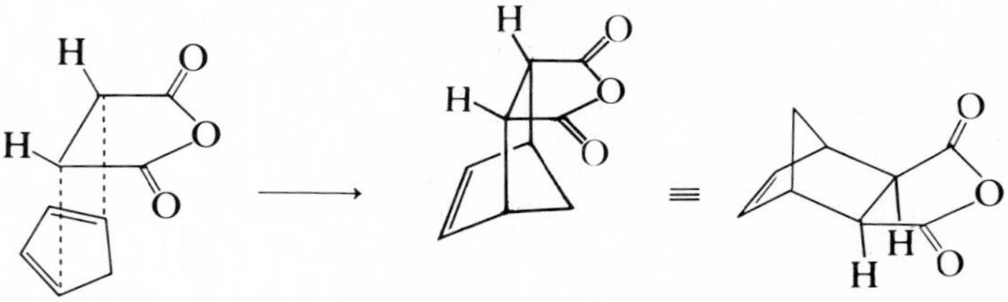

VI
Exo adduct

In the *endo*-adduct, the dienophile group is attached toward the *convex* side of the "boat" form of the six-membered ring.

Furan derivatives, provided they do *not* have an electron-withdrawing substituent ($-CH=O$, $-CO_2H$, $-C{\equiv}N$, etc.), undergo addition to maleic anhydride readily. The resulting adducts have in some instances been used to establish or to confirm the position of substituents in the furan ring. Upon heating with hydrobromic acid, the maleic adducts undergo opening of the oxygen bridge and furnish phthalic acid derivatives that can be identified readily.

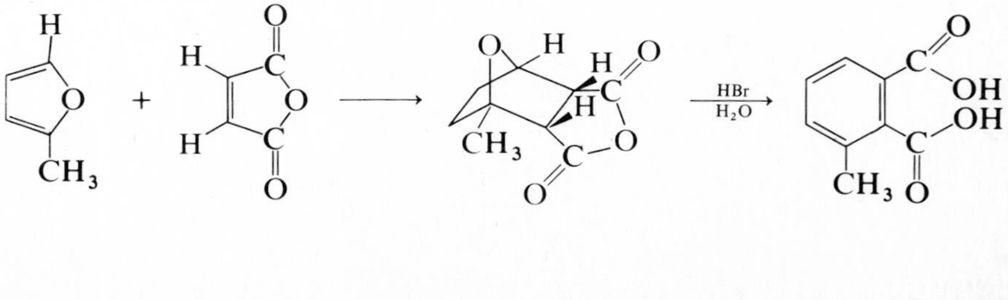

41.2 Preparations

(A) N-Phenylmaleimide

N-Phenylmaleamic acid. To a solution of 5 g (0.05 mole) of maleic anhydride in 40 mL of warm toluene (**Caution**—*flammable solvent!*) add slowly with swirling, a solution of 4.6 mL (4.7 g, 0.05 mole) of aniline in 10 mL of toluene. Reaction occurs rapidly and a white crystalline precipitate of the phenylmaleamic acid separates. After cooling the solution to 20°, collect the product on a suction filter and allow it to dry. The yield is 9–9.5 g; mp 200–201°(dec). It is unnecessary to purify it.

N-Phenylmaleimide (III). In a small Erlenmeyer flask prepare a slurry of 8 g (0.04 mole) of N-phenylmaleamic acid in 15 mL of acetic anhydride and add 1.5 g of anhydrous sodium acetate. Close the flask with a stopper bearing a drying tube and heat the mixture on a steam bath, swirling it occasionally, for 30 min. During this time the maleamic acid dissolves, and a deep red solution is formed. Pour the solution while stirring, into an ice–water mixture, and allow it to stand while the N-phenylmaleimide separates as a yellow flocculent mass. Wash the product with two 15-mL portions of cold water, then with one 10-mL portion of cyclohexane, and allow it to stand while the N-phenylmaleimide separates as a yellow flocculent mass. Wash the product with two 15-mL portions of cold water, then with one 10-mL portion of cyclohexane, and allow it to dry. The yield is 5–6 g; mp 86.5–88°.

The crude product may be used directly to form adducts with dienes. Recrystallization from 75% aqueous ethanol or from cyclohexane gives long yellow needles; mp 88–89°.

(B) N-Phenylmaleimide Adducts

N-Phenyl-4-cyclohexene-1,2-dicarboximide (IV). Place 3.5 g (0.03 mole) of 3-sulfolene, 3.5 g (0.02 mole) of N-phenylmaleimide, and 7 mL of technical mixed xylenes (**Caution**—*flammable solvent!*) in a 50-mL round-bottom flask. Attach a reflux condenser, and arrange a gas trap to take care of the sulfur dioxide evolved. Heat the mixture gently until the solids dissolve, and then heat under reflux for 45 min. Cool the solution below its boiling point, add about 10 mL of xylene, and transfer the solution to a 100-mL Erlenmeyer flask. Cool the solution to room temperature, and add 30 mL of petroleum ether (bp 60–90°). Collect the solid on a suction filter, and wash it with a small amount of cold methanol. The yield of off-white product is 2.5–3.5 g; mp 115–116°. Recrystallization from 70% aqueous methanol gives fine white crystals of the same melting point.

N-Phenyl-*endo*-norbornene-5,6-dicarboximide. Prepare about 5 g of cyclopentadiene from 20 g of cyclopentadiene dimer by the procedure described Chapter 43. This should be sufficient to make the adduct from phenylmaleimide and from maleic anhydride (see procedure C). The monomer must be kept cool and used promptly as it dimerizes on standing.

Dissolve 3.5 g (0.02 mole) of *N*-phenylmaleimide in 7 mL of mixed xylenes (**Caution**—*flammable solvent!*) in a 50-mL round-bottom flask and add 2 g (0.03 mole) of cyclopentadiene. Attach a reflux condenser and warm the mixture gently, or until the phenylmaleimide has dissolved. Cool the solution, add 15 mL of petroleum ether (bp 60–90°), and allow it to stand until the white crystals of the adduct have separated. The yield is 1.5–2 g; mp 139–142°; recorded mp 144–145°.

This is the *N*-phenylimide related to compound V.

(C) Maleic Anhydride Adducts

4-Cyclohexene-*cis*-1,2-dicarboxylic anhydride (II). Place in a 50-mL round-bottom flask, 3.5 g (0.03 mole) of 3-sulfolene, 2 g (0.02 mole) of finely powdered maleic anhydride and 5 mL of *dry* mixed technical xylenes (**Caution**—*flammable solvent!*). Attach a condenser, arrange a gas trap to remove the evolved sulfur dioxide, and warm the flask gently while swirling it until the solids dissolve. Heat the solution under reflux for 25–30 min and then cool to room temperature. Add about 10 mL of toluene and a little decolorizing carbon, and heat the material, swirling it, on a steam bath. Filter the hot solution through a folded filter (**Caution**—*flammable solvents!*) and add petroleum ether (bp 60–90°) in small portions until a turbidity develops. Allow the solution to cool and stand until crystallization is complete. The yield is 1–1.5 g; mp 101–102°. Recorded mp 103°.

***Endo*-Norbornene-*cis*-5,6,-dicarboxylic anhydride (V).** In a 50-mL round-bottom flask, dissolve 2 g (0.02 mole) of maleic anhydride in 7 mL of ethyl acetate by warming on a steam bath and add 7 mL of petroleum ether (bp 60–90°) (**Caution**—*flammable solvents!*). Cool the solution thoroughly in an ice bath, and add carefully 2 mL (1.6 g, 0.025 mole) of *dry* cyclopentadiene. Swirl the solution in the cooling bath until the initial exothermic reaction is over and the adduct has separated. Heat the flask on a steam bath until the product has dissolved, and then let the solution cool slowly and stand *undisturbed*. The yield of white crystals is 2–2.5 g; mp 163–164°.

The anhydride can be hydrolyzed to the *endo-cis*-dicarboxylic acid by boiling 2 g with 25–30 mL of distilled water until all of the solid and oil has dissolved and then letting the solution cool undisturbed. Usually it is necessary to induce crystallization by scratching with a glass rod or by adding a seed crystal. The yield is about 1.5 g, mp 178–180° (dec).

Questions

1. Write the structure of the Diels–Alder adduct formed by 4 + 2 cycloaddition of the following.
   (a) ethyl propiolate (H—C≡C—CO$_2$Et) and 2-ethoxy-1,3-butadiene
   (b) 2,5-dimethylfuran and acrylonitrile (CH$_2$=CH—C≡N)
   (c) 1,4-dimethoxy-1,3-butadiene and 1,4-naphthoquinone
2. The structure of the adducts formed in parts (a) and (b) of the preceding question might be established by conversion to known benzene derivatives. How could this be done?
3. Write the structure of the dimers formed through self-addition of the following, in a reaction of the Diels–Alder type (4 + 2 cycloaddition).
   (a) isoprene
   (b) cyclopentadiene
   (c) acrolein (CH$_2$=CH—CH=O)

42 Benzoquinone and Dihydroxytriptycene

42.1 Diels–Alder Reactions of Benzoquinone

Controlled oxidation of 1,2- and 1,4-dihydroxybenzene (catechol and hydroquinone) leads to *o*- and *p*-benzoquinone. The quinones possess a strong chromophore arising from extended conjugation of the carbonyl groups, for which the generic term *quinonoid structure* is used.

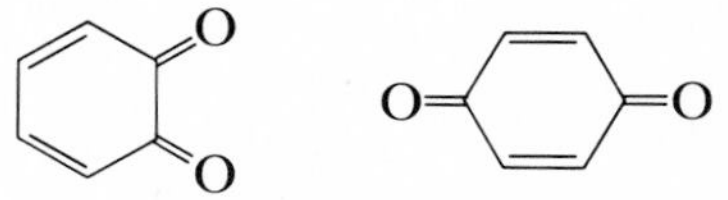

o-Benzoquinone
(red)

p-Benzoquinone
(yellow)

The ring system has lost its aromatic character; the quinones are highly active α,β-unsaturated ketones and their typical behavior involves 1,4-addition reactions. They are oxidizing (dehydrogenating) agents and can be reduced to their colorless, benzenoid precursors.

On a small scale, *p*-benzoquinone may be prepared conveniently by oxidation of hydroquinone with dichromate or potassium bromate in acid solution. Commercially, *p*-benzoquinone is made by oxidation of aniline with manganese dioxide and aqueous sulfuric acid. 2-Methylbenzoquinone is prepared in a similar way from *o*-toluidine.[1] The yields in the amine

[1] For synthesis of quinones by oxidation, see Cason, *Organic Reactions*, **4**, 305 (1948). Diels–Alder reactions of quinones are reviewed by Butz and Rytina, *ibid.*, **5**, 136 (1949).

oxidations are not high but the starting materials are readily available and cheap.

Benzoquinone acts as a dienophile in the Diels–Alder reaction, 4 + 2 cycloaddition (see Chapter 41). With one molecule of 1,3-butadiene it gives 5,8,9,10-tetrahydro-1,4-naphthoquinone; a second molecule of the diene can be added to obtain octahydro-9,10-anthraquinone.[2] Dehydrogenation of the latter affords a route to anthraquinone.

The central ring of anthracene (I) functions as a diene in Diels–Alder reactions even though its two double bonds participate formally in an aromatic structure. A striking example is the condensation of *p*-benzoquinone with anthracene to yield the bridged polycyclic system (II) containing two benzene rings and a cyclohexene-1,4-dione ring. The dione ring corresponds structurally to the unstable diketone tautomer of hydroquinone. Treatment of the adduct with either acids or bases transforms it

[2] Clar, *Ber.*, **64**, 1676 (1931).

rapidly to the more stable, aromatized hydroquinone form (III), which is dihydroxytriptycene. The hydroxyl functions of III can be removed by several means[3] to yield the highly symmetric triptycene (IV). This unusual name is based on the ancient Roman three-leaved book called a triptych.

In the Diels–Alder reaction of this procedure, p-benzoquinone is added to anthracene to produce dihydroxytriptycene (III), from which triptycene can be secured by a stepwise reduction process. In an alternate Diels–Alder reaction, benzyne, a synthetically efficient dienophile, can be used to obtain triptycene (IV) in one step by direct addition to anthracene.

Unfortunately, benzyne is extremely unstable and is formed slowly when conventional elimination reactions are used. Substances that yield benzyne at a practical rate for synthetic purposes often tend to be potentially dangerous. One source of benzyne is the decomposition of benzenediazonium-2-carboxylate (from anthranilic acid) under controlled conditions (42.1),[4] but this has occasionally caused explosions. Another method involves the thermal decomposition of diphenyliodonium-2-carboxylate (reaction 42.2).[5]

$$\text{anthranilic acid} \xrightarrow{RO-NO} \text{diazonium carboxylate} \longrightarrow \text{benzyne} + N\equiv N + CO_2 \qquad (1)$$

$$\text{o-iodobenzoic acid} \longrightarrow \text{diphenyliodonium-2-carboxylate} \longrightarrow \text{benzyne} + C_6H_5I + CO_2 \qquad (2)$$

42.2 Preparations and Reactions

(A) p-Benzoquinone

In a 125-mL Erlenmeyer flask place 5.5 g (0.05 mole) of hydroquinone and a warm solution of 3 mL of concentrated sulfuric acid in 50 mL of water. Dissolve the hydroquinone completely by gentle warming and then cool the solution at once in an ice bath, vigorously swirling it, to obtain a suspension of fine crystals.

➤ CAUTION Benzoquinone is extremely volatile and toxic. Its vapor is dangerously irritating to the eyes and the solid is a skin irritant. Manipulate quinone carefully to minimize contact with the vapor and the crystals.

[3] Bartlett, Ryan, and Cohen, *J. Amer. Chem. Soc.*, **64**, 2649 (1942); Craig and Wilcox, *J. Org. Chem.*, **24**, 1619 (1959).
[4] Friedman and Logullo, *J. Org. chem.*, **34**, 3091 (1969).
[5] LeGoff, *J. Am. Chem. Soc.*, **84**, 3786 (1962); Fieser and Williamson, *Organic Experiments*, 4th ed. (Lexington, MA. Heath, 1979).

Remove the flask from the cooling bath, and add in small portions, while swirling the flask, a solution of 7.1 g (0.024 mole) of sodium dichromate dihydrate in 10mL of water. Cool the reaction mixture as needed to maintain its temperature at 20–25°. During the early stages of oxidation, a greenish black precipitate of quinhydrone (a 1:1 complex of quinone and hydroquinone) separates. As the oxidation to quinone is completed, the color of the precipitate becomes yellowish green. About 20 min is required to complete the oxidation.

Cool the reaction mixture in an ice bath to 5–10° and collect the crystals of quinone with suction. Wash them with two or three 5-mL portions of ice-cold water, press firmly on the filter, and suck them as dry as possible. Purify the crude product by sublimation, using one of the methods shown in Figures 5.7 and 5.8. Transfer the sublimed crystals quickly to a dry, weighed sample tube. The yield is about 3 g. The bright yellow crystals of quinone become discolored upon standing.

Reactions. *p*-Benzoquinone undergoes 1,4-addition reactions with halogen acids, alcohols, and amines. The initial adducts undergo isomerization to form substituted hydroquinones; hydrogen chloride, for example, gives chlorohydroquinone. Acetic anhydride, in the presence of acid catalysts, adds in the same way and the intermediate diacetoxy compound is acetylated further to yield the triacetate of hydroxyhydroquinone (1,2,4-trihydroxybenzene).

The addition of alcohols and amines leads to alkoxy- and aminoquinones. The intermediate hydroquinones are dehydrogenated by benzoquinone, because it has a higher oxidation potential than the alkoxy- and aminoquinones. This oxidation does not occur with the halogenated hydroquinones, since the halogenated quinones are stronger oxidizing agents than benzoquinone.

Quinone forms a monoxime and a dioxime. The monoxime is converted rapidly into *p*-nitrosophenol by a simple proton shift. Oximation is done with hydroxylamine hydrochloride because a high concentration of free hydroxylamine reduces the quinone.

Hydroxyhydroquinone triacetate. In a test tube, place 1 g of benzoquinone and 3 mL of acetic anhydride. Mix the reactants thoroughly and add *one drop* of concentrated sulfuric acid. The temperature of the reaction mixture should not be allowed to rise above 50°. After shaking it for 5 min, pour the solution into 15 mL of cold water and rinse the tube with 4–5 mL of water. Collect the crystals with suction and crystallize them from 4–5 mL of ethanol. The yield is about 1 g. Hydroxyhydroquinone triacetate forms colorless crystals, mp 96–97°. The product becomes discolored when it stands exposed to the air.

(B) Dihydroxytriptycene

p-Benzoquinone–anthracene adduct. In a 100-mL round-bottom flask dissolve in 10 mL of xylene 1.8 g (0.01 mole) of pure anthracene and 1.1 g (0.01 mole) of *p*-benzoquinone. Attach a water-cooled reflux condenser and boil the solution for 45 min. Cool the solution to 15–20° and allow the adduct to crystallize. Collect the pale yellow crystals with suction and press them firmly on the filter. The yield of adduct is about 2.7 g. The crude product is sufficiently pure for conversion to dihydroxytriptycene. Do not attempt to determine its melting point.

Dihydroxytriptycene. Place the *p*-benzoquinone–anthracene adduct (about 2.7 g) in a 250-mL round-bottom flask and add a solution of 0.5 g of potassium hydroxide pellets in 50 mL of ethanol. Warm the flask on a steam bath for about 5 min, or until the adduct has dissolved. Dilute the ethanolic solution with 100 mL of water, cool the flask in an ice-water bath, and *carefully* neutralize the base by adding 20% hydrochloric acid drop by drop swirling the solution, until the liquid is acidic to litmus (about 2 mL of acid will be required). Collect the precipitated dihydroxytriptycene by suction filtration, wash it well with water, and spread it to dry in the air. The yield is about 2 g.

The crude product is almost colorless; it shows no carbonyl absorption in the 6-μm region of the infrared spectrum. To obtain pure dihydroxytriptycene, you may crystallize the crude material from ordinary ethanol (95%) or from ethanol–water mixtures. Do not attempt to determine its melting point.

Questions

1. Write equations for the reaction of benzoquinone with the following.
   (a) hydrogen chloride
   (b) sulfurous acid ($SO_2 + H_2O$)
   (c) hydroxylamine hydrochloride (one equivalent).
2. What is quinhydrone? Give examples of other addition compounds of a similar type.
3. Suggest a method, starting from *o*-toluidine, for the preparation of 2-methyl-1,4-naphthoquinone. What is the structural relationship of this quinone to the antihemorrhagic (blood-clotting) factors, vitamins K_1 and K_2?

4. What product(s) would you expect to obtain by treating hydroquinone diacetate with excess aluminum chloride (Fries rearrangement)?
5. Devise a synthesis of 3,6-dimethylphenanthrenequinone from *o*-benzoquinone by 4 + 2 cycloaddition.
6. 1,3,5-Cycloheptatriene, on treatment with a highly active organic cation, undergoes loss of hydride anion to form the 1,3,5-cycloheptadienyl (tropylium) cation. This proves to be a relatively stable organic cation and exhibits aromatic properties. How do you account for this (cf. Hückel's rule)?
7. What product would result from 4 + 2 addition of benzyne to furan? Predict its behavior on treatment with strongly acidic reagents. The ultimate product might be soluble in aqueous sodium hydroxide solution and have the same composition ($C_{10}H_8O$) as the original adduct.

43 Ferrocene and Acetylferrocene

43.1 Metallocenes

A fortunate accident in 1951 led to the discovery of an extraordinary organoiron compound,[1] dicyclopentadienyliron (I), which later became known as ferrocene. Studies of this remarkably stable and atypical organometallic compound led to its formulation as a combination of two cyclopentadienide anions with a ferrous cation.[2] The bonding involves the six π-electrons of each anion in a way that binds every carbon atom equally to the metal. In this rather flat "sandwich-type" structure (II), 12 electrons from 2 anions and 6 electrons from the ferrous cation lead to the stable electronic configuration of the inert gas krypton. Ferrocene is stable to 450° and is soluble in organic solvents and insoluble in water.

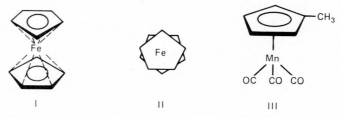

Metallocenes have been prepared with a great variety of metals as the central atom—ruthenium, osmium, plantinum, chromium, and many

[1] Kealey and Pauson, *Nature*, **168**, 1039 (1951).
[2] Wilkinson et al., *J. Amer. Chem. Soc.*, **74**, 2125 (1952).

others. Some of these have made significant contributions to theoretical chemistry, and some have proved to have highly useful catalytic properties for industrial chemical transformations. It is interesting to note that ferrocene is analogous electronically to dibenzenechromium, in which the central atom has zero valence; other metal derivatives of this type are known.

In addition to the metallocenes, there are many cyclopentadienyl derivatives of metal carbonyls, such as methylcyclopentadienylmanganese tricarbonyl (III). This compound is an effective antiknock additive for motor fuel and has been approved for use in nonleaded gasoline. Ferrocene itself also has antiknock properties.

The chemistry of ferrocene resembles that of benzene: it undergoes various electrophilic substitution reactions, similar to the Friedel–Crafts reaction, even *more easily* than benzene. It is somewhat limited in its synthetic applications by its susceptibility to oxidation (even by air), which leads to the blue ferricinium cation. In general it is desirable to carry out reactions of ferrocene with minimal exposure to air (and to moisture).

Among simple hydrocarbons cyclopentadiene is relatively acidic ($pK_a = 15.5$), but it is a *weak* acid. It needs special conditions for syntheses that require generation of the cyclopentadienide anion.[3] The choice of solvent media is important. Two special solvents of moderate polarity and good solvent activity are used in the present experiment: 1,2-dimethoxyethane (Glyme) and dimethyl sulfoxide (DMSO). Chromatographic methods are used to purify and identify the reaction products.

43.2 Preparations

(A) Ferrocene

Cyclopentadiene. Place 25 mL (27 g, 0.20 mole) of dicyclopentadiene in a 100-mL round-bottom flask and attach a fractionating column packed with glass beads.[4] Provide an efficient condenser and an ice-cooled receiver for the distillate. Heat the dimer gently until brisk refluxing occurs and the monomer begins to distill steadily in the range of 40–42°.

▰▶ *CAUTION* Cyclopentadiene is extremely volatile and flammable.

Continue the distillation at a rapid rate but do not allow the temperature of the distilling vapor to rise above 43–45°. This should afford about 11 mL (9 g) of cyclopentadiene in 40–50 min. Keep the distillate cooled to 0–5°; if

---

[3] *Organic Syntheses*, Collective Volume **IV**, 473, 476 (1963); **V**, 434, 578, 621 (1973).

[4] A simpler procedure that produces cyclopentadiene sufficiently pure for the Diels–Alder reaction is to replace the fractionating column with a Claisen adapter and carry out the distillation at a slow rate such that the head temperature does not exceed 55–60°.

it has become clouded by moisture, add 1–2 g of anhydrous calcium chloride. On standing, cyclopentadiene reverts to the dimer, which then polymerizes. It is important that the apparatus be disassembled and cleaned as soon as the distillation is completed.

While the distillation is in progress, apparatus and reagents for the next step may be assembled.

Ferrocene (dicyclopentadienyliron). Prepare an assembly consisting of a 250-mL round-bottom flask fitted with a Claisen adapter bearing a 125-mL separatory funnel, and place a stopper bearing a short pressure release tube in the side arm. *Use grease on all glass joints!*

In the flask place 40 g (0.70 mole) of potassium hydroxide (flakes or pellets), followed by 10 mL (8 g, 0.12 mole) of cyclopentadiene and 100 mL (87 g) of 1,2-dimethoxyethane (Glyme). Swirl the mixture occasionally as the black solution of potassium cyclopentadienide is formed.

Meanwhile, in a 125-mL Erlenmeyer flask prepare a solution of 12 g (0.074 mole) of ferrous chloride dihydrate in 60 mL (66 g) of dimethyl sulfoxide (DMSO; $CH_3-SO-CH_3$), by swirling the flask over a gentle steam bath for a few minutes (*not over 5 min*).

▶ *CAUTION*   Handle DMSO carefully. It can be absorbed directly through the skin; also, it is flammable when hot.

Cool the solution to 15–20° in an ice bath and transfer it to the separatory funnel of the assembled apparatus. Allow the ferrous chloride solution to flow drop by drop into the black solution of potassium cyclopentadienide, over a period of about 10 min, while swirling the mixture. Disconnect the reaction flask, swirl the liquid for 5 min longer, and then pour it into a beaker (600-mL or larger) containing 200 g of ice and 175 mL of $6N$ hydrochloric acid.[5] Rinse the flask with about 20 mL of water, pour this into the beaker, and stir with a glass rod until the ice has melted. *At this point, but not earlier, the experiment may be interrupted.*

Filter the reaction mixture with suction. Use a coarse filter paper in the Büchner funnel, and maintain liquid in the funnel until all of the mixture has been introduced. Wash the crude ferrocene on the filter with about 20 mL of water to remove the blue color from the crystals. This color is attributed to oxidation of ferrocene to the blue cation, $[Fe(C_5H_5)_2]^+$.

Dry the product by drawing air through the filter; the yield of crude ferrocene is 10–11 g. The product can be purified by sublimation, as described below (or by crystallization from hexane or methanol).

Sublimation. Weigh accurately 2.7–3.0 g of the crude ferrocene, place it in a dry suction filtering flask, and insert in the mouth of the flask a rubber stopper bearing a 16 × 150-mm test tube (see Figure 5.8). Attach the flask

---

[5] This is approximately 20% hydrochloric acid by weight (22 g/L, d 1.10 g/mL), that can be made from 90 mL of concentrated hydrochloric acid and 85 mL of water.

to an aspirator pump through a trap, and swirl the apparatus gently to distribute the solid evenly. Place the flask in a sand bath, apply suction, and warm it gently over a small flame to drive off the last traces of water. Fill the test tube with chipped ice and heat the base of the flask gently in the sand bath, with a micro burner. Sublimation begins in a few minutes and yellow crystals appear on the sides of the sublimation flask and on the cold test tube. Continue the heating until only a black carbonaceous residue remains (30–60 min). Allow the flask to cool and disconnect the suction line. Scrape the sublimed material from the test tube and the walls of the flask onto a clean, smooth paper and transfer it at once to a tared vial, which is kept tightly stoppered.[6]

Record the weight of sublimed ferrocene and calculate the percentage recovery after sublimation. Determine the melting point of the crude and the sublimed ferrocene; after the sample has been introduced into the capillary melting point tube, seal the upper end before taking the melting point. (Why?)

Determine appropriate conditions for TLC analysis and examine the crude and the sublimed product. Record and comment on your results.

The NMR of ferrocene is shown in Figure 43.1.

[6] Clean the suction flask thoroughly with acetone, then with soap and water, and finally with (in a hood) *aqua regia*, prepared by putting 12 mL of concentrated hydrochloric acid in the flask and adding carefully 4 mL of concentrated nitric acid (*noxious fumes!*).

FIGURE 43.1 NMR Spectrum of Ferrocene in Deuterochloroform

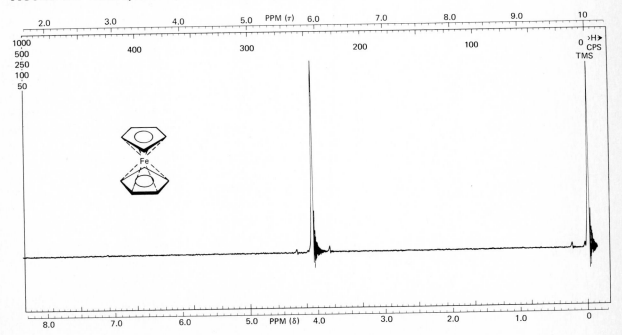

(B) Acetylferrocene

The conditions of this acetylation are designed to enhance conversion to acetylferrocene and diminish further reaction to give 1,1′-diacetylferrocene. Although an excess of acetylating agent is used, the extent of reaction is controlled by limiting the reaction time. Unchanged ferrocene and diacetyl derivative are removed later by column chromatography.

Fit a 100-mL round-bottom flask with a stopper bearing a drying tube tilled with anhydrous calcium chloride. In the flask place 1 g (0.0054 mole) of sublimed ferrocene and add 10 mL (0.105 mole) of acetic anhydride, measured carefully. To the reaction mixture add *carefully* 2 mL of 85% phosphoric acid while swirling the solution. Heat the flask, with drying tube attached, on a steam bath for about 10 min. After the reaction mixture has cooled, pour it onto about 40 g of ice in a 600-mL beaker. Add gradually, stirring thoroughly, small quantities of crushed solid sodium hydroxide until the solution is neutral. When the mixture has cooled to 20°, collect the product with suction and dry it by drawing air through the filter for about 15 min. Purify the crude acetylferrocene by column chromatography on alumina.

Column Chromatography of Acetylferrocene. Read the discussion of column chromatography in Section 7.5 and see Figure 7.3.

Clamp the clean column vertically on a ring stand using two clamps and make certain the drainage tube at the bottom is closed. Fill the column about two-thirds full of petroleum ether, bp 35–60° (**Caution**—*highly flammable!*). Drop a plug of glass wool onto the surface of the solvent and allow it to fill slowly with solvent to remove air bubbles. Use a glass rod or wooden dowel to push the plug to the bottom of the column and tamp it firmly to make it compact, so that adsorbent will not be lost from the column. Add a small amount of clean dry sand to form a 10-mm layer on top of the glass wool plug. Tap the column gently to give a smooth surface to the sand layer. Next introduce the dry alumina adsorbent in a *slow* stream at the top of the column so that it sinks slowly through the petroleum ether and makes an evenly packed column, about 10 cm high (about 11–13 g of alumina). Draw off solvent from the bottom of the tube, if necessary, to avoid overflow as the alumina is added. Avoid trapping air bubbles in the column as they will interfere with good resolution. When the desired length of column packing has been attained, drain off solvent until the level falls just to the upper surface of the packing. The column is now ready for operation.

Dissolve the crude acetylferrocene in a minimum volume of toluene (about 4 mL), ignoring any resinous insoluble by-products, and pipet the solution carefully onto the top of the column. Be especially careful to avoid disturbing the top layer. A small circle of filter paper placed on top of the alumina will prevent its surface being disturbed by further addition of solvent. Open the drainage clamp briefly to draw the liquid down to the top of the alumina layer. To complete the transfer of the crude acetylferrocene into the column, pour in carefully about 1 mL of benzene.

Do not allow the level of solvent to be drained below the upper surface of the column, as this will cause air bubbles to be drawn into the packing and disrupt the uniform flow.

Add a few milliliters of low-boiling petroleum ether carefully by pipet to the top of the column, and drain the solvent down to the level of the absorbent. Fill the column *very gently* with more petroleum ether, and begin to elute the column. Collect the fractions in 50-mL or 125-mL Erlenmeyer flasks; observe the unreacted ferrocene moving down the column as an orange-yellow band. This should be removed from the column in about the first 25 mL of solvent. Now elute the column with ~100 mL of a mixture of low-boiling petroleum ether (50 mL) and diethyl ether (50 mL). This more polar solvent will cause the orange-red band of acetylferrocene to move down the column. Collect the fractions containing the acetylferrocene, and then change the solvent to pure diethyl ether or ethyl acetate. Any diacetylferrocene will be eluted from the column with about 100 mL of these solvents.

Evaporate separately the solvent from the fractions containing the yellow-orange and the red-orange material, using a steam bath in a hood (**Caution**—*Flammable solvents!*). Obtain the weight, melting point, and TLC analysis of each of the two products. Recrystallize the acetylferrocene from petroleum ether, and check the melting point and TLC analysis. Calculate the yield of acetylferrocene. Check the material eluted by ether (or ethyl acetate) for diacetylferrocene using TLC.

The recorded melting points are ferrocene, 173°; acetylferrocene, 85°; 1,1′-diacetylferrocene, 130–131°.

Questions

1. Ferrocene enters into a number of electrophilic substitution reactions, often more readily than benzene. Why cannot ferrocene be nitrated successfully by the usual mixture of nitric and sulfuric acids?
2. Suggest a method for obtaining ferrocenecarboxylic acid.
3. In the formation of diacetylferrocene why does the second acetyl group not enter the same ring as the first?
4. If the stereoisomeric *syn* and *anti* forms of the oxime of acetylferrocene were subjected to a Beckmann rearrangement,
    (a) What products would you expect?
    (b) What products would be formed by hydrolysis of these products?

44 Dyes and Indicators

44.1 Diazonium-Coupling Reactions

Methyl orange belongs to a class of dyes known as "azo colors," which contain the —N=N— group linked to two aromatic nuclei. In addition to the azo group the dyes must contain salt-forming groups such as hydroxyl, amino, sulfonic acid, or carboxyl groups (auxochromes), which usually intensify the color and at the same time enable the molecule to attach itself to the fabric or combine with the mordant to form a lake. Two typical commercial azo dyes are Ponceau 2R and Chicago Blue.

Ponceau 2R
(a brilliant scarlet)

Chicago Blue

Azo dyes are formed by coupling a diazonium ion with a phenol or an aromatic amine.[1] Since many diazonium ions decompose rapidly in solution, it is desirable that the coupling reaction be completed quickly.

$$\text{C}_6\text{H}_5-\overset{+}{\text{N}}\equiv\text{N} + \text{H}-\text{C}_6\text{H}_4-\text{NR}_2 \longrightarrow \text{C}_6\text{H}_5-\text{N}=\text{N}-\text{C}_6\text{H}_4-\text{NR}_2$$

Diazonium ion   An arylamine (or a phenol)   Azo compound

The rate at which a diazonium ion couples with an aromatic amine is proportional to the product of the concentrations of the diazonium ion and the free (unprotonated) amine. At high pHs the diazonium ion is converted into the unreactive diazoate anion and at low pHs the free amine is converted into the unreactive ammonium salt. Only at intermediate pHs will there be a sufficient concentration of both required species to give a significant coupling rate.

$$\text{ArN}_2^+ + \text{H}_2\text{O} \underset{}{\overset{K_1}{\rightleftharpoons}} \text{ArN}_2\text{O}^- + 2\,\text{H}^+$$

$$\text{Ar}\overset{+}{\text{N}}\text{HR}_2 \underset{}{\overset{K_2}{\rightleftharpoons}} \text{ArNR}_2 + \text{H}^+$$

The pH dependence of the relative rate of coupling can be expressed quantitatively by solving the equilibrium equations for the concentrations of free amine and diazonium ion in terms of $[\text{H}^+]$ and K_1 and K_2. It is found that the rate is nearly constant for pHs between $\frac{1}{2}pK_1$ and pK_2 (pK is defined as $-\log K$ by analogy to pH) and diminishes sharply for pHs outside this range. There exists a broad plateau of maximum coupling rate that is of great practical significance, for it is within the range of pHs defining this plateau that the coupling reaction should be carried out. The plateau limits for the coupling of *p*-diazobenzenesulfonate and dimethylaniline to give methyl orange are about pH 4.4 (pK_2) and 10.7 ($\frac{1}{2}pK_1$). To ensure that the reaction solution is not outside this range the dimethylaniline may be converted into its acetate salt, which buffers the solution to about pH 4.6.

The influence of pH on the coupling rate of diazonium ions with phenols is qualitatively similar to the situation with amines. With phenols the *active* coupling species is the phenoxide anion, which, because of the greater availability of electrons, is more rapidly attacked than the free phenol (or than an aromatic amine). Because of the greater basicity of phenoxide anions compared to aromatic amines (the appropriate expression for pK_2 of phenols is ~ 7–10), the acceptable range of pHs for coupling with phenols is much narrower than with amines and more careful control of the pH is required.

---

[1] Reference books in this field are Sidgwick, *The Organic Chemistry of Nitrogen*, 3rd ed. revised by Millar and Springall (Oxford: Clarendon Press, 1966), ch. 16; Zollinger, *Diazo and Azo Chemistry*, trans. by Nursten, (New York: Interscience, 1961).

Benzenediazonium salts in a medium buffered with sodium acetate will undergo coupling reactions with certain primary and secondary amines to form diazoamino compounds. With aniline, diazoaminobenzene, $C_6H_5-N=N-NH-C_6H_5$, is formed. To avoid this type of reaction diazotization is carried out in a strongly acid medium, since it is the free amine that reacts to form the diazoamino compound.

Diazonium compounds in the presence of cuprous salts undergo loss of nitrogen, as in the Sandmeyer reaction, and bring about introduction of the aryl group to α,β-unsaturated carbonyl compounds. Thus, in the Meerwein arylation reaction, cinnamic acid can be converted to derivatives of stilbene, $C_6H_5-CH=CH-C_6H_5$.

Reduction of azo compounds leads to the corresponding primary aromatic amines. For example, reduction of methyl orange with sodium hydrosulfite, $Na_2S_2O_4$, in neutral or alkaline solution gives p-aminodimethylaniline (N,N-dimethylphenylenediamine) and sulfanilic acid.

$$HO_3S-\!\!\!\left\langle\;\right\rangle\!\!\!-N=N-\!\!\!\left\langle\;\right\rangle\!\!\!-N(CH_3)_2 \xrightarrow{4\,H^+} HO_3S-\!\!\!\left\langle\;\right\rangle\!\!\!-NH_2 + H_2N-\!\!\!\left\langle\;\right\rangle\!\!\!-N(CH_3)_2$$

Methyl orange

44.2 Preparations of Azo Dyes

(A) Methyl Orange

Diazotization. In a 400-mL beaker place 10 mL of a 5% solution of sodium carbonate, dilute it with water to about 25 mL, and add 2.6 g (0.015 mole) of anhydrous sulfanilic acid (or 2.9 g of the hydrate). Warm slightly on a water bath and if the sulfanilic acid does not dissolve completely add 2–3 mL more of 5% aqueous sodium carbonate (do not add more than 3 mL). If necessary, filter the solution with suction to remove any undissolved residue. Carefully weigh out 0.9 g (0.0125 mole) of sodium nitrite, dissolve it in about 5 mL of water, and add the nitrite solution to the sodium sulfanilate. Cool the solution in a slush of water and ice until the temperature is between 3 and 5°, then stir vigorously, and add, drop by drop, a solution of 2 mL of concentrated hydrochloric acid diluted with about 3 mL of water. Do not allow the resulting diazonium solution to stand any longer than necessary; proceed at once to the next step.[2]

---

[2] It is advisable to test for the presence of free nitrous acid (to ensure complete diazotization and to avoid a large excess of nitrous acid) by placing a drop of the solution on potassium iodide-starch test paper. In the presence of nitrous acid, iodine is liberated and the starch is colored blue *immediately*. The test paper may be prepared by dipping strips of filter paper into 1% aqueous potassium iodide, then into colloidal starch solution (prepared readily from "instant starch"), and allowing them to dry.

Sec. 44.2] PREPARATIONS OF AZO DYES

Coupling. To 1.6 mL (1.5 g, 0.0125 mole) of dimethylaniline in a test tube, add 0.8 mL (0.8 g, 0.014 mole) of glacial acetic acid and mix thoroughly. To the diazonium salt solution add quickly, with vigorous mixing, the dimethylaniline acetate and allow the mixture to stand with occasional stirring for 5–10 min. Finally make the solution alkaline by adding a solution of 1.8 g of solid sodium hydroxide in about 5 mL of water. This causes the deep red color to change to yellowish orange. The methyl orange separates at once; it can be made to precipitate more completely by adding about 5 g of clean salt. Collect the precipitate with suction, using a hardened filter paper, and crystallize the impure product from hot water. Usually about 5–6 mL of hot water will be required for each gram of material to be crystallized. Cool the hot filtered solution, filter the crystals with suction, and wash them with ethanol and finally with ether. The yield is 2–3 g of purified methyl orange. Do not attempt to determine the melting point of this substance.

Dissolve a little methyl orange in water, add a few drops of dilute hydrochloric acid, then make alkaline again with dilute sodium hydroxide solution. Observe the color changes. The effect of acids and alkalies is probably represented by the following structural changes.

$$^-O_3S-\text{C}_6H_4-N=N-\text{C}_6H_4-N(CH_3)_2 \underset{OH^-}{\overset{H^+}{\rightleftharpoons}} {}^-O_3S-\text{C}_6H_4-NH-N=\text{C}_6H_4=N(CH_3)_2{}^+$$

           Anion (alkaline solution)   Inner salt (acid solution)
                (yellow)   (red)

Many types of organic molecules can be used as indicators. All these have the property of undergoing practically instantaneous change (or changes) in structure in going from acid to alkaline solution, or the reverse, generally within a narrow pH range.

(B) Para Red

Para red (structure: 1-(4-nitrophenylazo)-2-naphthol)

In a 600-mL beaker dissolve 1 g (0.025 mole) of sodium hydroxide, 9.5 g (0.025 mole) of commercial trisodium phosphate ($Na_3PO_4 \cdot 12H_2O$), as a buffer, and finally 1.4 g (0.01 mole) of 2-naphthol in 200 mL of water, in the order given. Chill the solution by allowing it to stand in an ice bath for 5 min, stirring intermittently.

Meanwhile prepare a solution of *p*-nitrophenyldiazonium chloride. Dissolve 1.4 g (0.01 mole) of *p*-nitroaniline by warming with 3 mL of concentrated hydrochloric acid diluted with 3 mL of water. Pour the solution into a 100-mL Erlenmeyer flask containing about 10 g of chopped ice.

Swirl the mixture to obtain a fine suspension of the crystals of *p*-nitroaniline hydrochloride. While maintaining the mixture at 5–10° and agitating it vigorously, add as quickly as possible a cold solution of 0.8 g (0.011 mole) of sodium nitrite in 3–4 mL of water. Swirl the mixture until most of the hydrochloride has dissolved (2–3 min), allow it to stand a few minutes to complete the diazotization, and proceed at once to the next step.

Pour the diazonium solution all at once into the chilled alkaline solution of 2-naphthol. Stir the material vigorously for a few minutes to ensure complete reaction and then, after adding 5 mL of concentrated hydrochloric acid, raise the temperature to about 30° on a steam bath and stir for 30–40 min.

Collect the bright red dye with suction, using a hardened filter paper. Allow the solid to dry completely and then extract the inorganic salts by stirring with 100 mL of water. Collect the solid with suction and wash it with water until the filtrate is essentially free of chloride ion.[3] Wash the crystals finally with small portions of ethanol. After it is dried, the product weighs about 2.5 g. Do not attempt to determine its melting point.

44.3 Phthalein and Sulfonphthalein Indicators

Phthalic anhydride undergoes a stepwise condensation with two molecules of phenol, in the presence of acid catalysts such as sulfuric acid or zinc chloride, to form phenolphthalein. The phthaleins are derivatives of triphenylmethanol and are related structurally to the triphenylmethane dyes. In neutral or acidic solutions phenolphthalein exists in the colorless, lactone form (I). In basic solutions, in the range of pH 8.3–10, it is converted to the red dianion (II); in very strongly alkaline solutions, hydroxyl ion is taken

---

[3] For testing, acidify a 2-mL portion of the filtrate with 1 mL of concentrated nitric acid and add 3 mL of dilute (5–10%) aqueous silver nitrate solution.

up at the central carbon atom (destroying the quinonoid structure) and the resulting trianion (III) is colorless. Similar phthaleins can be obtained by using *o*-cresol or thymol instead of phenol. If phthalic anhydride is replaced by the anhydride of *o*-sulfobenzoic acid, sulfonphthaleins are produced. Both types of phthaleins are useful acid–base indicators.

$$\text{I} \underset{\text{H}^+}{\overset{\text{OH}^-}{\rightleftharpoons}} \text{II} \underset{\text{H}^+}{\overset{\text{OH}^-}{\rightleftharpoons}} \text{III}$$

Lactone (colorless)   II Dianion (red)   III Trianion (colorless)

The sulfonphthaleins are excellent pH indicators because they are moderately soluble in water and give brilliant color changes over narrow pH ranges. The parent compound, phenolsulfonphthalein, is formed by condensation of *o*-sulfobenzoic anhydride with phenol.

$$\text{o-sulfobenzoic anhydride} \xrightarrow[\text{(ZnCl}_2\text{)}]{3\ \text{C}_6\text{H}_5\text{OH},\ \text{heat}} \text{Phenolsulfonphthalein (Phenol red)}$$

The neutral form of phenol red is yellow, but above pH 8 the phenolic and sulfonic acid hydrogens dissociate to give the red dianion. In extremely strong acidic solutions the red cation is formed.

red (acidic solution) $\underset{\text{OH}^-}{\overset{\text{H}^+}{\rightleftharpoons}}$ yellow (neutral solution) $\rightleftharpoons$ red (basic solution)

TABLE 44.1 Properties of Sulfonphthalein Indicators

Sulfonphthalein indicators		pH at color change	Acid color	Base color
Cresol red	o-Cresolsulfonphthalein	1.0–2.0	Red	Yellow
Thymol blue	Thymolsulfonphthalein	1.2–2.8	Red	Yellow
Meta cresol purple	m-Cresolsulfonphthalein	1.2–2.8	Red	Yellow
Bromophenol blue	3′,3″,5′,5″-Tetrabromophenolsulfonphthalein	3.0–4.7	Yellow	Blue
Bromocresol green	3′,3″,5′,5″-Tetrabromo-m-cresolsulfonphthalein	3.8–5.4	Yellow	Blue
Bromocresol purple	5′,5″-Dibromo-o-cresol sulfonphthalein	5.2–6.8	Yellow	Purple
Bromothymol blue	3′,3″-Dibromothymol sulfonphthalein	6.0–7.6	Yellow	Blue
Phenol red	Phenolsulfonphthalein	6.6–8.0	Yellow	Red
Cresol red	o-Cresolsulfonphthalein	7.0–8.8	Yellow	Red
Meta cresol purple	m-Cresolsulfonphthalein	7.4–9.0	Yellow	Purple
Thymol blue	Thymolsulfonphthalein	8.0–9.6	Yellow	Blue

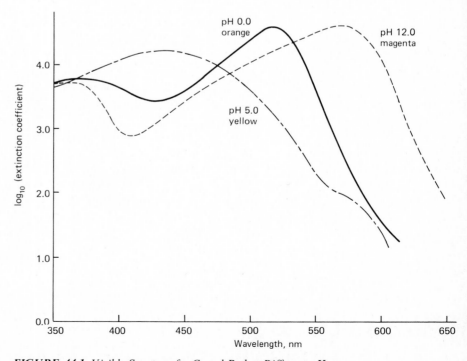

FIGURE 44.1 Visible Spectra of o-Cresol Red at Different pHs

If electron-withdrawing groups are added to the phenol rings both color changes occur at lower pHs; electron-releasing groups shift the color changes to higher pHs. The pH changes with different groups can be understood in terms of the relative stability of the cationic and anionic forms of the indicator. Table 44.1 lists the properties of a number of sulfonphthalein indicators. The visible spectra of *o*-cresol red at three different pHs are shown in Figure 44.1.

The general method for preparing sulfonphthaleins is to heat *o*-sulfobenzoic anhydride with the appropriate phenol in the presence of a Lewis acid. The preparation of *o*-cresolsulfonphthalein described below is a typical procedure. Bromine substituents can be introduced by direct bromination of the sulfonphthaleins in acetic acid solution.

An alternative procedure for preparing sulfonphthaleins is to use the imide of *o*-carboxybenzenesulfonic acid (insoluble saccharin) in place of *o*-sulfobenzoic anhydride. Substitution of the imide requires condensation catalysts such as sulfuric acid and more vigorous reaction conditions.

44.4 Preparation of *o*-Cresol Red

o-Cresol red

In a 15-mL test tube, place 1 g of freshly fused and pulverized zinc chloride, 1 g (0.0055 mole) of *o*-sulfobenzoic anhydride, and 1.6 g (0.015 mole) of *o*-cresol. Mix the reactants thoroughly with a stirring rod and then heat the mixture in a heating bath at 145° (bath temperature) for 1 hr. At the end of the heating period transfer the deep red mass to a 250-mL Erlenmeyer flask with the aid of several small portions of 10% aqueous sodium hydroxide (use a total of about 50 mL of solution). Acidify the resulting wine-colored solution with concentrated hydrochloric acid until the product precipitates as dark crystals (about 12 mL of acid). During the addition of acid, zinc salts will precipitate and then redissolve.

If the material in the flask remains liquid from unreacted *o*-cresol, boil the solution in the hood until the volume has been reduced by a half. The unreacted *o*-cresol will steam distill and the product should solidify. If any oily material remains add more water and repeat the steam distillation.

Collect the iridescent, green crystals with suction and allow them to air dry. The yield is about 1–1.5 g. Do not attempt to determine the melting point.

Questions

1. What effect will the following substituents have on the rate of coupling of the aryldiazonium ion, relative to the benzenediazonium ion?
   (a) *p*-nitro
   (b) *m*-nitro
   (c) *p*-methoxy
   (d) 2,4-dimethyl

2. What is a diazoamino compound? How may it be converted to an aminoaryl azo compound?

3. At pH 5–6, 7-amino-2-naphthol undergoes coupling at the 8-position, but at pH 10 the coupling occurs at the 1-position. Explain this behavior.

4. What compounds are formed by the reduction of methyl orange with strong reducing agents, such as sodium hydrosulfite ($Na_2S_2O_4$) or stannous chloride? Suggest a method for separating and characterizing the reduction products.

5. What is meant by (a) a chromophore group and (b) an auxochrome group? Give examples of each.

6. Define or explain the following terms.
   (a) direct or substantive dye
   (b) mordant or adjective dye
   (c) vat dye
   Cite an example of each.

7. Write equations showing a method of preparing a triphenylmethane dye, such as malachite green or crystal violet.

8. What is meant by the term *leuco base*? Write the structural formula of leuco-eosin (formed by the action of zinc dust on eosin in the presence of sodium hydroxide solution).

9. Explain the color change that occurs when a phthalein, such as phenolphthalein or fluorescein, is treated with sodium hydroxide solution.

10. Explain the fact that fluorescein diacetate does not give a similar color change when treated with sodium hydroxide solution.

11. Compare the structure of phenolphthalein with the structure of phenol red and comment on the colors of these two indicators at high pH (dianion forms).

12. Phenolphthalein is colorless at pH 6, but phenol red is yellow. Suggest a reason for this difference in color.

13. Why does bromophenol blue change color at lower pH than phenol red?

14. What color would be expected for the tetrabromophenolphthalein dianion (high pH form)?

45 Solvatochromic Dyes

45.1 Merocyanin Dyes

Dyes are highly colored compounds that usually contain polar functional groups, which serve both to help bind the dye to the fiber and to determine its color characteristics. The class of dyes known as the merocyanin dyes have the general structure represented by resonance structures Ia and Ib. Three other resonance structures of type I, involving different arrangements of the double bonds in the pyridine and benzene rings, can be drawn.

$$[\text{R–N}^+\text{=}\langle\ \rangle\text{–CH=CH–}\langle\ \rangle\text{–O}^- \longleftrightarrow \text{R–N}\langle\ \rangle\text{=CH–CH=}\langle\ \rangle\text{=O}]$$

Ia   Ib

The merocyanin dyes are neutral. Structures of type Ia contain two charges, but they are of opposite sign and cancel; structure Ib has no formal charges. As structure Ia implies, however, the merocyanin dyes possess a large dipole moment.

The merocyanin dyes are particularly interesting because their color is strongly dependent on the polarity and hydrogen bonding ability of the solvent. Depending on the solvent the color can be purple, blue, green, orange, red, or yellow. Such substances are called *solvatochromic* dyes. A proposed commercial use of these dyes is to detect adulteration of solvents.

In this preparation you will synthesize the merocyanin dye with R equal to methyl and examine the solvent dependence of its color.

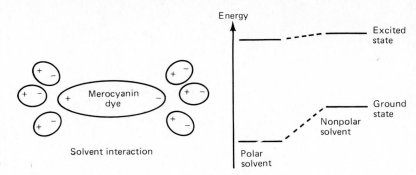

FIGURE 45.1
Interaction of Solvent with a Merocyanin Dye

45.2 Theoretical Basis for Solvatochromism

In the ground state of the merocyanin dyes, the electron density is higher at the oxygen end of the molecule than at the nitrogen end. The resulting dipole points in the direction indicated by structure Ia and is quite large because of the great distance between the charges. In a polar or hydrogen bonding solvent, molecules interact strongly with the merocyanin dipole, as shown at the left of Figure 45.1, and stabilize the ground state. Less polar solvents produce less stabilization.

On electronic excitation of a merocyanin molecule by absorption of visible light, electrons move from the negatively charged end toward the positively charged end. This reduces the merocyanin dipole moment and reduces the stabilization provided by the solvent. As a consequence (see Figure 45.1, right) the energy gap between the ground and excited state is greater in a polar solvent such as ethanol than in a nonpolar solvent such as acetone. From the Einstein relationship described in Chapter 10, $\Delta E = h\nu = hc/\lambda$, a larger excitation energy corresponds to a shorter wavelength for the absorbed light and longer wavelengths for the light that is *not* absorbed. In a nonpolar solvent the merocyanin is purple (absorption of low-energy long-wavelength yellow light), and in polar solvents it is yellow (absorption of high-energy short-wavelength purple light).

45.3 Synthesis of Merocyanin Dyes

The synthesis of the merocyanin dye with R = CH_3 proceeds in three steps.

1. $CH_3I\ +\ :N\!\!\diagup\!\!\diagdown\!\!-CH_3\ \longrightarrow\ CH_3\!-\!\overset{+}{N}\!\!\diagup\!\!\diagdown\!\!-CH_3\ +\ I^-$

2. $CH_3-\overset{+}{N}\!\!\diagup\!\!\diagdown\!\!-CH_3 + H-\overset{O}{\underset{}{C}}-\!\!\diagup\!\!\diagdown\!\!-OH \xrightarrow{\text{piperidine}} CH_3-\overset{+}{N}\!\!\diagup\!\!\diagdown\!\!-CH_2-\overset{OH}{\underset{H}{C}}-\!\!\diagup\!\!\diagdown\!\!-O^-$

red

3. $CH_3-\overset{+}{N}\!\!\diagup\!\!\diagdown\!\!-CH_2-\overset{OH}{\underset{H}{C}}-\!\!\diagup\!\!\diagdown\!\!-O^- \xrightarrow{KOH} CH_3-\overset{+}{N}\!\!\diagup\!\!\diagdown\!\!-CH=CH-\!\!\diagup\!\!\diagdown\!\!-O^-$

blue-red

In step 1 the tertiary amine, 4-methylpyridine, is quaternarized with methyl iodide in an S_N2 reaction. Other merocyanin dyes can be prepared by using different alkylating agents. With methyl iodide the reaction is rapid but proceeds somewhat more slowly with the larger alkyl iodides because of greater steric hindrance in the transition state. One general word of caution is that some alkylating agents, including methyl iodide, are suspected of causing cancer on prolonged exposure. All alkylating agents should be treated as hazardous substances and used only in the hood or other well-ventilated areas.

The second and third steps are the two parts of an aldol condensation. In step 2 the piperidine removes a proton from the ring methyl of the methyl 4-methylpyridinium salt to produce a low equilibrium concentration of the methyl anion. Normally piperidine is far too weak a base to remove protons from alkyl groups, but in this instance the acidity of the ring methyl protons is enhanced by the presence of the positive charge in the pyridine ring.

$CH_3-\overset{+}{N}\!\!\diagup\!\!\diagdown\!\!-CH_3 + \underset{\underset{H}{N}}{\bigcirc} \rightleftharpoons CH_3-\overset{+}{N}\!\!\diagup\!\!\diagdown\!\!-\overset{-}{C}H_2 + \underset{\underset{H\ \ \ H}{\overset{+}{N}}}{\bigcirc}$

Piperidine

The methyl anion adds to the carbonyl group of the 4-hydroxybenzaldehyde to produce, after an intramolecular proton transfer, the red alcohol shown in step 2.

In step 3 the red product is boiled briefly with dilute aqueous potassium hydroxide to complete the aldol reaction by the E2 elimination of water to yield the blue-red merocyanin dye.

Preparation and Measurements[1]

1,4-Dimethylpyridinium iodide. To a 50-mL round-bottom flask containing a solution of 4.8 mL (4.6 g, 0.05 mole) of 4-methylpyridine (**Caution**—*stench!* Use only in hood.) in 10 mL of 2-propanol, slowly add

[1] Adapted from the experiment described by Minch and Sadiq Shah, *J. Chem. Educ.*, **54**, 709 (1977).

3.1 mL (7.1 g, 0.05 mole) of methyl iodide (**Caution**—*cancer suspect agent!*). Attach a reflux condenser, add a boiling chip, and boil the solution gently for 30 min. Cool the solution in an ice-bath and collect the yellow crystals that separate on a Büchner funnel. Recrystallize the product from 1:1 95% ethanol:acetone (chill to maximize the recovery). The yield is 3.5–7.1 g (30–60% of theory), mp 144°. The product decomposes slowly on standing, and it is therefore best to proceed to the next step without delay.

Merocyanin dye.[2] In a 100-mL round-bottom flask dissolve the 7.1 g of 1,4-dimethylpyridinium iodide prepared above, 3.7 g of freshly recrystallized (from 1:3 water: 95% ethanol, then chilled) 4-hydroxybenzaldehyde (0.03 mole), and 2.5 mL of piperidine (0.025 mole) in 40 mL of *n*-propanol. Attach a reflux condenser and boil the solution *gently* for 1 hr. If less than 7.1 g of iodide was obtained in the first step, the quantities of other reactants and solvents should be scaled accordingly.

Cool the reaction mixture to room temperature and collect the red precipitate on a Büchner funnel. Suspend the solid in 175 mL of 0.2 M KOH and boil the solution gently for 30 min. Cool the solution and collect the blue-red crystals and recrystallize them several times from hot water. The yield is about 5.5 g (86% of theory), mp 220°.

Color experiments. Determine and record the color of solutions of small samples of the merocyanin dye in acetone, water, and ethanol. Pyridine and 4-methylpyridine give interesting colors although both solvents have unpleasant odors.

It is instructive to prepare an acetone solution of the dye and to determine the amount of ethanol or water required to produce a visible change in color.

If a UV–visible spectrophotometer is available, determine the absorption spectrum of a 2×10^{-5} M solution of the dye in water over the wavelength range of 300–700 nm.

Questions

1. If ethyl iodide were used in place of methyl iodide in the alkylation step, how would the rate of reaction change?
2. Normally the protons of methyl groups attached to benzene rings are not sufficiently acidic to be abstracted by a mild base such as piperidine. What is the enhanced acidity in the merocyanin dye preparation?
3. Predict the color of solutions of the merocyanin dye in the following solvents.
    (a) toluene
    (b) acetic acid
    (c) triethylamine
    (d) 2-propanol

---

[2] The proper name is 1-methyl-4-[(oxycyclohexadienylidene)ethylidene]-1,4-dihydropyridine.

4. The color of the merocyanin dyes do not change much as the alkyl side chain is varied from methyl to hexadecyl. Why?
5. Would you expect the solvent sensitivity of the merocyanin dyes to change if the phenoxide anion were converted to a phenyl ether by alkylation? Explain briefly.

46 Sugars

46.1 Introduction

Carbohydrates are one of the large and important groups of compounds found in nature. They include wood and cotton (cellulose), potato and corn starches, and honey and maple syrup (sugars). The simple sugars or monosaccharides, such as D-glucose, D-fructose, and D-galactose, are the units from which the more complex carbohydrates are built up by elimination of water (condensation polymerization). Low-molecular-weight polymers containing 2–10 units are called oligosaccharides (*oligo* means few). Examples of disaccharides are sucrose, maltose, cellobiose, and lactose. Raffinose is a trisaccharide with three units: galactose–glucose–fructose. Starches and cellulose are polysaccharides of high molecular weight, built up from thousands of glucose units.

The monosaccharides are polyhydroxyaldehydes (aldoses) or polyhydroxyketones (ketoses). In solution, they exist in an equilibrium involving an open-chain structure and cyclic hemiacetal forms. The latter may be five-membered rings (furanose) or six-membered rings (pyranose). The cyclic forms exist in two stereoisomeric configurations (called anomers), designated as α- and β-forms, which are *not* mirror-image forms. They arise because a new chiral carbon atom is created when the ring structure is formed, and it can exist in either of two configurations.

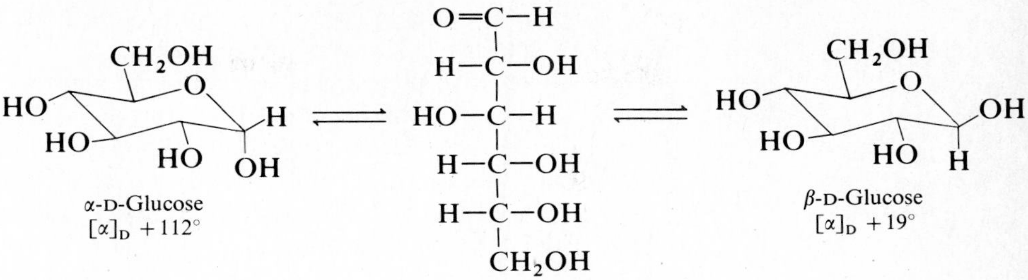

α-D-Glucose ⇌ (open chain) ⇌ β-D-Glucose

α-D-Glucose $[\alpha]_D +112°$

β-D-Glucose $[\alpha]_D +19°$

In aqueous solution either one of the individual α- or β-anomers will reach the same equilibrium mixture (mutarotation). For D-glucose, this equilibrium value is $+53°$. The anomers can be isolated in the form of derivatives such as α- and β-D-glucose pentaacetates, or α- and β-methyl-D-glucosides, when sugars are subjected to acylation or alkylation.

46.2 Monosaccharide and Disaccharide Tests

For the following tests prepare a 2% solution of the monosaccharides D-glucose (hydrate) ($C_6H_{12}O_6 \cdot H_2O$) and D-fructose ($C_6H_{12}O_6$) and of the disaccharides sucrose ($C_{12}H_{22}O_{11}$), maltose (hydrate) ($C_{12}H_{22}O_{11} \cdot H_2O$), and lactose (hydrate) ($C_{12}H_{22}O_{11} \cdot H_2O$) by dissolving 1 g of each pure sugar in 50 mL of distilled water. Use these solutions for tests (A), (B), and (C).[1]

Briefly describe and explain each result, using equations where possible. For comparison of the different sugars it is advantageous to record the results in a tabular form.

(A) Test for Reducing Sugars

Sugars that have a free aldehyde group (or can form one under the reaction conditions) are oxidized by cupric ion (see Chapter 33) to produce the carboxylic acid and cuprous ion.

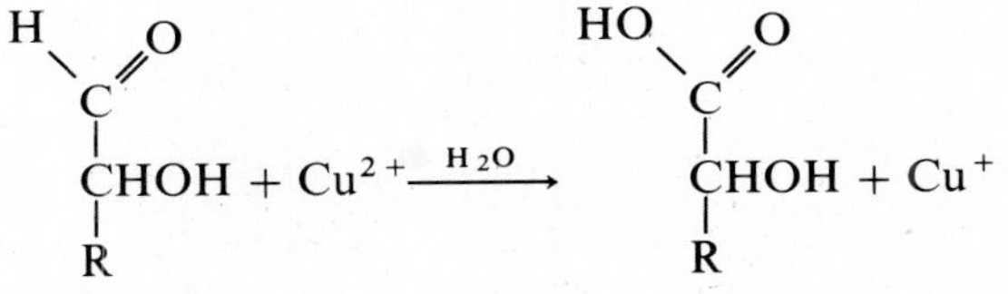

$$\underset{R}{\overset{H}{\underset{|}{C}}}\!\!=\!\!O \;\; \text{CHOH} + Cu^{2+} \xrightarrow{H_2O} \underset{R}{\overset{HO}{\underset{|}{C}}}\!\!=\!\!O \;\; \text{CHOH} + Cu^+$$

---

[1] If other monosaccharides are available, they may be tested similarly: D-galactose, $C_6H_{12}O_6$ (an aldohexose), D-mannose, $C_6H_{12}O_6$ (an aldohexose), D-xylose, $C_5H_{10}O_5$ (an aldopentose), L-arabinose, $C_5H_{10}O_5$ (an aldopentose).

Two reagents commonly used for this purpose are Fehling's solution[2] and Benedict's solution.[3] These are basic reagents in which the cupric ion is prevented from precipitating as the hydroxide by being complexed as either tartrate or citrate (both are blue). The cuprous ion does not form a tight complex and precipitates as the brick-red cuprous oxide, Cu_2O.

In a test tube place 5 mL of freshly mixed Fehling's solution (equal volumes of No. 1 and No. 2) or 5 mL of Benedict's solution.

Heat the solution to gentle boiling and add 2–3 drops of the glucose solution. Continue to boil gently and observe the results after a minute or two. Continue to add the glucose solution, 2–3 drops at a time, and heat for a short while after each addition, until the deep blue color just disappears. Fehling's solution contains a known amount of copper per milliliter of solution; suggest a method for determining the amount of glucose in a solution of unknown strength.

Repeat the test with the other sugar solutions. Discontinue the test if no reduction is observed after 5 or 6 drops of the sugar solution has been added.

Hydrolysis of sucrose. To 10 mL of the 2% solution of sucrose, add 1–2 mL of dilute hydrochloric acid and heat on a steam bath for 0.5 hr. Carefully *neutralize* with 10% aqueous sodium hydroxide and apply a test for reducing sugars.

(B) Osazone Test

α-Hydroxy carbonyl compounds, such as the reducing sugars, react with phenylhydrazine to produce bisphenylhydrazones, called osazones.

$$\begin{array}{c} H \\ \diagdown \\ C=O \\ | \\ CHOH \\ | \\ R \end{array} + 3\ C_6H_5NHNH_2 \longrightarrow \begin{array}{c} H \\ \diagdown \\ C=N-NHC_6H_5 \\ | \\ C=N-NHC_6H_5 \\ | \\ R \end{array} + C_6H_5NH_2 + NH_3 + H_2O$$

---

[2] **Preparation of Fehling's solution.** Fehling's solution is made by mixing together, at the moment it is to be used, equal volumes of a solution of copper sulfate (called Fehling's solution No. 1) and an alkaline solution of sodium potassium tartrate (called solution No. 2). For qualitative work these solutions may be prepared as follows: (No. 1) Dissolve 3.5 g of pure copper sulfate crystals ($CuSO_4 \cdot 5H_2O$) in 100 mL of water. (No. 2) Dissolve 17 g of sodium potassium tartrate crystals (Rochelle salt) in 15–20 mL of warm water, add a solution of 5 g of solid sodium hydroxide in 15–20 mL of water, cool, and dilute to a volume of 100 mL. The solutions are kept separately and not mixed until time for use, since the mixed solution deteriorates on standing.

[3] **Preparation of Benedict's solution.** Benedict modified Fehling's solution and devised a single test solution that does not deteriorate on standing. For qualitative work this solution may be prepared as follows: Dissolve 20 g of sodium citrate and 11.5 of anhydrous sodium carbonate in 100 mL of hot water in a 400-mL beaker. Add slowly to the citrate–carbonate solution, stirring continually, a solution of 2 g of copper sulfate crystals in 20 mL of water. The mixed solution should be perfectly clear (if not, pour it through a fluted filter).

The mechanism of this curious process involves initial formation of the *mono*phenylhydrazone, which undergoes a series of proton exchanges to yield a keto compound (I).

$$\underset{R}{\underset{|}{\overset{H}{\overset{|}{C}}=O}} \longrightarrow \underset{R}{\underset{|}{\overset{H}{\overset{|}{C}}=NNHC_6H_5}} \rightleftharpoons \underset{R}{\underset{|}{\overset{H}{\overset{|}{C}}-NHNHC_6H_5}} \rightleftharpoons \underset{R}{\underset{|}{\overset{HH}{\overset{||}{C}}-NHNHC_6H_5}}$$

$$\text{CHOH} \text{CHOH} \text{C}-\text{OH} \text{C}=O$$

I

The keto compound reacts with a second molecule of phenylhydrazine to give a new hydrazone (II).

$$\underset{R}{\underset{|}{\overset{HH}{\overset{||}{C}}-NHNHC_6H_5}} \longrightarrow \underset{R}{\underset{|}{\overset{HH}{\overset{||}{C}}-NHNHC_6H_5}}$$
$$\text{C}=O \text{C}=\text{N}-\text{NHC}_6\text{H}_5$$

I → II

This hydrazone then undergoes an oxidative proton abstraction and N—N bond cleavage to produce the imine hydrazone (III) and aniline. The imine hydrazone (III) then undergoes an addition-elimination reaction with a third molecule of phenylhydrazine to form the observed osazone.

$$C_6H_5NHNH_2$$

$$\underset{R}{\underset{|}{H-C-NH-NHC_6H_5}} \longrightarrow \underset{R}{\underset{|}{\overset{H}{\overset{|}{C}}=NH}} + NH_2C_6H_5$$
$$\text{C}=\text{N}-\text{NHC}_6\text{H}_5 \text{C}=\text{NNHC}_6\text{H}_5$$

II → III

$$\downarrow C_6H_5NHNH_2$$

$$\underset{R}{\underset{|}{\overset{H}{\overset{|}{C}}=N-NHC_6H_5}} + NH_3$$
$$\text{C}=\text{N}-\text{NHC}_6\text{H}_5$$

Osazone

In applying the osazone test for comparison of various sugars, it is advisable to perform all the tests simultaneously. Place 5 mL of each sugar solution separately in large test tubes, add to each 3 mL of *freshly prepared* phenylhydrazine reagent (**Caution**—*phenylhydrazine is a suspected carcinogen*),[4] and 2 or 3 drops of saturated aqueous sodium bisulfite (to avoid oxidation). Mix thoroughly, and heat in a beaker of boiling water. Note the time of immersion and record the time in which the osazones[5] precipitate. Shake the tubes from time to time to avoid forming supersaturated solutions of the osazones. Continue heating them in the bath of boiling water for 15 or 20 min and allow them to cool slowly.

From each tube in which crystals have formed, transfer a small quantity of material to a microscope slide, and examine it under the microscope. Filter the remainder of the crystals at once with suction, wash with a little water, and let them dry. The osazones discolor on standing. The melting points of the common osazones lie close together (~ 200–$206°$) and are not useful for identification.

(C) Acetylation of Glucose

In a porcelain mortar, pulverize finely 2.7 g (0.015 mole) of anhydrous glucose, and mix it thoroughly with 1.5 g of anhydrous sodium acetate. Transfer the mixture to a 100-mL round-bottom flask and add 14 mL (15 g, 0.15 mole) of acetic anhydride. Heat the mixture under a reflux condenser on a steam bath for 2.5 hr with occasional vigorous agitation. Pour the warm solution in a thin stream, with vigorous stirring, into 125 mL of ice-cold water in a beaker. Disintegrate any lumps of the crystalline precipitate and allow the finely divided material to stand in contact with the water, occasionally stirring it, until the excess acetic anhydride has been hydrolyzed. This will require about 2 hr. It is desirable to permit the reaction mixture to stand overnight or longer. Collect the crystals with suction and press them as dry as possible with a clean cork or flat glass stopper. Transfer the crystals to a beaker, mix thoroughly with about 125 mL of water, and allow to stand with occasional stirring for about 2 hr longer. Collect the crystals with suction and press them as dry as possible on the filter. Recrystallize the crude β-D-glucose pentaacetate from about 15 mL of

---

[4] **Preparation of phenylhydrazine reagent.** Dissolve 4 g of phenylhydrazine hydrochloride, $C_6H_5-NH-NH_2 \cdot HCl$, in 36 mL of water, and add 6 g of sodium acetate crystals and a drop of glacial acetic acid. If the resulting solution is turbid, add a small pinch of decolorizing charcoal, shake vigorously, and filter the solution. The reagent deteriorates rapidly.

A satisfactory reagent may also be made by dissolving 4 g (4 mL) of phenylhydrazine (the free base is a liquid) in a solution of 4 g (4 mL) of glacial acetic acid in 36 mL of water, and clarifying it as before with a pinch of decolorizing charcoal. This procedure is less desirable owing to the **poisonous character of phenylhydrazine** (vapor or liquid). If the liquid comes in contact with the skin, it must be washed off at once, using 5% acetic acid, followed by soap and water.

[5] Consult Shriner, Fuson, Curtin, and Morrill, *The Systematic Identification of Organic Compounds*, 6th ed. (New York: Wiley, 1980).

The precipitate formed with D-mannose (0.5–1 min) is the sparingly soluble phenylhydrazone. Sucrose must undergo hydrolysis before precipitation of glucosazone occurs.

methanol. The yield of purified product is 3–4 g. The recorded melting point of the purified compound is 135°.

α-D-Glucose pentaacetate can be prepared by acetylation with acidic catalysts. Acids effect the interconversion of α- and β-D-glucose *and* of the pentaacetates, and at *equilibrium* in acetic anhydride, the α-pentaacetate constitutes 90% of the product. With sodium acetate, a basic catalyst, the β-pentaacetate predominates because sodium acetate catalyzes interconversion of α- and β-D-glucose but not the pentaacetates, and the rate of acetylation of β-D-glucose is much faster than that of the α-isomer.

(D) Benzoylation of Glucose

In a 125-mL Erlenmeyer flask, place 25 mL of the 2% glucose solution, 14 mL of 10% aqueous sodium hydroxide, and 2 mL of benzoyl chloride (**Caution**—*irritating vapor!*). Shake the flask vigorously at frequent intervals until the odor of benzoyl chloride no longer can be detected (test cautiously); at least 15 min will be required. If necessary add a little more alkali to keep the solution alkaline. When the product has crystallized, collect the crystals by suction filtration and wash them thoroughly with water. If the product tends to remain gummy, it may be allowed to stand in contact with the alkaline solution until the following laboratory period.

Recrystallize the crude product from ethanol and determine the melting point. The recorded melting point of D-glucose pentabenzoate is 179°.

Questions

1. What is the precipitate that forms 46.2(A)? Compare these results with those obtained in previous tests of simple aldehydes and ketones in Chapter 9.
2. What is meant by each term?
    (a) aldohexose
    (b) ketopentose
3. Of what value is Fehling's test in classifying an unknown sugar?
4. Compare the melting points and rates of precipitation of the osazones of the more common sugars and explain how these values might be used in the identification of an unknown sugar.[5]
5. Explain the fact that glucose and fructose give the *same* osazone.
6. What is the significance of the fact that five acetyl or benzoyl groups can be introduced into the molecule of glucose or fructose?
7. What is the significance of the fact that only eight acetyl or benzoyl groups (and not ten) can be introduced into the molecule of the disaccharide sucrose?
8. Give an accurate definition of the term *carbohydrate*.
9. What sugars are present in the solution after hydrolysis of sucrose? Will the sugars form an osazone with phenylhydrazine?
10. Why is the hydrolysis of sucrose sometimes referred to as the inversion of sucrose? What is invert sugar?
11. What agents other than acids promote the hydrolysis of complex carbohydrates?

47 Biosynthesis of Alcohols

47.1 Fermentation of Sugars

The fermentation processes involved in bread making, wine making, and brewing are among the oldest chemical arts. For many years it was believed that the transformation of sugar by yeasts into ethanol and carbon dioxide was inseparably connected with the life process of the yeast cell. This view was abandoned when Eduard Buchner[1] (Nobel laureate, 1907) demonstrated that yeast juice will bring about alcoholic fermentation in the absence of any live yeast cells. The fermenting activity of yeast is due to a number of remarkably active catalysts of biochemical origin. It is now recognized that most of the chemical transformations that go on in living cells of plants and animals are brought about by enzymes.

The overall expression for fermentation of sucrose is typical.

$$\text{Sucrose} \longrightarrow \text{glucose} + \text{fructose} \longrightarrow 4\ CH_3CH_2OH + 4\ CO_2$$
$$C_{12}H_{22}O_{11} \qquad C_6H_{12}O_6 \quad C_6H_{12}O_6$$

The first step in the fermentation of disaccharides, such as sucrose or maltose, is a simple hydrolysis to monosaccharides: hexoses like glucose and fructose. These are then converted to their 6-phosphate esters, then to fructose 1,6-diphosphate, and to phosphate esters of the trioses: dihydroxyacetone (a ketotriose) and glyceraldehyde (an aldotriose). The next step

---

[1] Eduard Buchner is widely confused with Ernst Büchner (note difference in spelling) who invented (1888) the funnel bearing his name.

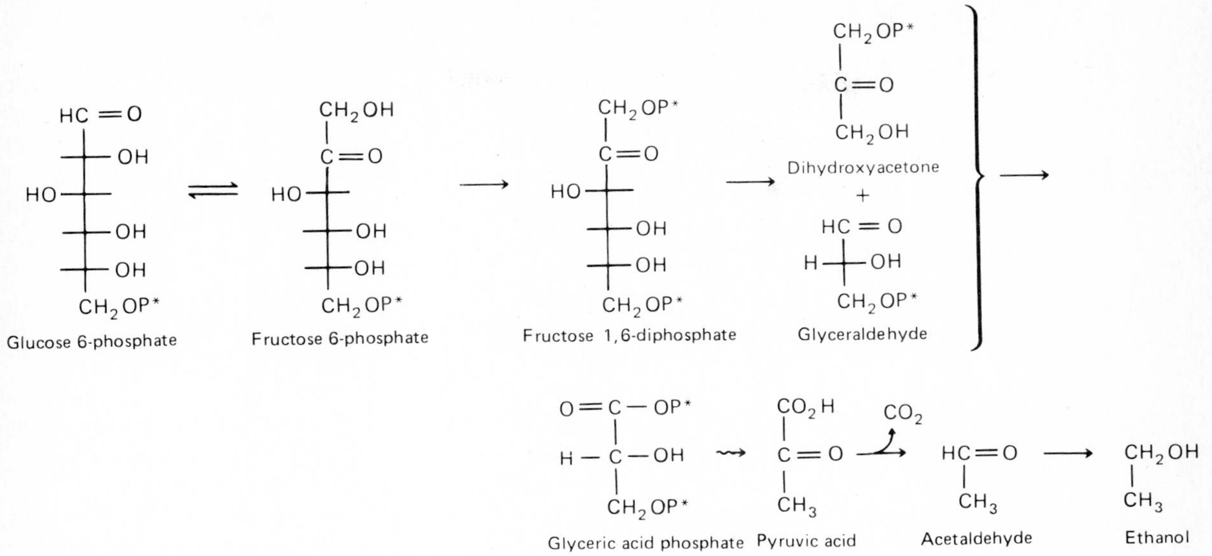

FIGURE 47.1 *General Outline for Steps in Alcoholic Fermentation (P* = phosphate ester)*

involves oxidation to the phosphoglyceric acids, which lead to pyruvic acid; this acid loses carbon dioxide to form acetaldehyde, and the aldehyde is reduced to ethanol (Figure 47.1).

Enzymes show an extraordinary selectivity—each step of the fermentation requires a particular enzyme as catalyst. Inorganic salts are also important: Pasteur found that salts such as magnesium sulfate and various phosphates were needed to promote yeast growth and fermentation. The chief industrial sources of sugars for fermentation are molasses residues and starches from grains (corn, rye) and potatoes. The starches are high-molecular-weight polysaccharides that are converted by the enzyme diastase into the disaccharide maltose, which furnishes two molecules of glucose (see Chapter 46).

After fermentation has been completed, fractional distillation produces an ethanol–water azeotrope (bp 78.15°) containing 95.6% alcohol by weight (97.2% by volume). To obtain anhdrous (absolute) ethanol commercially, water is removed from the azeotrope by distillation with benzene to furnish the pure product (bp 78.37°). For tax purposes, beverage alcohol is rated in terms of proof spirt; 100 proof (U.S.A.) is 50% ethanol by volume, 42.5% by weight. To avoid payment of the beverage tax, alcohol can be *denatured* by addition of substances that make it unfit to drink.

Higher alcohols (C_3–C_5), called fusel oil, are obtained in small amounts by fractionation of the fermented liquor. These alcohols do not come from the sugars, but arise from enzyme action on amino acids derived from the raw materials used and from yeast cells. The alcohols are all primary, mainly *n*-propyl, isobutyl, isopentyl (3-methyl-1-butanol), and optically active 2-methyl-1-butanol.[2]

Yeasts, molds, and bacteria are used commercially for the large-scale production of various organic compounds. An important example, in addition to ethanol, is the anaerobic fermentation of starch by certain bacteria to yield *n*-butyl alcohol, acetone, ethanol, carbon dioxide, and hydrogen.

In the United States, industrial ethanol is manufactured mainly from ethylene, a product of the "cracking" of petroleum hydrocarbons. By reaction with concentrated sulfuric acid, ethylene is converted to ethyl hydrogen sulfate, which is hydrolyzed to ethanol by dilution with water. Isopropyl, *sec*-butyl, *t*-butyl, and higher secondary and tertiary alcohols also are produced on a large scale from alkenes derived from the cracking process.

Mixtures of C_5 alcohols are manufactured by hydrolysis of C_5 alkyl chlorides obtained by chlorination of *n*-pentane and isopentane (from petroleum). Higher alcohols, C_{12}–C_{18}, are manufactured by catalytic reduction of fatty esters.

Laboratory syntheses of alcohols often make use of the reaction of an organomagnesium halide (Grignard reagent) with an aldehyde, ketone, or ester; examples are given in Chapters 17 and 35. They can also be prepared by hydration of an alkene as described in Chapter 18.

47.2 Ethanol by Fermentation

Place 40 g (0.12 mole) of sucrose (common granulated sugar) in a 500-mL round-bottom flask, add 350 mL of water, 35 mL of Pasteur's salts solution,[3] and one-half envelop of dry yeast or one-half cake of compressed yeast, rubbed to a thin paste with 15–20 mL of water. Shake vigorously and close the flask with a rubber stopper holding a delivery tube arranged so that any gas evolved must bubble through about 10 mL of water in a test tube. Allow the mixture to stand at a temperature of 25–35° until fermentation is complete, as indicated by the cessation of gas evolution. (About a week is required.)

Without stirring up the yeast any more than is necessary, decant the liquid through a plug of cotton or glass wool into a 500-mL round-bottom flask. Attach a fractionating column (preferably one providing about three

---

[2] The C_5 alcohols formerly were called amyl alcohols but are now generally designated as pentyl. The term *active* denotes optical activity, referring to action on plane polarized light; optical isomers are enantiomorphs, which are structures having nonsuperposable mirror-image spatial configurations.

[3] Pasteur's salts solution consists of 2.0 g potassium phosphate, 0.20 g calcium phosphate, 0.20 g magnesium sulfate, and 10.0 g ammonium tartrate in 860 mL of water. If Pasteur's salts solution is not available, 0.25 g of disodium hydrogen phosphate may be substituted. The fermentation can be carried on without any added salts but is slower and yields a more difficultly separated yeast residue.

theoretical plates) and connect it to a condenser set downward for distillation. Add two tiny boiling chips and distill about 60 mL of liquid into a weighed 100-mL graduated cylinder (discard the yeast residue in the distilling flask). Record the volume of the distillate and determine its weight by the difference. If a hydrometer is available, a more accurate measure of the specific gravity of the distillate can be obtained.[4] Calculate the weight of distillate, and from the table of densities of aqueous solutions of ethanol given in the Appendix, calculate the weight of alcohol in the distillate.[5] Calculate the percentage yield of ethanol in the fermentation.

Transfer the distillate to a 100-mL round-bottom flask and distill *slowly* using the same fractionating apparatus. Collect the following fractions: A, 78–82°; B, 82–88°; C, 88–95°. Discard the residue containing the fusel oil. If fraction A amounts to less than 15 mL, redistill fractions A, B, and C.

Carry out the *iodoform test* and *esterification test* with small portions of fraction A as described in Chapter 9. The NMR spectrum of ethanol is shown in Figure 10.24.

Questions

1. Using as many systems as seem appropriate, name all of the alcohols having the carbon skeleton of
   (a) 3-methylpentane
   (b) neohexane
   (c) methylcyclopentane

2. When benzene is used to dry 95% (by weight) ethanol, two azeotropes are involved: a ternary azeotrope (bp 65°) containing 74% by weight benzene, 18.5% ethanol, and 7.5% water, and a binary azeotrope (bp 68°) consisting of 68% benzene, and 32% ethanol. Assuming a safety factor of 10% excess, calculate the amount of benzene needed to dry 1 kg of 95% ethanol. How much absolute ethanol would be obtained?

3. How could one ascertain whether the alcohol or the carboxylic acid furnishes the hydroxyl group that appears in the molecule of water produced in formation of an ester?

---

[4] The hydrometer should be the short form for use with 50 mL or less of liquid, with a range of 0.900–1.000 in graduations of 0.002 or less.

[5] The density, d_4^{20}, of a liquid is the measured specific gravity multiplied by the density of water at 20°.

48 Peptides

48.1 Structure

Peptides are polymers derived from amino acids and have molecular weights less than 5000–10,000; similar polymers with molecular weights greater than this are called proteins. The component amino acids are joined by amide bonds from the amino group of one unit to the carboxyl group of the next.

$$H_2N-\underset{R}{CH}-\underset{\parallel}{C}-\left(\underset{H}{N}-\underset{R}{CH}-\underset{\parallel}{C}\right)_n-\underset{H}{N}-\underset{R}{CH}-CO_2H$$

N-terminal amino acid — C-terminal amino acid

Polypeptide

In writing structures of polypeptides, the convention is to place the N-terminal amino acid on the left and the C-terminal amino acid on the right. Then the amino acids are named left to right using either their common names or their standardized three-letter codes.

serine — phenylalanine — glycine

Serylphenylalaninylglycine (Ser-Phe-Gly)

48.2 Biological Function

Peptides and proteins serve a number of functions in biological systems. In addition to the role of the high-molecular-weight polyamides as structural materials (muscle, tendon, hair, skin), the lower molecular weight polyamides serve as essential biological regulators. Enzymes, which have complex structure and typical molecular weights in excess of 10,000, selectively catalyze the many organic reactions of living systems. Frequently, the catalytic activity of an enzyme is controlled by the concentration of the product, with the result that the product concentration stays fairly constant although the starting material concentrations change markedly.

48.3 Synthesis of Polypeptides

The standard strategy for forming amide bonds is to prepare a chemically active component and allow it to react with an amine. For example, a carboxylic acid can be converted to an acyl chloride, which, on treatment with an amine, gives an amide. This reaction is the basis for both carboxylic acid and amine identification (see Chapter 9).

$$R-\underset{\underset{O}{\|}}{C}-OH \xrightarrow{\text{activation}} R-\underset{\underset{O}{\|}}{C}-Cl \xrightarrow{NH_2R} R-\underset{\underset{O}{\|}}{C}-NHR + HCl$$

This simple strategy, so effective for monofunctional group components, fails miserably when applied to making amide bonds between amino acids. With such bifunctional group compounds, the initial activation step produces an intermediate that reacts with itself to produce a polymer before the amine component is added. For example, the simple amino acid glycine, on conversion to glycyl chloride, polymerizes to polyglycine and other condensation products.

$$NH_2-CH_2-CO_2H \longrightarrow NH_2-CH_3-\underset{\underset{O}{\|}}{C}Cl \longrightarrow$$

$$NH_2-CH_2-\underset{\underset{O}{\|}}{C}\left(NH-CH_2-\underset{\underset{O}{\|}}{C}\right)_n NH-CH_2-CO_2H + \text{other products}$$

To avoid this unwanted reaction it is necessary to temporarily "protect" the amino group, that is, to convert the NH_2 to some other unreactive functional group. After the desired coupling has occurred, the protecting group can be removed.

The repeated sequence of protection, activation, and reaction terminated by a final deprotection step is time-consuming and fraught with many subtle difficulties. However, the ability to synthesize any arbitrary amino acid polymer is so important to modern biochemical research that a great effort has been invested in making the process practical. A brilliant achievement in this area was Merrifield's step-by-step synthesis of the enzyme bovine pancreatic ribonuclease, a protein with a chain of 124 amino acid units. A total of 369 consecutive chemical reactions was achieved with an overall yield of 17% which corresponds to an average yield of better than 99% for each step.

In the peptide synthesis described here, we will tackle a much simpler example, glycylglycine. The chemical sequence to be followed is outlined below.

1. Phthalic anhydride + Glycine → Phthaloylglycine

2. Phthaloylglycine + SOCl$_2$ (Thionyl chloride) → Phthaloylglycyl chloride

3. Phthaloylglycyl chloride + Glycine → Phthaloylglycylglycine

4. Phthaloylglycylglycine + NH$_2$NH$_2$ (Hydrazine) → Phthalhydrazide + Glycylglycine

In step 1 the amino group of a glycine is protected by reaction with phthalic anhydride to produce phthaloylglycine, in which the original NH_2 group has been transformed into an imide. Because the nitrogen of an imide is conjugated with a pair of carbonyl groups, it is not longer basic or nucleophilic. Because the nitrogen bears no hydrogens, it cannot even hydrogen-bond to other bases. The amino group has been chemically deactivated and one can now proceed to activate the carboxylic acid without fear of having it react with the amino group of a glycine unit.

In step 2 the amine-protected glycine is converted into an acid chloride with thionyl chloride. In this reaction an equivalent of hydrogen chloride is liberated. If the amino group had not been protected it would have been converted into a hydrochloride salt, which would have slowed the reaction with thionyl chloride.

In step 3 the acid chloride is allowed to react with another equivalent of glycine to produce the phthaloylglycylglycine. In principle, one could convert this material into an acid chloride, as in step 2, and use that to add a third amino acid and so on. We will stop at the dipeptide stage.

The final step, after all the desired amino acids have been joined, is to remove the protecting phthaloyl unit. This can be accomplished by heating the phthaloylglycylglycine with hydrazine, which does two consecutive nucleophilic displacements on the imide to yield glycylglycine and the very stable phthalhydrazide. Directions for this last step are not given in the procedure section because hydrazine is quite toxic and hazardous to work with.

48.4 Preparation of Phthaloylglycylglycine

Phthaloylglycine. Place a well-pulverized mixture of 3 g (0.02 mole) of phthalic anhydride and 1.5 g (0.02 mole) of glycine in a 100-mL beaker. Clamp the beaker and insert a 250° thermometer into the solids so that the bulb is covered. Heat the tube with a burner until the solids melt, stir them gently with a glass rod so that they mix, and then heat the molten mass at 150–190° for 15 min. Allow the reaction mixture to cool and then recrystallize it from 50 mL of water. The yield of phthaloylglycine is about 3.9 g, mp 196–198°.

If IR and NMR instruments are available, determine the spectra of the product.

Phthaloylglycyl chloride. In a small round-bottom flask place 2.1 g (0.01 mole) of recrystallized phthaloylglycine and 8 mL (13.1 g, 0.11 mole) of thionyl chloride (**Caution**—*vapors and liquid are strongly irritating to the skin, nose, and eyes*). Attach a condenser bearing a drying tube and heat the mixture at a gentle reflux for 1 hr.

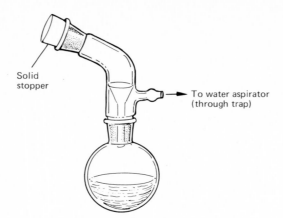

FIGURE 48.1 *Apparatus for Removal of Excess Thionyl Chloride*

Remove the condenser and evaporate the excess thionyl chloride from the reaction mixture under *reduced pressure* using the vacuum adapter, as illustrated in Figure 48.1. Recrystallize the residue from 5:1 acetone–petroleum ether (bp 60–80°). The yield is about 2.2 g, mp 74–82°.

Phthaloylglycylglycine. In a 125-mL Erlenmeyer flask prepare a suspension of 0.7 g of glycine and 0.5 g of magnesium oxide in 30 mL of water and cool it to 5° in an ice bath. Add to the cooled suspension, *drop by drop*, a solution of 1.8 g (0.008 mole) of phthaloylglycyl chloride in 10 mL of tetrahydrofuran. The reaction mixture should be maintained at 5° during the addition. When the addition is complete, stir the reaction mixture for 10–15 min at room temperature to complete the reaction. Acidify the mixture (litmus paper) with hydrochloric acid. Chill the acidic mixture in an ice bath for 20 min and collect on a Büchner funnel the crystalline product that separates. Recrystallize the product from about 35 mL of water. The yield is about 1.5 g, mp 224–226°.

If IR and NMR instruments are available, determine the spectra of the product and compare them with the spectra of phthaloylglycine.

Questions

1. Explain why the preparative procedure used to synthesize glycylglycine would fail if you attempted to introduce threonine, tyrosine, or glutamic acid.
2. The pK_1 values for glycine, glycylglycine, and glycylglycylglycine are 2.35, 3.14, and 3.23, respectively. How can one account for the large difference between glycine and the peptides derived from it?
3. Explain why a benzoyl group could not be used as a N-protecting group for peptide synthesis.

Appendix

Tables of Physical Data

Vapor Pressures of Organic Substances and of Water at 30–120°C

(In millimeters of mercury)

t, °C	Water	Benzene	Bromo-benzene	p-Dibromo-benzene	Toluene	Ethanol
30	31.8	118	6	—	37	78
40	55.3	181	10	—	60	135
50	92.5	269	17	—	93	223
60	149.2	389	28	—	138	353
70	233.8	547	43	4	203	543
80	355.5	754	66	7	288	813
90	526.0	1016	98	12	402	1187
95	634.0	1171	118	15	472	1415
100	760.0	1344	141	18	557	1695
110	1074.0	1748	199	28	747	2364
120	1489.0	2238	275	42	965	—

Vapor Pressures of Organic Substances[a] at 100°C

(In millimeters of mercury)

Acetophenone	26.5	p-Dichlorobenzene	73.1
Aniline	45.7	Dimethylaniline	37.9
Benzaldehyde	60.5	Ethyl acetoacetate	80
Benzoic acid	1.8	Ethylbenzene	307
Bromobenzene	141	Ethyl benzoate	17.4
Biphenyl	9	2-Furaldehyde	110
n-Butyl acetate	325	Nitrobenzene	20.8
n-Butyl alcohol	390	p-Nitrobenzoic acid	0.006
o-Chlorotoluene	133	p-Nitrotoluene	7.9
p-Chlorotoluene	123	Salicylic acid	0.86
o-Cresol	13.6	Toluene	557
p-Dibromobenzene	18	Triphenylmethanol	0.01

[a] The vapor pressures of salts, such as aniline hydrochloride and sodium acetate, are generally so low at 100°C that they can be considered to be nil.

Vapor Pressures of Water and Organic Substances at 90–100°C[a]

(In millimeters of mercury)

t, °C	Water	α-Pinene	p-Dichlorobenzene	o-Chlorotoluene
90.0	525.8	99.6	49.5	84.1
90.5	525.8	101.5	50.5	85.6
91.0	546.0	103.3	51.5	87.3
91.5	556.4	105.2	52.5	88.9
92.0	566.9	107.2	53.6	90.6
92.5	577.6	109.1	54.6	92.2
93.0	588.5	111.1	55.7	94.0
93.5	599.5	113.1	56.8	95.7
94.0	610.7	115.2	58.0	97.5
94.5	622.1	117.3	59.1	99.3
95.0	633.7	119.4	60.3	101.1
95.5	645.5	121.5	61.5	103.0
96.0	657.4	123.7	62.7	104.8
96.5	669.6	125.9	63.9	106.7
97.0	681.9	128.2	65.2	108.7
97.5	694.4	130.5	66.5	110.7
98.0	707.1	132.8	67.7	112.7
98.5	720.1	135.1	69.1	114.7
99.0	733.2	137.5	70.4	116.8
99.5	746.5	139.9	71.8	118.9
100.0	760.0	142.4	73.1	121.0

[a] Based on data presented in Weast, Ed., *Handbook of Chemistry and Physics* (Boca Raton, FL: CRC, Inc., 1983).

Density and Vapor Pressure of Water at 0–35°C

t, °C	Vapor pressure, Hg	Density $d_{4°}^{t°}$	t, °C	Vapor pressure, Hg	Density $d_{4°}^{t°}$	t, °C	Vapor pressure, Hg	Density $d_{4°}^{t°}$
0	4.58	0.99987	12	10.48	0.99952	24	22.18	0.99733
1	4.92	0.99993	13	11.19	0.99940	25	23.54	0.99708
2	5.29	0.99997	14	11.94	0.99927	26	24.99	0.99682
3	5.68	0.99999	15	12.73	0.99913	27	26.50	0.99655
4	6.09	1.00000	16	13.56	0.99897	28	28.10	0.99627
5	6.53	0.99999	17	14.45	0.99880	29	29.78	0.99597
6	7.00	0.99997	18	15.38	0.99862	30	31.55	0.99568
7	7.49	0.99993	19	16.37	0.99843	31	33.42	0.99537
8	8.02	0.99988	20	17.41	0.99823	32	35.37	0.99505
9	8.58	0.99981	21	18.50	0.99802	33	37.43	0.99473
10	9.18	0.99973	22	19.66	0.99780	34	39.59	0.99440
11	9.81	0.99963	23	20.88	0.99757	35	41.85	0.99406

Aqueous Ethanol

Density $d_{4°}^{20°}$	C_2H_5OH % by weight	C_2H_5OH % by volume	g/100 mL	Density $d_{4°}^{20°}$	C_2H_5OH % by weight	C_2H_5OH % by volume	g/100 mL
0.98938	5	6.2	4.9	0.85564	75	81.3	64.2
0.98187	10	12.4	9.8	0.84344	80	85.5	67.5
0.97514	15	18.5	14.6	0.83095	85	89.5	70.6
0.96864	20	24.5	19.4	0.81797	90	93.3	73.6
0.96168	25	30.4	24.0	0.81529	91	94.0	74.2
0.95382	30	36.2	28.6	0.81257	92	94.7	74.8
0.94494	35	41.8	33.1	0.80983	93	95.4	75.4
0.93518	40	47.3	37.4	0.80705	94	96.1	75.9
0.92472	45	52.7	41.6	0.80424	95	96.8	76.4
0.91384	50	57.8	45.7	0.80138	96	97.5	76.9
0.90258	55	62.8	49.6	0.79846	97	98.1	77.4
0.89113	60	67.7	53.5	0.79547	98	98.8	77.9
0.87948	65	72.4	57.1	0.79243	99	99.4	78.4
0.86766	70	76.9	60.7	0.78934	100	100	78.9

Acids

Aqueous hydrochloric acid			Aqueous nitric acid			Aqueous sulfuric acid		
Density $d_{4°}^{15°}$	HCl, % by wt.	HCl, g/100 mL	Density $d_{4°}^{20°}$	HNO_3, % by wt.	HNO_3, g/100 mL	Density $d_{4°}^{20°}$	H_2SO_4, % by wt.	H_2SO_4 g/100 mL
1.010	2.14	2.2	1.0256	5	5.1	1.0317	5	5.2
1.015	3.12	3.2	1.0543	10	10.5	1.0661	10	10.7
1.020	4.13	4.2	1.0842	15	16.3	1.1020	15	16.5
1.025	5.15	5.3	1.1150	20	22.3	1.1394	20	22.8
1.030	6.15	6.4	1.1469	25	28.7	1.1783	25	29.5
1.035	7.15	7.4	1.1800	30	35.4	1.2185	30	36.6
1.040	8.16	8.5	1.2140	35	42.5	1.2599	35	43.8
1.045	9.16	9.6	1.2463	40	49.9	1.3028	40	52.1
1.050	10.17	10.7	1.2783	45	57.5	1.3476	45	60.6
1.055	11.18	11.8	1.3100	50	65.5	1.3951	50	69.6
1.060	12.19	12.9	1.3393	55	73.7	1.4453	55	79.5
1.065	13.19	14.1	1.3667	60	82.0	1.4983	60	89.6
1.070	14.17	15.2	1.3913	65	90.4	1.5533	65	101.0
1.075	15.16	16.3	1.4134	70	98.9	1.6105	70	112.7
1.080	16.15	17.4	1.4337	75	107.5	1.6692	75	125.2
1.085	17.13	18.6	1.4521	80	116.2	1.7272	80	138.2
1.090	18.11	19.7	1.4686	85	124.8	1.7786	85	151.2
1.095	19.06	20.9	1.4826	90	133.5	1.8144	90	164.3
1.100	20.01	22.0	1.4850	91	135.1	1.8195	91	165.6
1.105	20.97	23.2	1.4873	92	136.8	1.8240	92	167.8
1.110	21.92	24.3	1.4892	93	138.5	1.8279	93	170.0
1.115	22.86	25.5	1.4912	94	140.2	1.8312	94	172.1
1.120	23.82	26.7	1.4932	95	141.9	1.8337	95	174.2
1.125	24.78	27.8	1.4952	96	143.5	1.8355	96	176.3
1.130	25.75	29.1	1.4974	97	145.2	1.8364	97	178.1
1.135	26.70	30.3	1.5008	98	147.1	1.8361	98	179.9
1.140	27.66	31.5	1.5056	99	149.0	1.8342	99	181.6
1.145	28.61	32.8	1.5129	100	151.3	1.8305	100	183.0
1.150	29.57	34.0						
1.155	30.55	35.3						
1.160	31.52	36.6						
1.165	32.49	37.9						
1.170	33.46	39.2						
1.175	34.42	40.4						
1.180	35.39	41.8						
1.185	36.31	43.0						
1.190	37.23	44.3						
1.195	38.16	45.6						
1.200	39.11	46.9						

Acids (continued)

Aqueous acetic acid			Aqueous hydrobromic acid			Fuming sulfuric acid		
Density $d_{4°}^{20°}$	CH_3CO_2H, % by wt.	CH_3CO_2H, g/100 mL	Density $d_{4°}^{20°}$	HBr, % by wt.	HBr, g/100 mL	Sp. gr. $d_{20°}^{20°}$	Free SO_3, % by wt.	Free SO_3, g/100 mL
1.0125	10	10.1	1.0723	10	10.7	1.860	1.54	2.8
1.0263	20	20.5	1.1579	20	23.2	1.865	2.66	5.0
1.0384	30	31.1	1.2580	30	37.7	1.870	4.28	8.0
1.0488	40	42.0	1.3150	35	46.0	1.875	5.44	10.2
1.0575	50	52.9	1.3772	40	56.1	1.880	6.42	12.1
1.0642	60	63.8	1.4446	45	65.0	1.885	7.29	13.7
1.0685	70	74.8	1.5173	50	75.8	1.890	8.16	15.4
1.0700	80	85.6	1.5953	55	87.7	1.895	9.43	17.7
1.0661	90	96.0	1.6787	60	100.7	1.900	10.07	19.1
1.0498	100	105.0	1.7675	65	114.9	1.905	10.56	20.1
						1.910	11.43	21.8
						1.915	13.33	25.5
						1.920	15.95	30.6
						1.925	18.67	35.9
						1.930	21.34	41.2
						1.935	25.65	49.6

Bases

Aqueous sodium hydroxide			Aqueous ammonia			Aqueous sodium carbonate		
Density $d_{4°}^{20°}$	NaOH, % by wt.	NaOH, g/100 mL	Density $d_{4°}^{20°}$	NH_3, % by wt.	NH_3, g/100 mL	Na_2CO_3, % by wt.	Density $d_{4°}^{20°}$	Na_2CO_3, g/100 mL
1.0207	2	2.0	0.9939	1	1.0	2	1.0190	2.04
1.0428	4	4.2	0.9895	2	2.0	4	1.0398	4.16
1.0648	6	6.4	0.9811	4	3.9	6	1.0606	6.36
1.0869	8	8.7	0.9730	6	5.8	8	1.0816	8.65
1.1089	10	11.1	0.9651	8	7.7	10	1.1029	11.03
1.1309	12	13.6	0.9575	10	9.6	12	1.1244	13.50
1.1530	14	16.1	0.9501	12	11.4	14	1.1463	16.05
1.1751	16	18.8	0.9430	14	13.2	16	1.1682	18.50
1.1972	18	21.5	0.9362	16	15.0	18	1.1905	21.33
1.2191	20	24.4	0.9295	18	16.8	20	1.2132	24.26
1.2738	25	31.8	0.9229	20	18.5			
1.3279	30	39.8	0.9164	22	20.2			
1.3798	35	48.3	0.9101	24	21.8			
1.4300	40	57.2	0.9040	26	23.5			
1.4779	45	66.5	0.8980	28	25.1			
1.5253	50	76.3	0.8920	30	26.8			

In Case of Accident

Always call or notify a laboratory instructor as soon as possible.

Fire

Burning Reagents: Immediately extinguish any gas burners in the vicinity. Fire extinguishers, charged with carbon dioxide or monoammonium phosphate powder under pressure, are available in various parts of the laboratory.

For burning oil use powdered sodium bicarbonate.

Burning Clothing: Avoid running (which fans the flame) and take great care not to inhale the flame. Rolling on the floor is often the quickest and best method for extinguishing a fire on one's own clothing.

Smother the fire as quickly as possible using wet towels, laboratory coats, heavy (fire) blankets, or carbon dioxide extinguisher.

Treatment of Small Burns: Submerge the burned area in cold water until the pain subsides. Blot the area dry, gently, with a sterile gauze and apply a dry gauze as a protective bandage. In small second or third degree burns in which blisters have formed or broken, or in which deep burns are encountered, see a physician as soon as possible.

Extensive Burns: These require special treatment to avoid serious or fatal outcome—*summon medical treatment at once*. Combat the effects of shock by keeping the patient warm and quiet.

Injuries and Chemical Burns

Reagents in the Eye: Wash *immediately* with a large amount of water, using the ordinary sink hose, eye-wash fountain, or eye-wash bottle—*do not touch the eye*. After the eye has been washed thoroughly for 15 min, if *any* discomfort remains, see a physician.

Reagents on the Skin: *Acids*—Wash immediately with a large amount of water, then soak the burned part in sodium bicarbonate solution. Cover the burned area with a dressing bandage and see a physician.

Alkali—Wash immediately with a large amount of water, then soak the burned area in 1% boric acid solution to neutralize the alkali. Cover the burned area with a dressing bandage and see a physician.

Bromine—Wash *immediately* with a large amount of water, then soak the burned area in 10% sodium thiosulfate, or cover with a *wet* sodium thiosulfate dressing, for at least 3 hr. and see a physician.

Organic Substances—Most organic substances can be removed from the skin by washing immediately with ordinary ethanol, followed by washing with soap and warm water. If the skin is burned (as by phenol), soak the injured part in water for at least 3 hr. and see a physician.

Cuts: Wash the wound with sterile gauze, soap, and water. Cover with a sterile dressing and keep dry.

Index

Numbers in **boldface** refer to preparations or test procedures.

Absorbance, definition, 212
p-Acetamidobenzenesulfonamide, **340**
p-Acetamidobenzenesulfonyl chloride, **339**
p-Acetamidobenzoic acid, **337**
p-Acetotoluidide, **336**
α-Acetoxydiphenylacetic acid, **361**
Acetylation
  of aniline, **325**
  of ferrocene, **406**
  of glucose, **426**
  of salicylic acid, **332**
Acetylbenzilic acid, **361**
4-Acetylbiphenyl, **315**
2-Acetylcyclohexanone, **386**
Acetylferrocene, **406**
Acetylsalicylic acid, **332**
Acid amides, **168**
Acid chlorides, **168**
Acylation
  of amines, **175–76**
  of aromatic compounds, 313
Adsorbants for liquid–solid
   chromatography, 94
Adsorption coefficient, 92
AIBN, free radical initiator, 226
Alcoholic silver nitrate, test for halides, **153**
Alcohols
  conversion to alkyl halides, 233
  derivatives of, 155, 158
  oxidation of, 276
  tests for, 140
Aldehydes, 281, 291
  derivatives of, 157, 161–62
  tests for, 142
Aldol condensations, 345
Alkenes
  hydration of, 267
  preparation of, 250
Alkylation of benzene, 312
Alkyl halides
  from alcohols, 233
  test for, **153**
  table of, 180
Alkynes
  coupling of, 272
  hydration of, 267
Amines
  derivatives of, 173–75
  test for, 147
p-Aminobenzenesulfonamide, **340**
p-Aminobenzoic acid, **337**
Analgesics, analysis by TLC, 112
Anthraquinone, **380**
Anthrone, **381**
Apparatus
  boiling point, micro sample, 129
  column chromatography, 102
  common, 4

443

Apparatus (*cont.*)
  extraction, 80, 86
  filtration, 73
  fractional distillation, 33
  GLC, 106
  for introduction of air, 274
  melting point, 66
  reflux, 119
  removal of excess thionyl chloride, 436
  simple distillation, 31
  sodium fusion, 132
  steam distillation, 58
  sublimation, 74
  suction filtration, 73
  TLC, 101
  vacuum distillation, 48
  water separator, 384
Aromatic hydrocarbons, test for, 152
Aryl halides, test for, 153
Aryloxyacetic acids, **177**
Aspirin, **332**
Atomic weights, inside back cover
Azeotropes, 22
Azeotropic mixtures, 25

Baeyer test for unsaturation, **151**
Beer–Lambert law, 212
Benedict's solution, **424**
Benzil, **353**
  monohydrazone, **356**
  monosemicarbazone, **356**
  α-monoxime, **356**
Benzilic acid, **359**
Benzilic acid rearrangement, 358
Benzoic acid, **293**, 300
Benzoin, **351**
  acetate, **352**
  α-oxime, **352**
Benzoin condensation, 348
Benzophenone, **361**
Benzopinacol, **375**
Benzopinacolone, **376**
p-Benzoquinone, **398**
  anthracene adduct, **400**
Benzoylation of glucose, **427**
Benzyl alcohol, **292**
S-Benzylthiouronium chloride, **171**
S-Benzylthiouronium salts, **171**
Biosynthesis of alcohols, **428**
Boiling chips, 18, 51
Boiling point, 19
  estimation at reduced pressure, 45
  of micro samples, 128
Boiling point diagrams
  carbon tetrachloride–toluene, 21
  ethanol–water, 24
  methanol–water, 23
Bromination of phenols, **178**
Bromine test for unsaturation, **151**
Bromo derivatives of phenols, 178
p-Bromonitrobenzene, **321**
Bubble plate column, 25
Büchner funnel, 73
Butesin, **337**
n-Butyl acetate, **299**
n-Butyl bromide, **236**, 261
sec-Butyl bromide, **237**
t-Butyl chloride, **238**
  solvolysis of, 245
n-Butyl iodide, **243**

Caffeine, extraction from tea, 90
Camphene, 388
Camphor, **388**
Cannizzaro reaction, 291
Carbonyl addition reactions, 281
Carboxamides
  preparation of, **168**
  test for, 147
Carboxylic acid chlorides, **168**
Carboxylic acids
  derivatives of, 166, 169
  ionization constants of, 303
  neutralization equivalent of, 167
  test for, 146
Carcinogens, 2
Chemical communication, 369
Chemical kinetics, 245
Chlorination
  of 2,3-dimethylbutane, **230**
  free radical, 225
  substituent effects, 229, **231**
Chromatography, 92
  column, 101
  flash, 95, 103
  gas–liquid, 97, 106
  ion-exchange, 95
  liquid–liquid, 96
  liquid–solid, 92
  thin-layer, 93, 100
Chromic acid oxidation test, **140**
Cis-trans isomerism of alkenes, 253
Claisen–Schmidt condensation, 346
Classification tests, 138
Cleaning and drying glassware, 5
Collins' reagent, 276
Column chromatography, 93, 101
Contact lenses, 2
Cooling baths, 118

Copper acetate
  catalyst for alkyne coupling, 272
  catalyst for benzoin oxidation, 354
o-Cresol red, **415**
Crystallization, 68
Cuprous chloride, **275**
Cycloadditions, 390
Cyclohexanone, **279**
Cyclohexene, **255**
4-Cyclohexene-cis-1,2-dicarboxylic anhydride, **394**
N-1-Cyclohexenylpyrrolidine, **386**
Cyclopentadiene, **403**

Dalton's law, 20
Density
  of aqueous acids and bases, 440–41
  of water, 439
Depression of melting points, 62
Derivatives
  of alcohols, 158
  of aldehydes, 161–62
  of amines, 174–75
  of carboxylic acids, 169
  of ketones, 162
  of phenols, 178
  of sulfonic acids, 170
  table of, 139
Derivatization of functional groups, 154
Detergents and wetting agents, 299
Diazonium-coupling reactions, 408
Dibenzalacetone, **347**
Dicyclopentadienyliron, **404**
Diels–Alder reaction, 390, 396
Diethyl benzylphosphonate, **289**
N,N-Diethyl-m-toluamide, **371**
Dihydroxytriptycene, **400**
Dilantin, **357**
Dimerization of isobutylene, 252
1,4-Dimethylpyridinium iodide, **419**
m-Dinitrobenzene, **320**
3,5-Dinitrobenzoate derivatives, **156**
3,5-Dinitrobenzoates of phenols, **177**
2,4-Dinitrophenylhydrazones, **161**
2,4-Dinitrophenylhydrazone test, **143**
5,5-Diphenylhydantoin, **357**
Diphenylmethanol, **284**
Disaccharide tests, 423
Distillation, 19
  fractional, 25
  laboratory procedure, **31**
  relative efficiency, 28
  steam, 55
  vacuum, 45
Distribution coefficient for extraction, 81

Drying agents, 116
Dulcin, **334**
Dyes and indicators, 408

E1 and E2 reactions, 251
Electromagnetic spectrum, 186
Element detection, 129
  halogens, 131, **133**
  nitrogen, 130, **133**
  sulfur, 130, **133**
Enamine reaction, 383
Equilibria and rates of carbonyl reactions, 283
Ester hydrolysis, **166**
Esters, 295
  derivatives of, 165
  table of, 164
  test for, 145
Ethanol by fermentation, **430**
p-Ethoxyacetanilide, **333**
p-Ethoxyphenylurea, **334**
Ethyl alcohol, see Ethanol
Ethynylcyclohexanol, oxidative coupling of, **273**
Eutectic mixtures, 68
Extinction coefficient, 212
Extraction, 79
Extraction funnels, use of, 86

Fats and fatty oils, 298
Fehling's solution, **424**
Fermentation of sugars, 428
Ferric complex, test for phenols, **150**
Ferrocene, **404**
Fieser's reagent, 277
Finkelstein reaction, 235, 241
Fire hazards, 2
First aid, 442
First-order kinetics, 245
Flash chromatography, 95, 103
Fluted filter preparation, 71
Fractional distillation, 25
  apparatus, 33
Fractionating columns, 25
  efficiency, 28
Free radical halogenation, 225
Friedel–Crafts reaction, 312
Friedel–Crafts test for aromatics, **153**
Fries reaction, 314
Fuming nitric acid, 318
Functional group identification, see individual group

Gas absorption traps, 120
Gas burners, 9

Gas–liquid chromatography, 97, 106
Geometric *(cis-trans)* isomers of
  alkenes, 253
Glaser–Eglinton–Hayes reaction, 272
Glucose pentaacetate, **426**
Glucose pentabenzoate, **427**
Glyceryl esters, 298
Greasing joints, 5
Grignard reaction, 259–61, 363
Ground-glass joints, care of, 5

Half-life, definition, 247
Halogenated hydrocarbons, table of, 180
Halogenation, free radical, 225
Halogens, test for, 131, **133**
Heating mantles, 10
Heat sources, 9
Henderson–Hasselbalch equation, 304
2-Heptanone, **270**
HETP, 28
  in gas chromatography, 99, 109
1-Hexene, oxymercuration of, **270**
Hinsberg test, **149**
Holdup, 27, 30
Hot plates, 10
Hydration of alkenes and alkynes, 267
Hydrocarbons
  derivatives of, 179
  tests for, 150
Hydrolysis of nitriles and amides, **173**
Hydroxamate test
  for amides and nitriles, **147**
  for esters, **146**
Hydroxyhydroquinone triacetate, **399**

Ideal solutions, 19
Identification
  by chemical methods, 125
  by spectrometric methods, 182
Ignition test, 127
Indicators and dyes, 408
Infrared spectra
  acetamide, 191
  acetic acid, 191
  aniline, 195
  benzaldehyde, 194
  benzophenone, 284
  n-butanol, 193
  n-butyl acetate, 192
  cyclohexanone, 193
  cyclohexene, 195
  diphenylmethanol, 285
Infrared spectroscopy, 186
  alcohols, 192
  aldehydes, 193

amines, 194
assignment of functional groups,
  189–90
hydrocarbons, 194
ketones, 193
sampling techniques, 196
sulfonic acids and derivatives, 195
table of frequencies, 188
Ink pigments, separation, **111**
Insect repellents, 370
Interconversion of weight and volume, 8
Iodoform test, **141**
Ion-exchange chromatography, 95
Ionization of carboxylic acids, 303
Isoborneol, **389**
Isobornyl acetate, **388**

Jones' reagent, **389**

Ketones, 281
  derivatives of, 157, 162
  tests for, 142
Kinetic control of products, 313

Labels, information on, 223
Laboratory notes, 12, 217–19
Liquid–liquid chromatography, 96
Liquid–solid chromatography, 92
Lucas test, **141**
Lumière–Barbier acetylation, **325**

MEL-TEMP, 66
Magnetic stirring, 121
Maleic anhydride adducts, 394
Manometers, 50
Mass spectrometry, 183
Mass spectrum, ethyl acetate, 184
Mechanical stirring, 121
Melting point, 62
  determination, 65
Menshutkin reaction, 241
Mercuric acetate, as hydration catalyst,
  270
Merocyanin dye, 417, **420**
Mesionic compounds, 342
Metallocenes, 402
Methone derivatives, **161**
p-Methoxystilbene, **289**
Methyl benzilate, **361**
Methyl benzoate, **300**
2-Methyl-2-hexanol, **262**
2-Methylhexenes, **263**
Methyl *m*-nitrobenzoate, **322**
Methyl orange, **410**
Methylpentenes, **254**

Methyl *n*-propyl ketone, **278**
Mixed melting points, 64
Molecular formula by mass spectrometry, 185
Molecular weight determination, 63
Monosaccharide tests, 423
Multiple extraction, 82

Neutralization equivalent of an acid, 170
Nitration
   of anilines, 324
   of aromatics, 318
Nitriles
   derivatives of, 172
   test for, **147**
p-Nitroacetanilide, **328**
p-Nitroaniline, **328**
p-Nitrobenzoate derivative, **156**
o-Nitrobenzoic acid, **311**
p-Nitrobenzoic acid, **310**
Nitrogen, test for, 130, **133**
N-Nitroso-*N*-phenylglycine, **343**
NMR spectra
   acetaldehyde, 209
   [18]annulene, 202
   n-butyl bromide, 238
   t-butyl chloride, 239
   ethanol, 210
   ethylbenzene, 208
   ferrocene, 405
   methyl benzoate, 301
   2-methyl-1-hexene, 264
   2-methyl-2-hexene, 264
   α-methylstyrene, 208
   phenylacetylene, 209
   3-phenylsydnone, 344
   propiophenone-^{13}C, 207
   propiophenone-^{1}H, 205
   1,2,2-trichloropropane, 201
NMR spectroscopy, 197
   chemical shift, 198
   ^{13}C chemical shift table, 206
   ^{1}H chemical shift table, 199
   exchange, 201
   nuclei other than protons, 205
   spin-spin coupling constants, 204
   spin-spin interactions, 202
Nomograph for boiling points, 46
Nonideal solutions, 22
endo-Norbornene-*cis*-5,6-dicarboxylic anhydride, **394**
Notebooks, 12, 217
   sample page, 219
Nuclear magnetic resonance, *see* NMR spectroscopy

Off!, 371
Osazone test, **424**
Oxidation of alcohols, 276
Oxime derivatives, **163**
Oxymercuration-demercuration of alkenes, 268, **270**

PABA (*p*-aminobenzoic acid), **337**
Para red, **411**
2-Pentanone, **278**
n-Pentyl bromide, **236**
Peptides, 432
Permanganate test for unsaturation, **151**
Phenacetin, **333**
Phenanthraquinone, **381**
Phenols
   derivatives of, 176, 178
   test for, 149
Phenylcarbamate derivatives, formation of, **157**
N-Phenyl-4-cyclohexene-1,2-dicarboximide, **393**
N-Phenylglycine, **343**
Phenylhydrazine reagent, **426**
Phenylmagnesium bromide, **365**
N-Phenylmaleamic acid, **393**
N-Phenylmaleimide, **393**
N-Phenyl-*endo*-norbornene-5,6-dicarboximide, **394**
3-Phenylsydnone, **343**
Pheromones and insect repellents, 369
Photochemical reduction, **375**
Phthalein dyes, 412
Phthaloylglycine, **435**
Phthaloylglycyl chloride, **435**
Phthaloylglycylglycine, **436**
Pinacol–pinacolone rearrangement, 373
Plant pigment separation, 111
p*K* determination of carboxylic acids, 307
Polypeptides, 432
n-Propyl bromide, **236**

Quaternary ammonium salts, **176**
Quinones, 378, 396

Raoult's law, 19
Rast molecular weight method, 63
Rate constant, definition, 245
Reducing sugars, test for, **423**
Reduction of carbonyl compounds, 282
Reflux ratio, 30
Refluxing, 118
Retention times for different liquid phases, 98

R_f value defined, 93
Rotary evaporation, 122

Safety glasses, 1
Salts, effect on solubility, 83
Saponification, 295
Saponification equivalent, **165**
Schiff's fuchsin test, **144**
Schotten–Baumann acylation, **176**
Semicarbazone derivatives, **163**
Side chain oxidation, 309
Silver mirror test, **144**
Simple distillation, 17
 apparatus for, 31
S_N1 and S_N2 reactions, 233, 241, 245
Sodium borohydride, 282, **284**
Sodium fusion, **131**
Sodium iodide, test for reactive halides, **154**
Sodium methoxide, **289**
Solubility
 classification scheme, 135
 effect of salts, 83
 generalizations, 84, 133
Solvatochromic dyes, 417
Solvents
 for crystallization, 69
 for liquid–solid chromatography, 94
Solvolysis of t-butyl chloride, 245
Soxhlet extractor, 80
Spectrometric methods of identification, 182
Standard separation, 29
Steam baths, 10
Steam distillation, 55
 composition of distillate, 56
Stem correction, 38
Stirring, 11, 121
Stork enamine reaction, 383
Sublimation, 62, 74
Sugars, 422
Sulfanilamide, **338**
Sulfonamides, **171**
Sulfonic acids
 derivatives of, 170
 test for, 146
Sulfonphthalein indicators, 412
Sulfur, test for, 130, **133**

Sulfuryl chloride, 226

Theoretical plate, 28
Thermodynamic control of products, 283, 313
Thermometer calibration, 38
Thiele melting point tube, 65
Thin-layer chromatography, 93, 100
Thionyl chloride, 168, 371
Titration curves of acid neutralizations, 305
Tollen's test, **144**
p-Toluidides, preparation of, **168**
m-Toluyl chloride, **371**
Toxic chemicals, 2
Triarylmethanols, 363
2,4,4-Trimethyl-1- and -2-pentenes, **256**
Triphenylmethane, **376**
Triphenylmethanol, **365**
β,β,β-Triphenylpropionic acid, **367**
Tylenol, 333

Ultraviolet absorptions, characteristic, 213
Ultraviolet spectroscopy, 211
Ultraviolet–visible spectra
 cresol red, 414
 methyl benzoate, 212
Unsaturation, test for, **151**
Urethan derivatives, **157**

Vacuum distillation, 45
Vacuum pumps, 47
Van Deemter equation, 109
Vapor pressure, 18
 of organic substances, 437–38
 of water, 437–39
Variac, 11
Visible spectroscopy, 211
Vitamin B_1 catalysis, 350

Wagner–Meerwein rearrangements, 387
Water separator, 384
Weighing reagents, 7
Williamson ether synthesis, 241
Wittig reaction, 287
Wurtz–Fittig reaction, 364

Yields, calculation of, 220

Atomic Weights

Aluminum	Al	26.98	Magnesium	Mg	24.31
Antimony	Sb	121.75	Manganese	Mn	54.94
Arsenic	As	74.92	Mercury	Hg	200.59
Barium	Ba	137.34	Nickel	Ni	58.71
Boron	B	10.81	Nitrogen	N	14.01
Bromine	Br	79.91	Oxygen	O	16.00
Calcium	Ca	40.08	Palladium	Pd	106.40
Carbon	C	12.01	Phosphorus	P	30.97
Chlorine	Cl	35.45	Platinum	Pt	195.09
Chromium	Cr	52.00	Potassium	K	39.10
Copper	Cu	63.55	Silicon	Si	28.09
Fluorine	F	19.00	Silver	Ag	107.87
Hydrogen	H	1.01	Sodium	Na	22.99
Iodine	I	126.90	Sulfur	S	32.06
Iron	Fe	55.85	Tin	Sn	118.69
Lead	Pb	207.19	Vanadium	V	50.94
Lithium	Li	6.94	Zinc	Zn	65.37